Terahertz Sources and Systems

NATO Science Series

A Series presenting the results of scientific meetings supported under the NATO Science Programme.

The Series is published by IOS Press, Amsterdam, and Kluwer Academic Publishers in conjunction with the NATO Scientific Affairs Division

Sub-Series

I. Life and Behavioural Sciences	IOS Press
II. Mathematics, Physics and Chemistry	Kluwer Academic Publishers
III. Computer and Systems Science	IOS Press
IV. Earth and Environmental Sciences	Kluwer Academic Publishers

The NATO Science Series continues the series of books published formerly as the NATO ASI Series.

The NATO Science Programme offers support for collaboration in civil science between scientists of countries of the Euro-Atlantic Partnership Council. The types of scientific meeting generally supported are “Advanced Study Institutes” and “Advanced Research Workshops”, and the NATO Science Series collects together the results of these meetings. The meetings are co-organized bij scientists from NATO countries and scientists from NATO's Partner countries – countries of the CIS and Central and Eastern Europe.

Advanced Study Institutes are high-level tutorial courses offering in-depth study of latest advances in a field.
Advanced Research Workshops are expert meetings aimed at critical assessment of a field, and identification of directions for future action.

As a consequence of the restructuring of the NATO Science Programme in 1999, the NATO Science Series was re-organized to the four sub-series noted above. Please consult the following web sites for information on previous volumes published in the Series.

http://www.nato.int/science
http://www.wkap.nl
http://www.iospress.nl
http://www.wtv-books.de/nato-pco.htm

Series II: Mathematics, Physics and Chemistry – Vol. 27

Terahertz Sources and Systems

edited by

R. E. Miles

P. Harrison

University of Leeds,
Leeds, United Kingdom

and

D. Lippens

Université des Sciences et Technologies de Lille,
Lille, France

Kluwer Academic Publishers

Dordrecht / Boston / London

Published in cooperation with NATO Scientific Affairs Division

Proceedings of the NATO Advanced Research Workshop on
Terahertz Sources and Systems
Chateau de Bonas, France
22–27 June, 2000

A C.I.P. Catalogue record for this book is available from the Library of Congress.

ISBN 0-7923-7096-1

Published by Kluwer Academic Publishers,
P.O. Box 17, 3300 AA Dordrecht, The Netherlands.

Sold and distributed in North, Central and South America
by Kluwer Academic Publishers,
101 Philip Drive, Norwell, MA 02061, U.S.A.

In all other countries, sold and distributed
by Kluwer Academic Publishers,
P.O. Box 322, 3300 AH Dordrecht, The Netherlands.

Printed on acid-free paper

Contents

ORGANISING COMMITTEE

Co-Directors : D Lippens (Université des Sciences de Lille)
N N Zinovev (University of Leeds

R E Miles (Leeds)	Secretary
P Harrison (Leeds)	Treasurer
J M Chamberlain (Leeds)	Proceedings
M C Quirke (Leeds)	Workbook

SPONSORS

We wish to thank the following for their contribution to the success of this conference:

European Office of Aerospace Research and Development, Air Force Office of Scientific Research, United States Air Force Research Laboratory

Agilent Technologies UK Limited

Marconi Materials Technology

Spectra-Physics Lasers

C R L Ltd

PREFACE

This ARW on "Terahertz Sources and Systems" is a follow up to the ASI on "New Directions in Terahertz Technology" held in 1996, also at the Chateau de Bonas. The contributions to this volume show how, during the intervening years, Terahertz Technology has moved on from being a useful but expensive circuit technique, applied largely to astronomy and space science, to a subject in its own right with important applications. If one thing stands out in particular it must be Terahertz Imaging - an application of Terahertz Technology that was not even mentioned during those pioneering sessions back in 1996. It is perhaps a measure of the "coming of age" of Terahertz Technology that there was hardly a mention of space science and astronomy in the 2000 ARW!

The first two days of the Workshop concentrated on terahertz sources, discussing the state of the art in frequency conversion (i.e. *up* from millimetre waves and *down* from infra-red) and direct generation. As far as the latter is concerned the "terahertz gap" is still with us but tremendous advances have been made in down conversion, both pulsed and CW. Work on the Heterostructure Barrier Varactor (HBV) for electronic upconversion continues but the generation of power above about 0.25 THz by this method still eludes us.

The driving force in Terahertz Technology is currently in imaging and spectroscopy. We now have the means to obtain images and chemical information in this frequency band which, until recently, has been difficult to access. The images shown at the workshop (and reproduced in this volume) are striking and not surprisingly the clinical and analytical uses are already the subject of intense activity. Imaging systems must surely be the first commercially viable terahertz products.

Many of us who participated in the 1996 ASI were of the opinion that terahertz systems would have their first impact in the area of communications where their wide bandwidth potential would be an advantage. In practice, communications applications have not yet materialised, mainly because we still do not have a complete range of active terahertz electronic components: sources and amplifiers being particularly conspicuous by their absence. However, an encouraging conclusion of the Workshop was that, in the foreseeable future, terahertz frequency electronics will be necessary in communications systems. This could be in the form of direct transmission at terahertz frequencies for a local area network or terahertz modulation on optical fibre. It is quite likely that spin-off from the imaging technology will in due course provide the necessary components.

The organisers would like to thank all of the speakers, every one of whom captured the spirit of the Workshop to create, once again, the uniquely stimulating atmosphere that is a NATO ARW at the Chateau de Bonas. Suggestions that this may also have had something to do with Patrice and his staff's hospitality should not be discounted. Thanks also to the participants for their contributions to the discussions and poster session. We also thank the NATO Science Committee for their support, without which this ARW would not have been possible.

We look forward to an exciting future for terahertz technology.

Theme 1

Physics of Terahertz Devices and Materials

TERAHERTZ DYNAMICS IN QUANTUM STRUCTURES:

TOWARDS A FUNDAMENTAL TERAHERTZ OSCILLATOR

S.J. ALLEN, J. S. SCOTT, M.C. WANKE
Institute for Quantum Engineering Science and Technology
Center for Terahertz Science and Technology

K. MARANOWSKI, A.C. GOSSARD
Materials Department

M.J.W. RODWELL
Electrical and Computer Engineering Department
University of California at Santa Barbara
Santa Barbara, Californa 93106, USA

D.H. CHOW
HRL Laboratories
Malibu, California

The terahertz part of the electromagnetic spectrum marks a decade wide gap in semiconductor technology between transport-based electronics and quantum transition based photonics. As such, it appears that potential, future solid-state oscillators, that will fill this gap, will marry transport and quantum transitions in semiconductor quantum structures. Fundamental limits of quantum transport devices like resonant tunneling diodes are discussed. Monolithic power combining leads naturally to multi-quantum well superlattices in uniform electric fields, which are recognized as Bloch oscillators. Multi-quantum well superlattices are integrated into quasi-optical arrays and loaded into terahertz cavities driven by radiation from the UCSB free-electron lasers in order to measure the change in terahertz dynamical conductance under electrical bias. Electric field domain formation prevents the material from presenting the anticipated terahertz gain required for an oscillator.

1 Introduction

From the perspective of semiconductor electronics, the terahertz part of the electromagnetic spectrum defines the cutting edge of high-speed electronics. But the compromise or trade off between speed and complexity of high-speed semiconductor electronics better defines this frontier. This is depicted in Figure 1, where the most complex semiconductor electronics, the microprocessor unit, is currently limited to gigahertz clock speeds while the simplest device, a bolometer, in principle knows no limits. While a somewhat simplified view of the world of semiconductor technology, it does highlight the fact that the terahertz regime is the frontier for a broad class of electronic devices.

R.E. Miles et al. (eds.), Terahertz Sources and Systems, 3–14.

Two terminal fundamental oscillators are a key technology that can open the terahertz regime and a closer look at this class of device exposes a more fundamental issue. Figure 2 shows the operating regimes of *two-terminal* electronics and photonic devices [1]. Com-mercial devices like Impatt and Gunn diodes define the electronic part of the spectrum and operate up to several hundred gigahertz. Commercial photonic devices operate at wavelengths as long as 30 μm or 10 terahertz. Between several hundred gigahertz and 10 terahertz there is a gap in solid state terahertz technology.

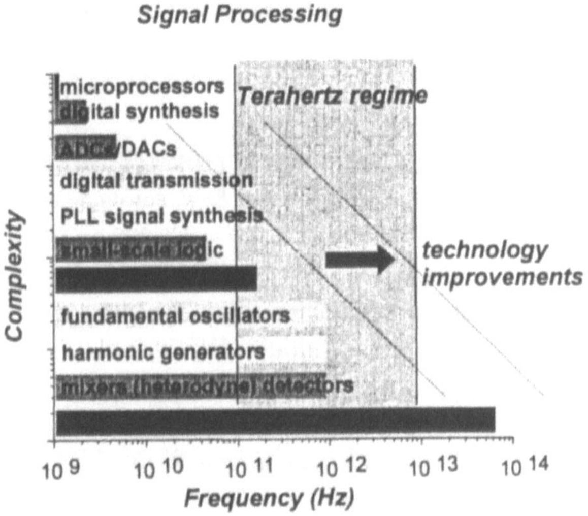

Figure 1. A Schematic plot of complexity versus speed exposes the electronics frontier at terahertz frequencies.

Whereas quantum physics defines the basic properties of materials used in electronic devices, the device physics is essentially described by quasi-classical transport. Photonic devices, on the other hand are essentially described by quantum transitions. From this perspective the terahertz regime is seen as a transition in device physics from classical transport to quantum transition. It seems likely that a successful solid state terahertz device will embrace both transport and quantum transitions,a "quantum transport device".

Research two terminal solid-state sources are also shown in Figure 2 [2]. With the exception of systems based on harmonic generation or photomixing, they all involve some form of quantum transport. On the photonic side of the terahertz technology gap are quantum cascade lasers, generating 10's of milliwatts at wavelengths as long as 17 μm. On the electronic side are various resonant tunneling devices. Quantum transport exhibits negative differential conductance and is used at frequencies controlled by the external circuit. While these approaches have made inroads into the terahertz gap they display rapidly falling power and their extension to several terahertz

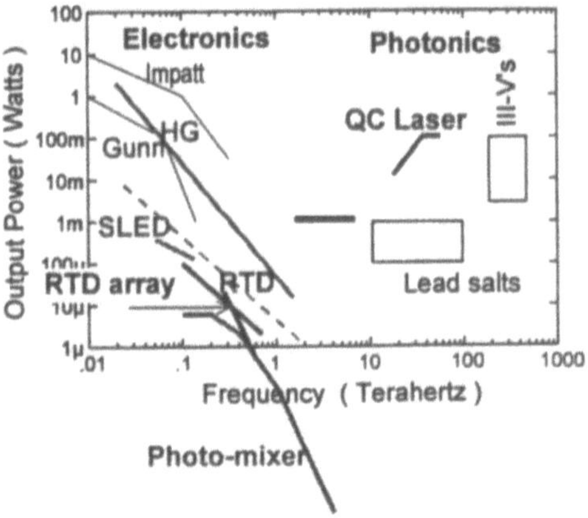

Figure 2. The terahertz regime marks a transition in fundamental oscillator device physics from transport to quantum transition.

does not look promising despite the fact that the essential physics does capture both transport and quantum transitions.

2 Fundamental Limits

The apparent roll off shown in Figure 2 can be understood following discussion by H. Grönqvist et al., [3], where the maximum power available from the resonant tunneling diode is

$$P = \frac{(J \cdot A)^2}{4\pi} \cdot R_L \quad (1)$$

Here J is the current density swing at the negative differential conductance, A is the area and R_L the effective load resistance. See Figure 3.

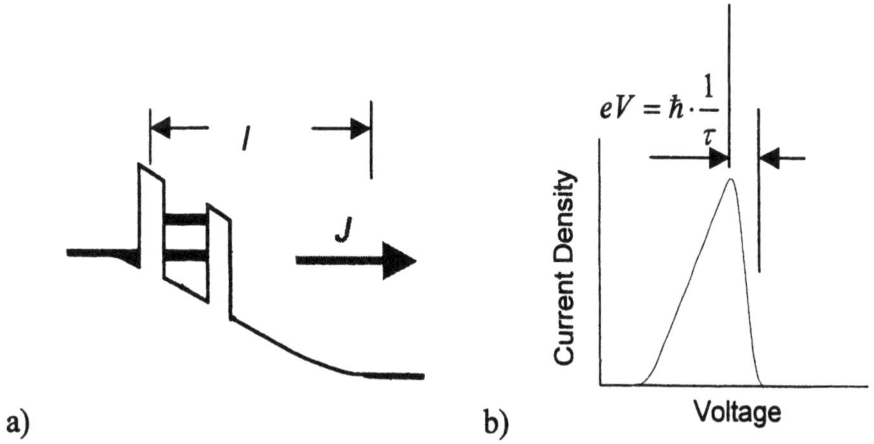

Figure 3. a) A schematic of a resonant tunneling diode with maximum current density J. b) Schematic current density vs applied voltage. The intrinsic width of the negative resistance region is determined by the tunneling rate out of the quantum well, the intrinsic width of the transmission resonance.

To operate at a frequency ω the "RC" time constant should satisfy

$$2\omega R_L C = 1, \text{ where } C = \frac{A\varepsilon}{l} \quad (2)$$

R_L is the load and C the resonant tunneling diode capacitance defined by its area A, the dielectric permittivity, ε, and l, the effective diode thickness. From (2), the area, A, scales inversely with the frequency, ω, and the maximum power, P in (1), scales inversely with the frequency squared.

Assuming a maximum current density of 10^5 A/cm^2 and a load impedance of 50 Ω we estimate maximum delivered power from a resonant tunneling diode indicated by the dashed line in Figure 3. It seems clear that single lumped elements, be they classical devices or quantum transport devices, will deliver limited power at terahertz frequencies and beyond.

There is a feature of the resonant tunneling diode that is appealing.

Conceptually, the resonant tunneling diode captures both a quasi-classical device at low frequencies and a quantum transport device at high frequencies. The gain or loss exhibited by the resonant tunneling diode with the *I-V* characteristic shown in Figure 3b is expressed by the differential conductance $G = {dI}/{dV}$. But *G* expressed as a finite difference if the frequency of the probing field is greater than the tunneling rate, 1/τ, which controls the width of the negative resistance region. That is to say, for $\omega\tau > 1$ [4]

$$G = \frac{I_{DC}(V + \hbar\omega/e) - I_{DC}(V - \hbar\omega/e)}{2\hbar\omega/e}. \tag{3}$$

For the resonant tunneling diode used to estimate the maximum power versus frequency in Figure 2, $\omega\tau = 1$ at approximately 1 terahertz.

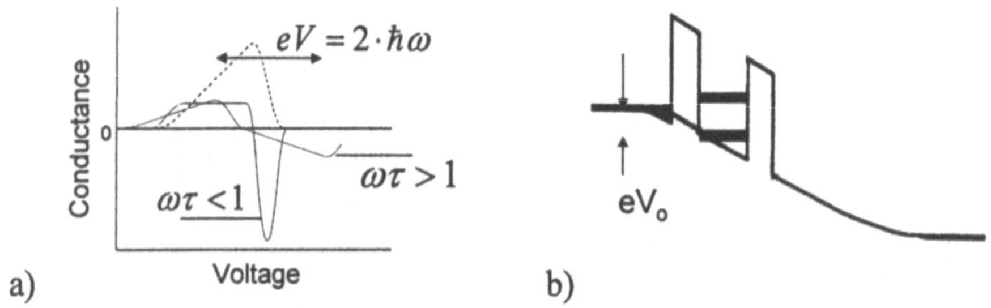

Figure 4. a) The differential conductance is the derivative of the I-V characteristic for ωτ < 1 and the finite difference for ωτ > 1. b) For ωτ > 1, the resonant tunneling diode appears as a laser diode with inversion between the emitter and quantum well ground state.

In Figure 4. a) the differential conductance for the schematic *I-V* is depicted for low and high frequency. Most important, at high frequency, the differential conductance exhibits gain at voltages beyond the features in the static *I-V*. But this can be understood by referring to Figure 4. b) where the gain appears as a quantum transition between the emitter and the ground, quantum well state. In this manner, the resonant tunneling diode conceptually captures the negative resistance, transport, device, and a quantum transition laser diode. But, it cannot overcome the intrinsic power limits discussed above.

3 Power combining

To overcome the inherent power limitations some form of power combining is necessary. Integrating the gain medium in an array or using a wave guiding structure can accomplish this. Further power combining can be achieved by fabricating the device as a multi-quantum well superlattice. The quantum cascade laser takes advantage of both of these forms of power combining. They are fabricated as ridge wave-guides and they consist of a multi-quantum well superlattice with an engineered "unit cell" that enhances inversion and the requisite gain.

3.1 SUPERLATTICE DEVICES

A simple superlattice can support gain without inversion (Figure 5.). If the number of quantum wells is sufficiently large we can ignore the contact or emitter and focus on the periodic quantum well structure. Further, if the level broadening caused by momentum and energy relaxation is less than the Stark splitting then we expect that the differential conductance of the electrically biased structure will exhibit gain for frequencies below the Stark splitting and loss for frequencies above.

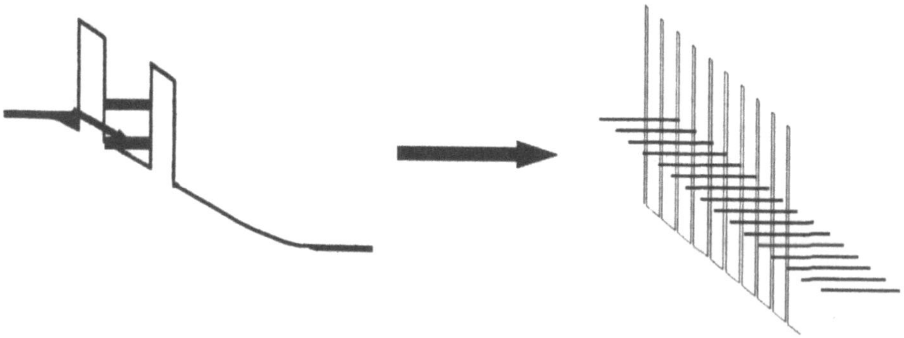

Figure 5. An electrically biased superlattice is a logical extension of the resonant tunneling diode laser, but it supports no inversion.

Analytic expressions for the differential conductivity can be obtained by recognizing that the Stark ladder is the quantum model of the classic Bloch oscillator in a strong electric field. The spacing between the rungs of the ladder is $\hbar\omega_B$, where ω_B is the Bloch frequency, $\hbar\omega_B = eEa$, and a is the superlattice period. Following Ktitorov et al. [5], we have for the differential conductivity

$$\sigma(\omega) = \sigma_0 \frac{1-\omega_B^2\tau_p\tau_e - i\omega\tau_e}{\left(\omega_B^2-\omega^2\right)\tau_p\tau_e + 1 - i\omega\left(\tau_p+\tau_e\right)}, \tag{4}$$

where τ_p and τ_e are the momentum and energy relation times respectively. σ_0 is the conductivity in the absence of an applied electric field.

The dynamical response is marked by negative conductance (gain) from DC to a frequency corresponding to the Stark Splitting provided that $\omega_B\tau_p > 1$. (Figure 6). In this respect it behaves like the Gunn effect with the important distinction that the gain region is not limited by the energy relaxation time but rather the Bloch frequency. Equally important, the

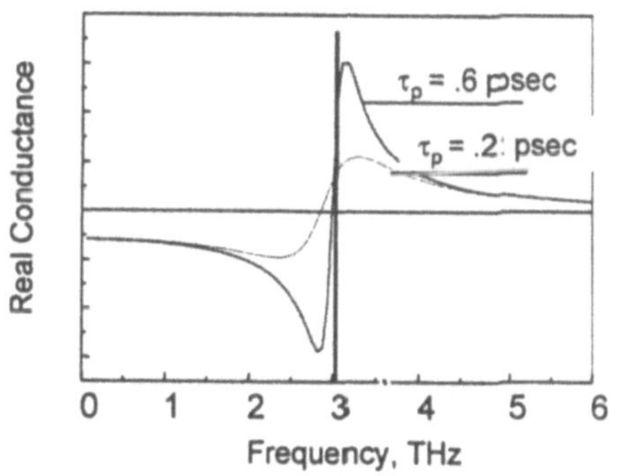

Figure 6. The differential conductance for an electrically biased superlattice with Stark splitting of 3 THz and energy relaxation time of 1 psec.

gain is resonant at frequencies just below the Bloch frequency. The superlattice under electrical bias provides a gain medium that can be tuned with the applied electric field.

It is worth noting that if $\omega_B \tau_p < 1$, then the gain region is limited to frequencies below the energy relaxation rate, $1/\tau_e$, and the behavior is indistinguishable from the Gunn effect.

3.2 QUASI-OPTICAL ARRAYS

While the superlattice effectively combines many quantum transport devices in series, power combining can continue by arranging the devices in a two dimensional quasi-optical array [6]. (Figure 7.)

If the wavelength of the radiation is greater than the periodicity of the array then the array presents a two-dimensional sheet admittance defined by the conductance of the array elements and all of the parasitic elements.

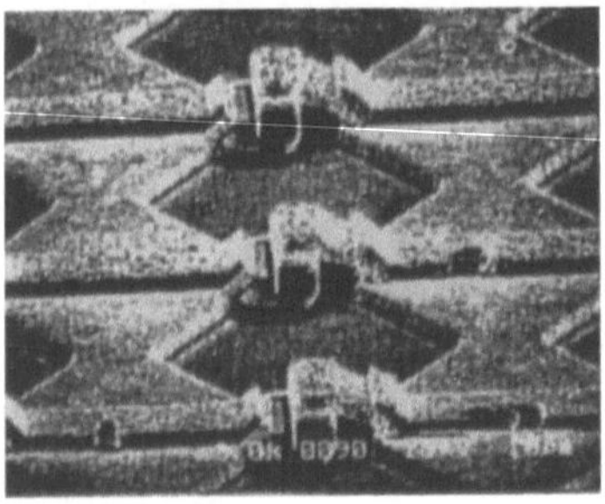

Figure 7. Scanning electron micrograph of elements of a quasi-optical array of superlattices connected "head to foot".

4 Experimental Approach

To experimentally assess the potential of electrically biased semiconductor superlattices as sources of terahertz radiation, we have fabricated a prototype oscillator system and measured its terahertz response as the embedded superlattice is subject to electrical bias. While the cavity will eventually form the oscillator, it is used in these early stages as a test structure to measure terahertz loss and gain in electrically biased superlattices.

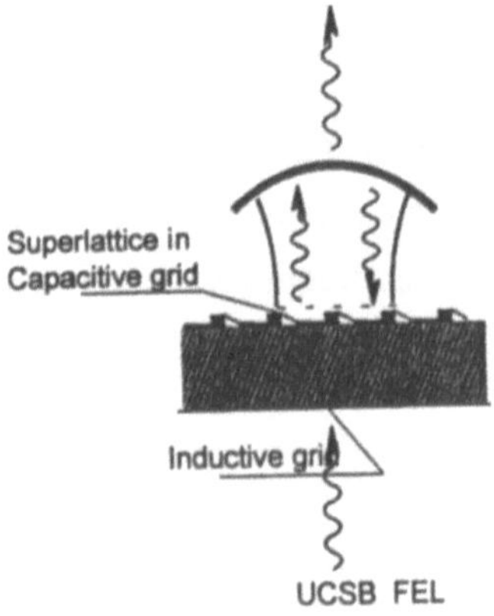

Figure 8. A terahertz cavity is formed by an inductive grid on the backside of the substrate and a cylindrical metallic reflector. The superlattice is embedded in a quasi-optical array formed by a capacitive grid

4.1 TERAHERTZ CAVITY SYSTEM

The prototype system is shown in Figure 8. The cavity is defined by a highly reflecting inductive mesh on the back of the substrate and a spherical reflector that can be moved relative to the substrate. The active element, the superlattice loaded quasi-optical array, is located on the front of the substrate. In the experiments described below the quasi-optical array is actually a capacitive grid with a 20 μm period Figure 9.

4.2 UCSB FREE-ELECTRON LASERS

The source of radiation in these experiments was the UCSB free-electron lasers that deliver kilowatts of tunable radiation from 120 GHz to 4.8 THz [7]. Kilowatts is more than sufficient power to drive the superlattice in the quasi-optical array into a non-linear regime, but here care was taken to attenuate the output to assure that we measured the linear response of the system. Under these conditions, we take full advantage of the tunability but use the high output power only to ensure that we always can recover a signal.

Figure 9. Scanning electron micrograph of a section of a 2 mm x 2 mm quasi-optical array loaded with GaAs /AlGaAs superlattices mesas.

4.3 MEASUREMENTS

Experiments were performed by first measuring the transmission through the substrate with quasi-optical array and inductive grid but without the external spherical reflector. The terahertz radiation from the UCSB free-electron lasers is tuned through the transmission Fabry-Perot resonance. At this frequency the external spherical reflector is inserted and the overall transmission maximized by tuning its separation from the substrate. Applying a voltage to the entire quasi-optical array then modulated the transmission through the tuned cavity system. The modulated fractional change in transmission was projected on models that included the dynamical terahertz conductance of the superlattice mesas, the inherent parasitic elements of the quasi-optical array, and the elements of the terahertz cavity.

5 **Experimental Results**

Experiments were carried at out at six different frequencies from 330 GHz to 2.55 THz. We discuss only the extremes here. All experiments are carried out at room temperature.

5.1 ZENER BREAKDOWN

Before discussing the terahertz dynamics of the system under electrical bias we need to address the issue of Zener breakdown and electric field domain formation. The current voltage characteristic of the quasi-optical array is displayed in Figure 10. Since there are 100 elements in the array, the voltage drop per element is is 0.01 times that displayed on the voltage axis. Like wise, since the period of the array is 20 μm the

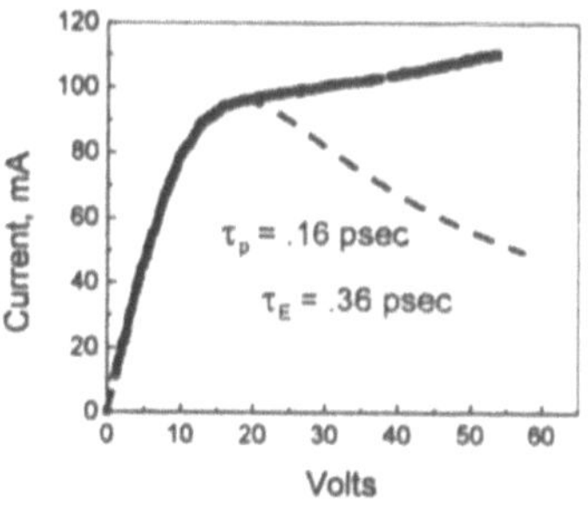

Figure 10. Measured I-V characteristic, solid line. Modeled I-V characteristic assuming uniform electric fields, dashed line.

current through an element of the array, mesa cross sectional area of 2x20 μm^2, is 0.01 times the current axis values. At large electric fields the current is expected to drop [8] and display the negative conductance evidenced at zero frequency in Figure 6. The I-V characteristic is predicted to follow

$$J = \frac{\sigma_{00} E}{1 + \omega_B^2 \tau_p \tau_e}. \tag{5}$$

The experimental I-V characteristic does not.

It is well understood that materials that exhibit bulk differential negative resistance are unstable against the formation of domains[9,10]. For a voltage controlled multi valued I-V these domains are electric field domains. In superlattices the high field domain is thought to be resonant tunneling from the ground state of a quantum well into the excited state of the neighboring well, the low field domain to be resonant tunneling between successive ground states. See Figure 11.

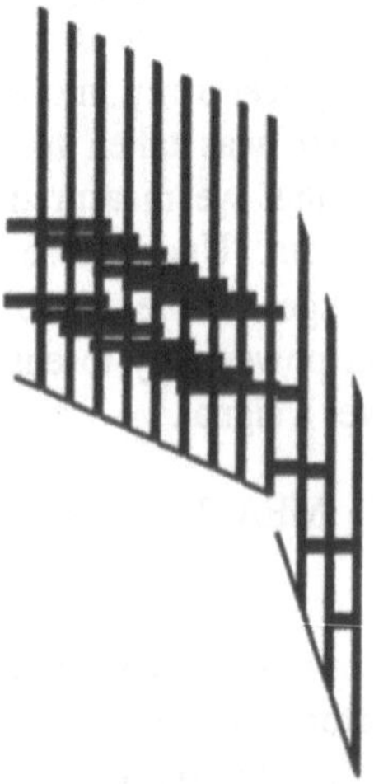

Figure 11. At high voltages the superlattuce is assumed to break into a low and high field domain marked by ground state to ground state tunneling and ground state to excited state tuneeling.

The dashed curve in Figure 10. is fit to eq. (5). We independently determine τ_p by measuring the transverse magneto-resistance with an applied field along the current direction of the quasi-optical array (perpendicular to the current flow in the superlattice itself). τ_e is then extracted from the voltage at which the current saturates. At this point the current should remain roughly constant as the current is limited by the maximum current that can flow through the miniband and the additional voltage is dropped across the high voltage domain defined by ground to excited state tunneling. The experimental current is not precisely constant but is nearly so. At this point we assume that the low voltage domain will shrink but present a terahertz conductance given by the model just as the superlattice develops zero conductance. The high voltage domain is assumed to be a terahertz short circuit.

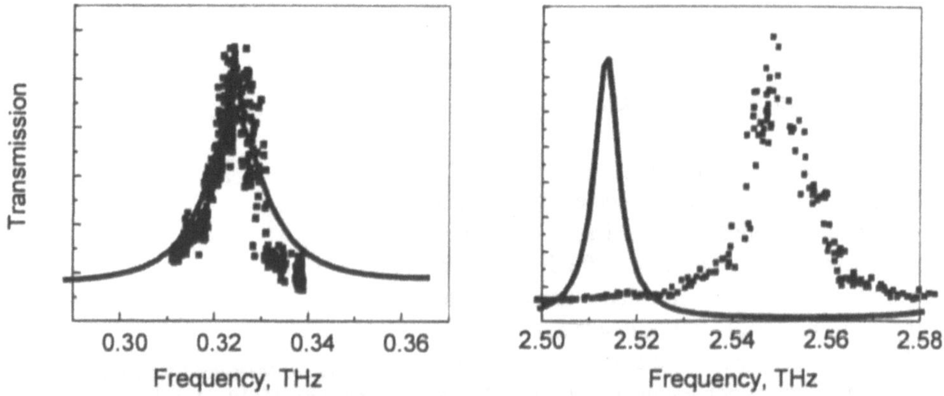

Figure 12. Resonant transmission through the quasi-optical array / substrate / inductive grid Fabry-Perot. The solid line is the model transmission.

5.2 TERAHERTZ CAVITY TRANSMISSION

The resonant transmission through the substrate without the external spherical reflector is shown in Figure 12 . The model reproduces the measured transmission quite well. The position of the resonance is a very sensitive function of the actual substrate thickness and effective real dielectric constant. The failure to exactly lie on top of the measured resonance is of no consequence. The width is more important and the model reproduces the measured values to within a factor of two.

With the free-electron laser tuned to the substrate resonance the external spherical mirror is then moved and the transmission through the cavity system measured. See Figure 13. . Here we model the transmission through the spherical reflector, which is caused by diffraction around the periphery by a reflection coefficient. It is used to model the response of the cavity at each frequency.

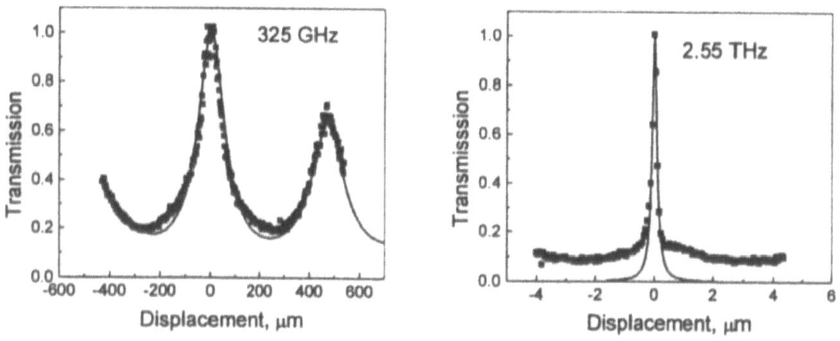

Figure 13. Transmission through terahertz cavity as a function of displacement of spherical reflector. The solid lines are modeled by assuming an effective reflectivity for the spherical reflector.

5.3 MODULATED TRANSMISSION

With the cavity and free-electron laser so tuned, the quasi-optical array is biased with an applied voltage and the change in transmission recorded. The results are shown in Figure 14. At low frequencies large, ~100%, changes in transmission are recorded while at the highest frequency the modulated transmission is only ~ 2%.

We model these results as discussed above In the first model we assume that the applied voltage produces a uniform field (dashed lines in Figure 14.). In the second we assume that the superlattice breaks into high and low field domains at the point that the DC dynamical conductance is zero. The high field domain is assumed to present a terahertz short circuit.

Including the domain formation produces rough overall agreement reproducing the magnitude of the changes in transmission and the break in modulated transmission at the point where domains are formed.

It is instructive to examine the predicted conductivity versus frequency for the

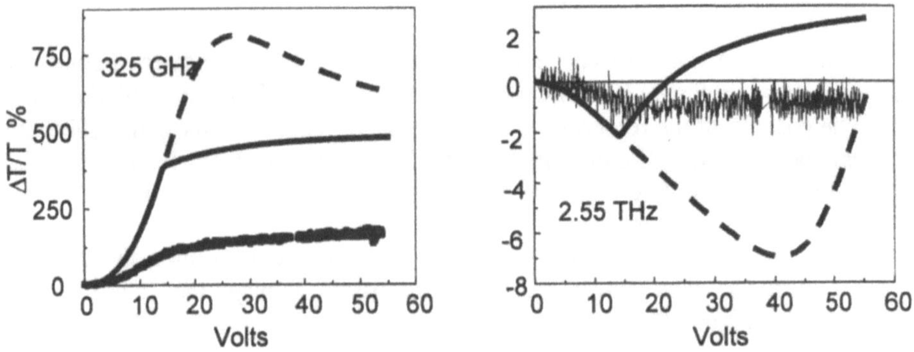

Figure 14. Modulated cavity transmission at 325 GHz and 2.55THz. The dashed lines are modeled by assuming no electric field formation (uniform fields). The solid lines assume domain formation with dynamics as discussed in the text.

superlattice, assuming uniform electric field (Figure 15). As the applied voltage exceeds that required for negative DC conductance, we expect relatively little change. At this point the superlattice will break into two domains; one with an electric field at the critical point for development of negative resistance, the other with a field that will support positive conductance, presumably due to alignment of the staircase structure shown in Figure 11. Below this critical field or voltage large changes in the low frequency transmission can be induced, but at high frequencies very small changes are expected, as shown experimentally in Figure 14.

There remains the fact that the fractional change in transmission at low frequencies differs by more than a factor of 2 from that predicted by the model. We believe that this discrepancy may be accounted for by including finite contact resistance between the metalization in the quasi-optical array and the superlattice itself. Models including the effect of parasitics are in progress.

6 Conclusions

We have modeled the terahertz dynamical response of electrically biased semiconductor superlattices and shown how the material can provide voltage tunable gain at terahertz frequencies. The multi-quantum well superlattice is a logical extension of a resonant tunneling diode but provides intrinsic power combing, an essential feature of any solid-state terahertz device that will deliver useful power.

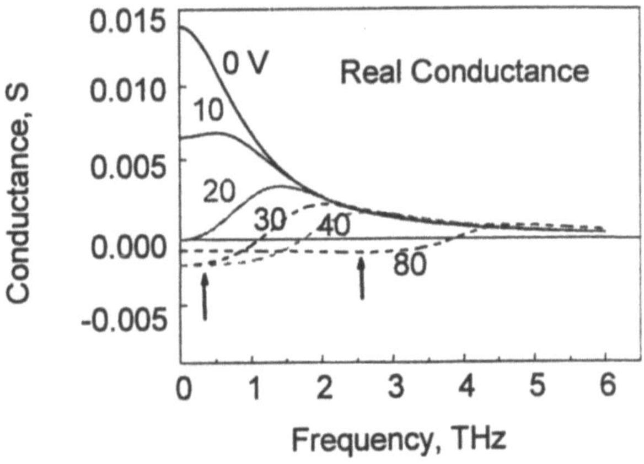

Figure 15. Predicted changes in the conductance or admittance of the superlattice loaded quasi-optical array. The two arrays indicate the frequencies where data is displayed in *Figure 14.*

We propose to use this voltage tunable gain in an oscillator by integrating superlattice mesa defined diodes into a quasi-optical array. This provides further power combining. The quasi-optical array is then inserted into a terahertz cavity defined by an inductive grid on the back of the substrate and mechanically tuned, millimeter size spherical reflector.

Current experiments use the tunable terahertz radiation from the UCSB Free-electron lasers to measure the transmission through the terahertz cavity, loaded with the active quasi-optical array, as voltage is applied to the superlattice. Large changes in transmission are measured at low frequencies while substantially smaller changes are recorded above two terahertz. A critical feature and a major obstacle to success is the formation of electrical field domains just as the superlattice system begins to develop useful gain. Suppressing domain formation is essential. At present we are able to satisfactorily model the terahertz dynamical response of the complete system if we include electric field domain formation.

Acknowledgments: We would like to acknowledge the invaluable technical assistance provided by David Enyeart and Gerald Ramian at the Center for Terahertz Science and Technology.

7 References

1. We ignore here the developments in high-speed transistors. For example, Rodwell, M., Lee, Q., Mensa, D., Guthrie, J., Martin, S.C., Smith, R.P., Pullela, R., Agarwal, B., Jaganathan, S., Mathew, T., and Long, S, (1999) Transferred-substrate HBT integrated circuits, *Solid-State Electr.* **43**, 1489-1495.
2. *Quantum cascade lasers*: Tredicucci, A., Gmachl, C., Capasso, F., Sivco, D.L., Hutchinson, A.L., and Cho, A.Y. (1999) Long wavelength superlattice quantum cascade lasers at lambda approximately=17 μm, *Appl. Phys. Lett.,* **74**, 638-40; Capasso, F., Gmachi, C., Kohler, R., Paiella, R., Tredicucci, A., Hutchinson, A.L., Sivco, D.L., Baillargeon, J.N., Cho, A.Y., and Liu, H.C., (2000) High performance quantum cascade lasers for the lambda =4 to 17 μm region and their chemical sensing applications in *Conference Proceedings. 2000 International Conference on Indium Phosphide and Related Materials*, IEEE, Piscataway, NJ, USA, pp.262-265.
   Photo-mixer: Verghese, S., McIntosh, K.A., and Brown, E.R. (1997) Optical and terahertz power limits in the low-temperature-grown GaAs photomixers, *Appl. Phys. Lett.*, 2743-2745.
   Harmonic generation: T. Crowe, this volume.
   Resonant tunneling diodes (RTD's): Brown, E.R., Soderstrom, J.R., Parker, C.D., Mahoney, L.J., Molvar, K.M., and McGill, T.C. (1991) Oscillations up to 712 GHz in InAs/AlSb resonant-tunneling diodes, *Appl. Phys. Lett.*, **58**, 2291-2293.
   RTD arrays: Reddy, M.; Martin, S.C.; Molnar, A.C.; Muller, R.E.; Smith, R.P.; Siegel, P.H.; Mondry, M.J., Rodwell, M.J.W., Kroemer, H., and Allen, S.J., Jr.

(1997) Monolithic Schottky-collector resonant tunnel diode oscillator arrays to 650 GHz, *IEEE Electr. Dev. Lett.* **18**, 218-221.
Superlattice electron device (SLED): Schomburg, E., Scheuerer, R., Brandl, S., Renk, K.F., Pavel'ev, D.G., Koschurinov, Yu., Ustinov, V., Zhukov, A., Kovsh, A., Kop'ev, P.S. (1999) InGaAs/InAlAs superlattice oscillator at 147 GHz, *Electr. Lett.* **35**, 1491-1492.

3. Gronqvist, H., Kollberg, E., and Rydberg, A. (1991) Quantum-well and quantum-barrier diodes for generating submillimeter wave power, *Micro. and Opt. Tech. Lett.* **4**, 33-38.
4. Tucker, J.R. (1979) Quantum limited detection in tunnel junction mixers. *IEEE J. Quant. Electr.* **QE-15**, 1234-1258.
5. Ktitorov, S.A., Simin, G.S., and Sindalovski, V.Y. (1971) Bragg reflections and the high-frequency conductivity of an electronic solid-state plasma, *Fizika Tverdogo Tela* **13**, 2230-2233 [Soviet physics – Solid State **13**, 1872-1874 (1972)]
6. Neikirk, D.P., Rutledge, D.B., Muha, M.S., Park, H., and Chang-Xuan Yu (1982) Far-infrared imaging antenna arrays, *Appl. Phys. Lett.* **40**, 203-205.
7. Allen, S.J., Craig, K., Felix, C.L., Guimaraes, P., Heyman, J.N., Kaminski, J.P., Keay, B.J., Markelz, A.G., Ramian, G., Scott, J.S., Sherwin, M.S., Campman, K.L., Hopkins, P.F., Gossard, A.C., Chow, D., Lui, M., and Liu, T.Y. (1994) Probing tetrahertz dynamics in semiconductor nanostructures with the UCSB free-electron lasers, *J. Luminescence* **60-61**, 250-255.
8. Esaki, L and Tsu, R., (1970) Superlattice and negative differential conductivity in semiconductors, *IBM J. Res. Dev.* **40**, 61-65.
9. Shockley, W., (1954) Negative resistance arising from transit time in semiconductor diodes, *Bell Syst. Tech. J.* **33**, 799-826.
10. Kroemer, H., (1970) Generalized Proof of Shockley's Positive Conductance Theorem, *Proc. IEEE* **58**, 1844-1845.

SEMICONDUCTOR TERAHERTZ EMITTERS

N. N. ZINOV'EV
Institute of Microwaves and Photonics,
School of Electronic and Electrical Engineering,
University of Leeds, Leeds LS2 9JT, UK

AND

Ioffe Physical Technical Institute,
194021 St Petersburg, Russia

1. Introduction

Far-Infared (FIR) spectroscopy has, for many years, been a powerful tool for the study of the chemical and structural properties of Condensed Matter. Fourier Transform Far-Infrared (FTIR) spectroscopy has provided clear proof of the importance of Terahertz frequency radiation in detecting the spectral signatures of elementary excitations in solids, liquids, gases and even in more complex materials such as DNA and human tissue. The terahertz sources used in this work to date have all been based on blackbody radiation. Other sources such the Free Electron Laser (FEL), optically pumped FIR gas lasers, and semiconducting p-Ge lasers have become available but their widespread use in THz technology is limited by problems of large size, tunability or the necessity of low (liquid He) operating temperatures. These drawbacks have seriously limited the use of these sources, not only in FTIR, but also in new and important imaging and spectroscopical applications in the biomedical sciences, surveillance technologies, industrial process control and communications. The lack of a compact, powerful and tunable THz source has significantly restricted applications in this area of optics. The absence of a compact THz source has created what has come to be known as the "Terahertz Gap" viz, the band of frequencies between 0.1 and 100 THz where it is difficult to generate power.

The past decade has seen a surge in research into the physics of THz sources, mainly for use in imaging systems. This has been facilitated by recent advances in ultra-fast and cw-lasers that have made possible the

R.E. Miles et al. (eds.), Terahertz Sources and Systems, 15–24.

realization of non-linear optical approaches to the down-conversion of visible or near-infrared radiation into the THz Gap. Optical rectification in a non-linear crystal or the pulsed illumination of photoconductive antennas produces a freely propagating transient electric field[2, 3, 4]. The resulting availability of coherent broad band THz radiation has had an enormous impact on developments in both THz spectroscopy and THz imaging. These novel techniques have renewed the interest in developing a compact condensed matter based source of THz emission. Similarly, the quest for a solid state THz source has also been motivated by the development of mid-infrared quantum cascade lasers [7] based on the pioneering ideas of [5] to tailor the inter-subband transitions in a superlattice or multi-quantum well (QW) heterostructure to a particular emission band.

The paper comprises the following parts. The first part discusses some key problems of Semiconducting THz Emitters. The second part addresses non-linear optical generation techniques and the related issues of the detection of THz transients. This is followed by a comparison of the achievements of the two approaches. In the final part, two important examples of the applications of THz sources to (a) spectroscopic imaging and (b) micro-imaging using near field optics are considered.

2. Electrically Pumped Semiconducting Terahertz Emitters

The original concept of the inter-subband optical emitter was suggested by Kazarinov and Suris and published in their seminal paper[5] soon after the discovery of superlattices (SL) by Esaki and Tsu[6]. Their concept

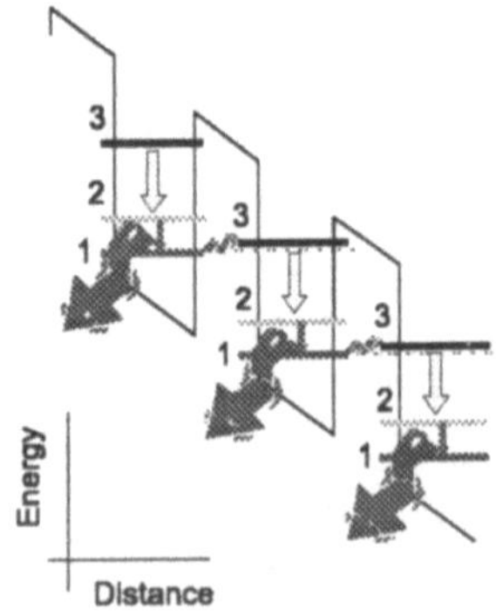

Figure 1. Transitions in a multi quantum well/superlattice structure accordingly to [5].

relied upon the following key features (Fig. 1): the energy levels of electrons in quantum wells (1,2,3) depend strongly on layer thickness; photons are created as electrons jump down to lower quantized levels in the QW, e.g.

level 2 to 1; many photons are created by a single electron cascading down through the quantum wells.

This novel unipolar semiconducting laser was first suggested by [5]. In order to obtain effective emission from a Quantum Cascade Structures (QCS), a few crucial problems have been solved[7]. These are summarized in the following Fig. 2:

1. Under operating conditions each "active region", be it quantum well or barrier, must have no net charge. Otherwise space charge accumulation associated with current flow will either detune or even destroy the scheme of the energy levels. The introduction of a reservoir of positive fixed charge to compensate for the space charge of flowing electrons was a key innovative feature that led to the successful implementation of the unipolar laser.
2. This reservoir is also used to implement two other important functions: (a)To suppress electron leakage out of the topmost level into the reservoir, an electronic Bragg filter formed by a doped SL reduces unwanted tunneling from the upper subband into the reservoir. (b) The doped SL region makes for more efficient de-population of the ground state of the "active region". A digitally doped or graded structure can creat a built-in electric field which, under certain conditions of applied bias, flattens the potential distribution: this makes for very efficient de-population of the ground states by means of miniband transport.

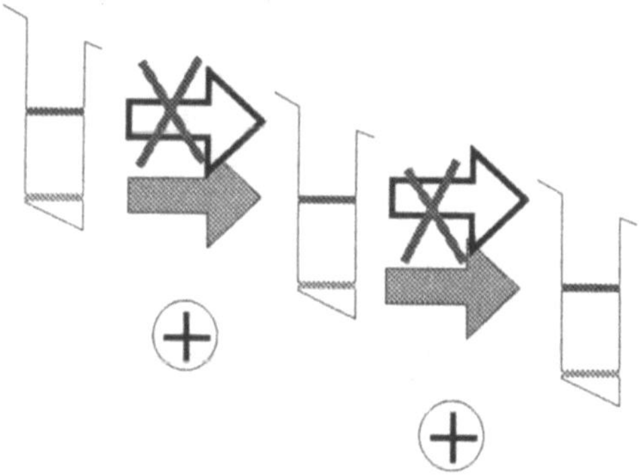

Figure 2. The scheme of quantum cascade structure.

Practical realisations of the QCS [7] have resulted in semiconducting lasers operating at wavelengths up to 20μm i.e. 15 THz [8]. Calculations suggest that a properly engineered QCS can be made to generate radiation in region of $\sim 1-15$ THz frequencies. An estimate of the optical gain, $g(\omega)$, obtained from inter-subband transitions can be made using the following relationship

[9]:

$$g(\omega) = \frac{(n_2 - n_1)}{d} \frac{c^2}{\epsilon_\infty \tau_R \omega^2} \frac{\tau_{21}}{[1 + (\omega - \omega_0)^2 \tau_{21}^2]} \quad (1)$$

where n_1 and n_2 are the 2D-populations per unit area of subbands 1 and 2 respectively, d is the thickness of the QW, τ_R is the spontaneous radiative lifetime for optical transition between levels 2 and 1, $\omega_0 = (E_2 - E_1)/\hbar$ and τ_{21} is the inter-subband relaxation (non-radiative) lifetime. Using (1) the peak optical gain, $g(\omega_0)$, is estimated to be in the region of $10^3 - 10^4 \text{cm}^{-1}$ for the AlGaAs/GaAs heterostructure whose $\tau_{21} \sim 10^{-13}$s, $\tau_R \sim 10^{-4}$s. The values of optical gain achievable will however depend strongly on intra- and inter-subband electron kinetics and (1) is not bound to any particular scattering mechanism responsible for τ_{21}.

The first successful attempt at THz emission generated by current pumping was realized in a SL sample [10]. These experiments went far beyond simple ohmic heating of the quantized system and gave clear evidence of current pumping of electron states, though no stimulated emission was achieved. In a recent paper[11] spontaneous THz emission has been reported for the first time in a AlGaAs/GaAs QCS at a power level of several pico-Watts.

Another type of QCS, not yet demonstrated as a THz emitter, relies on type-II Quantum Wells[12]. In spite of the fact that the growth of these systems is far more difficult than for conventional GaAs/AlGaAs structures, they could address some of the drawbacks of the type-I QCS. Firstly, τ_R is higher for conduction-to-valence band radiative emission. Secondly, the Auger recombination rate is much lower in type-II QWs[13, 14] because of the reduced in-plane hole effective mass and the removal of the resonance between the energy gap and inter-valence band transitions. Thirdly, LO-phonon relaxation is totally eliminated. Further advantages over type-I QCS include larger conduction and valence bands offsets and larger densities of states (DOS) for both electrons and holes.

In spite of the encouraging achievements in obtaining THz emission from electrically pumped semiconducting structures [10, 11, 15] the low value of external quantum yield remains as a key problem. It would appear that this relates to subtle peculiarities of electron kinetics in QCS designed for THz operation. In order to assess the problems of THz QCS implementation, consider a two level system describing nonequilibrium (radiative and nonradiative) processes. Current flow, J, results in a non-equilibrium population of the topmost level, n_2. This excess carrier population relaxes predominantly via non-radiative channels (Auger electron-electron scattering) but a very small amount occurs via the emission of photons. Tunneling into the collector contact depopulates the lowest level at a rate given by $1/\tau_{tun}$ and maintains the population inversion in the lasing regime. The kinetics of carriers in the two-level system are governed by the rate equations

$$\frac{dn_2}{dt} = J - \frac{n_2}{\tau_{21}} - \frac{n_2}{\tau_R} \tag{2}$$

$$\frac{dn_1}{dt} = \frac{n_2}{\tau_{21}} + \frac{n_2}{\tau_R} - \frac{n_1}{\tau_{tun}} \tag{3}$$

The steady state electron population of the first subband is determined by the relationship: $n_1 = J\tau_{tun}$. This basic relationship holds for all wavelengths. The ratio of the populations for the two levels is obtained from

$$\frac{n_1}{n_2} = \frac{\dfrac{\tau_{tun}}{\tau_{21}}}{1 - \dfrac{n_2}{\tau_R J}} \tag{4}$$

Eq. (4) shows that the ratio n_1/n_2 is a complex, implicit function of the injection current J. τ_{21} may also depend on J if the mechanism of intersubband relaxation is due to electron-electron scattering. In a structure designed for THz emission this factor becomes very important. In the THz frequency range, the characteristic energies are less than the optical phonon energy. Thus scattering due to optical phonons between the bottoms of the subbands is greatly reduced. However another important factor operating against a high nonequilibrium population appears at this point. The rate of electron-electron scattering increases in the THz range because of the increased likelihood of transfer of inter subband excess energy into electron momentum, according to the calculations of [13, 14, 16, 17, 18]. In spontaneous regime the parameter $n_2/\tau_R \ll J$. When lasing occurs n_2/τ_R is transformed into, it is transformed into $(g(\omega)c/\epsilon_\infty)\Phi_{ph}$, with Φ_{ph} as the photon flux. Below the laser threshold, the population factor (n_1/n_2) is determined solely by the ratio of τ_{tun}/τ_{21}. Hence to achieve a good working QCS the hierarchy of the lifetimes must satisfy the following relationship - $\tau_{tun} \ll \tau_{21}$, irrespective of the value of the radiative lifetime, τ_R, is. However, the threshold for lasing, does depend on τ_R through Eq. (1) when the gain achieved at a threshold current, g_{th}, compensates the losses occurring in the structure. Applying the standard laser threshold equation we obtain

$$N\frac{d}{t_m}g_{th} = N\frac{d}{t_m}\alpha_{ar} + \frac{t_{cont}}{t_m}\alpha_{cont} + \alpha_{ph} + \frac{1}{L}ln\left(\frac{1}{R}\right) \tag{5}$$

Here N is the number of active regions/periods, $t_m = \lambda/2\overline{n}$ is the effective mode thickness, $\overline{n}$ is the average refractive index of the QCS, α_{ar}, α_{cont}, α_{ph} are the absorption coefficients for the active region, contacts and phonon absorption respectively, L is the length of the optical resonator, R is the reflectivity of the external facets. It is seen from Eq. (4) and (5) that although the quest to design a structure with a high τ_R is well justified the development of a QCS with the least possible value of τ_{tun}/τ_{21} is even more important as it is the corner-stone of an efficient THz emitter. Much effort must still be invested in the optimization of the complete structure that is, developing the THz resonator, waveguide and optical coupler to the active region. The emitter-collector contacts also require serious attention as free carrier absorption in these regions is one of the main loss factors due to $\propto \lambda^2$.

3. Non-Linear Optical Conversion - Coherent THz Transient Spectroscopy

THz emission can be obtained from a non-linear optical process which involves two frequencies ω_1 and ω_2 lying in the visible or near-infrared region of the spectrum. All possible mixing mechanisms are included in the non-linear susceptibility tensor, $\hat{\chi}$.

$$P = \hat{\chi}\mathbf{E} = \hat{\chi}^{(1)}\mathbf{E} + \hat{\chi}^{(2)}\mathbf{EE} + \hat{\chi}^{(3)}\mathbf{EEE} + ... \tag{6}$$

(6) can be reduced to the lowest order of non-linearity to give

$$\mathbf{P}^{(2)} = \hat{\chi}^{(2)}[\mathbf{E}(\omega_1) + \mathbf{E}(\omega_2)]^2. \tag{7}$$

(7)contains the second harmonics at $2\omega_1$ and $2\omega_2$, the sum frequency at $\omega_1 + \omega_2$ and the difference frequency, $\omega_1 - \omega_2$. The last term describes the generation of THz waves via optical mixing when ($\omega_1 \neq \omega_2$) or optical rectification when ($\omega_1 = \omega_2$). Although the theories of non-linear mixing and rectification were proposed in the 1960s by [19], the first experimental demonstration was not published until the 1970s[2], and only relatively recently at the end of the 1980s [3, 4] has it become possible to generate sufficient THz power for the important applications of imaging and spectroscopy. Over the last 20 years two main techniques to generate Terahertz frequencies have emerged.

One of the two techniques relies upon the optical rectification of ultra-short laser pulses. If the amplitude of the incident light wave varies rapidly (i.e. an ultra-short laser pulse) the the non-linear polarization also changes rapidly and generates a few cycles of a THz electric field[2, 4]. The magnitude of THz electric field can, neglecting spatial effects, be estimated from $\mathbf{E} \propto -\frac{\partial^2 \mathbf{P}}{\partial t^2}$. In this technique a broad spectrum, typically tens of THz, is radiated. Conversely the Pockels Effect can be used for free space electro-optic detection using a THz electric field as the bias [4, 22].

The second generation technique uses photoconductive dipole antennas[3]. By shining light onto the gap of a Hertzian dipole antenna filled with a photoconductor a transient current, $\mathbf{j}(t)$, is induced if the antenna is dc-biased. The transient current produces a THz frequency oscillating electric field $\mathbf{E} \propto \frac{\partial \mathbf{j}}{\partial t}$. The generation of THz light with photoconductive antennas (PCA) normally produces much higher output intensity but the bandwidth of the emitted THz radiation is much narrower than for optical rectification. A PCA can also be used in reverse as a sensitive THz detector as shown in Fig. 3. The spectrum of THz waves emitted or detected by a PCA is typically confined to below 5THz. However, with the use of very short pulses (e.g. $\sim$ 10fs) and proper design of the PCA, this technique can also deliver frequencies up to 20THz[20].

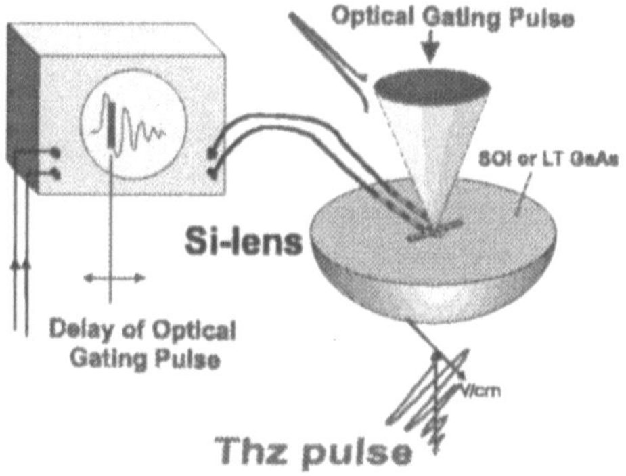

Figure 3. Photoconductive dipole antenna used as a detector.

THz radiation can be obtained by mixing two coherent laser beams of slightly different frequency. In this case a coherent THz beam at the difference frequency, $\omega_1 - \omega_2$ is created (Eq. 7). Coherent mixing requires the interacting laser beams to be of very narrow bandwidth with stringent requirements on amplitude and frequency stability. These conditions are more demanding than those required in the time domain mode of THz generation [21]. Photomixing techniques are therefore more suitable for the use with cw-laser beams.

The techniques of THz generation and detection mentioned above are widely used in THz imaging systems - Fig. 4. The incident radiation, delivered by a femtosecond Ti:sapphire laser is split into two beams to implement a *pump-probe* technique. One beam (the pump beam) is focused onto a photoconductive antenna used as the THz emitter. The emitted THz radiation, after interacting (transmission, reflection) with the object to be imaged is incident on the second antenna used as the THz detector. The THz electric field biases the PCA detector. The second or probe, beam, is used to gate the detector antenna by creating photocarriers. Time resolution is obtained by varying the optical path difference between the pump and probe beams. X-Y scanning of the object allows for spatial mapping of THz properties which can be in transmission (as shown in Fig. 4) or reflection depending on the experimental arrangement. The THz bandwidth can be widened by replacing the PCA-emitter with a piezo-material such as ZnTe. A flatter frequency responses obtained by replacing the PCA-detector with free space electro-optic polarization sensitive detection sampling[22]. The advantages

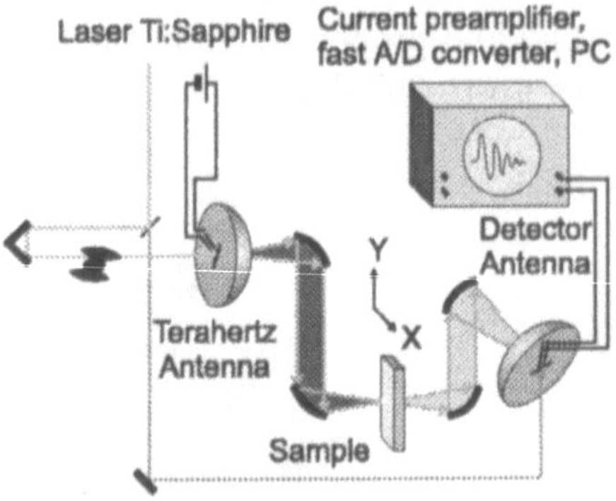

Figure 4. Imaging scheme with photoconductive antennas used as the emitter and detector

of the non-linear optical schemes lie in: coherent detection; ultra-short time resolution; image formation dynamics; high dynamic range and sensitivity. The reviews cited in [22, 23, 24] consider THz imaging applications more specifically and reveal the huge potential of this field.

4. THz Microscopy - THz Near Field Imaging

It is normally assumed that the diffraction limit of $\frac{\lambda}{2.3}$ is the fundamental limit to the resolution of any optical imaging system. There are, however, ways to obtain images without the use of conventional optical elements which allow this limit to be circumvented. Near field techniques allow us to break through the diffraction limit - see Fig. 5. To date a few attempts have been made to realise THz imaging beyond the diffraction limit[25, 26, 27]. However, the best resolution achieved was at a level of $\sim 100\mu$m i.e. $\geq \lambda/3$. Although it is not yet clear that the diffraction limit has been overcome the potential advantages of near field THz imaging, especially of cells and biological materials, means that this technique is now being pursued actively. Although a very low limit of resolution can be achieved

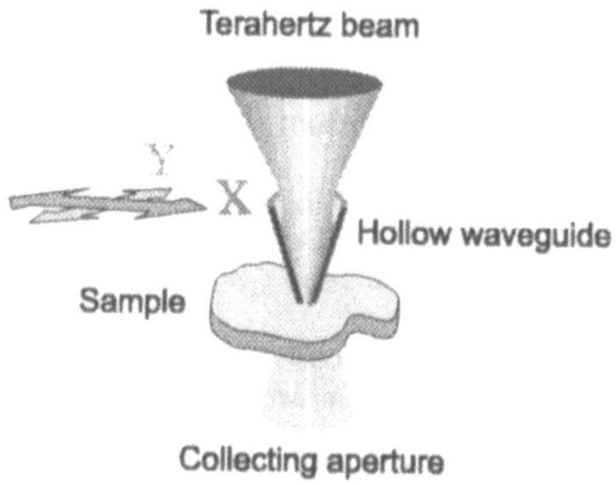

Figure 5. Near field optical setup with a hollow fibre

with the near field approach, there is a trade-off between the far- and near-field techniques. Near field intensities vary as $\propto R^{-3}$ whereas, for the far field, the intensity varies as $\propto R^{-1}$). High spatial resolution will therefore almost certainly come at the cost of image acquisition time.

5. Conclusions

To summarize we have reviewed the perspectives of semicinducting THz emitters taking the most characteristic examples: electrically pumped injection sources, coherent THz emitters and microscopic obbjectives. It is clearly seen that in spite of different levels of current development each approach complements the others in fulfilling the needs of various applications. Although THz coherent sources seem to be currently better suited to developing the applications concerning imaging and microscopy, the adventure of electrically pumped semicoinducting THz lasers and light emitting diods presents a prevailing challenge in portability and tunability which would lead to breakthroughs in a variety of THz applications.

References

1. Peter R. Griffiths and James A. de Haseth, *Fourier Transform Infrared Spectroscopy* in A Series of Monographs on Analytical Chemistry and Its Applications: Chemical Analysis, ed. by P.J. Elving, J.D. Winefordner, and I.M.Kolthoff, **83**, John Willey and Sons.
2. K. H. Yang, P. L. Richards, and Y. R. Shen, Appl. Phys. Lett. **19**, 320 (1971).
3. P. R. Smith, D. H. Auston and M. C. Nuss, *"Subpicosecond Photoconducting Dipole Antennas"*, IEEE J. Quantum Electronics, **24**, 255 (1988).
4. D. H. Auston and M. C. Nuss *Electrooptic Generation and Detection of Femtosecond Electrical Transients"*, IEEE J. Quantum Electronics, **24**, 184 (1988).
5. R. F. Kazarinov and R. A. Suris Fiz. Tekh. Poluprov. *"On a Possibility of Electromagnetic Waves Amplification in a Semiconductor with Superlattice"* **5**, 797 (1971); [Sov.Phys. - Semicond. **5**, 675 (1971)]Fiz. Tekh. Poluprov. **6**, 148 (1972)[Sov.Phys.

- Semicond. **6**, 120 (1972)].
6. L. Esaki and R. Tsu IBM J. Res. Dev. **14**, 61 (1970).
7. J. Faist, F. Capasso, D. L. Sivco, C. Sirtori, A. L. Hutchinson, A. F. Cho *"Quantum Cascade Laser"* Science, **264**, 553 (1994).
8. Alessandro Traveducci, this volume.
9. W. Heitler. The Quantum Theory of Radiation.
10. M. Helm, E. Colas, P. England, F. DeRosa, and S. J. Allen,Jr. *"Observation of Grating Induced Intersubband Emission from GaAs/AlGaAs Superlattices"* Appl. Phys. Lett. **53**, 1714 (1988).
11. M. Rochat, J. Faist, M. Beck, U. Oesterle and M. Ilegems *"Far-Infrared ($\lambda = 88\mu m$) Electroluminescence in a Quntum Cascade Structure"* Appl. Phys. Lett. **73**, 3724 (1998);J. Faist, this book.
12. J. R. Meyer, I. Vurgaftman, R. Q. Yang, and L. R. RamMohan *"Type-II and type-I intersubband lasers"* Electronic Letters, **32**, 45 (1996)
13. A. S. Polkovnikov and G. G. Zegrya *"Auger Recombination in Semiconductor Quantum Wells"* Phys. Rev. B **58**, 4039 (1998).
14. G. G. Zegrya and A. D. Andreev *"Mechanism of Suppression of Auger Recombination Processes in Type-II Heterostructures"* Appl. Phys. Lett. **67**, 2681 (1995).
15. J. Ulrich, R. Zobl, W. Schrenk, G. Strasser, K. Unterrainer, and E. Gornik, *"Terahertz quantum cascade structures: Intra- versus interwell transition"* Appl. Phys. Lett. **77**, 1928 (2000); K. Unterrainer, this book.
16. P. Hyldgaard and J. W. Wilkins, *"Electron-Electron Scattering in Far-Infrared Quantum Cascade Lasers"*, Phys. Rev. B **53**, 6889 (1996).
17. P. Harrison, R. W. Kelsall, K. Donovan, and P. Kinsler, *"Solid State Terahertz sources Using Quantum Well Intersubband Transitions"* IEEE Transactions on Microwave Theory and Technology **48**, 645 (2000).
18. P. Kinsler, P. Harrison, R. W. Kelsall, *"Intersubband Electron-Electron Scattering in Assymetric Quantum Wells Designed for Far Infrared Emission"* Phys. Rev. B **58**, 4774 (1998).
19. G. A. Askaryan *"Cherenkov Radiation and Transition Radiation from Electromagnetic Waves"* Zh. Eksp. Teor. Fiz. **42**, 1360 (1962)(Sov. Phys. JETP **15**, 943 (1962).
20. S. Kono, M. Tani, P. Gu, and K. Sakai *"Ultra Broadband Detection of Terahertz Radiation up to 20 Thz with a LT-GaAs Photoconductive Antenna Gated by 15fs Laser Pulse"* Proceedings of the 8th Int. Conf. on Terahertz Electronics, 28 Sep 2000, Darmstadt, p.87.
21. M. Tani, P. Gu, M. Hyodo, K. Sakai, and T. Hidaka *"Generation of Coherent Terahertz Radiation by Photmixing of Dual-Mode Lasers"* Optics and Quantum Electronics **32**, 503 (2000).
22. Q. Wu and X.-C. Zhang *"Free Space Electro-Optic Sampling of Terahertz Beams"* Appl. Phys. Lett. **67**, 3523 (1995); X.-C. Zhang, this volume.
23. D. Arnone, this volume.
24. P Haring, this volume.
25. S. Hunsche, M. Koch, I. Brener, and M. C. Nuss *"Thz Near-Field Imaging"* Opt. Commun. **150**, 22 (1998).
26. Q. Chen, Z. Jiang, G. X. Xu, and X.-C. Zhang *"Near-Field Terahertz Imaging with a Dynamic Aperture"* Opt. Letters **25**, 1122 (2000).
27. O. Mitrofanov, I. Brener, M. C, Wanke, R. R. Ruel, J. D. Wynn, A. J. Bruce, and J. Federici *"Near-Field Microscope Probe for Far Infrared Time Domain Measurements"* Appl. Phys. Lett. **77**, 591 (2000).

Theme 2

Generation of Terahertz by Multiplication

HIGH EFFICIENCY HBV MULTIPLIERS FOR MILLIMETRE WAVE GENERATION

L. DILLNER, M. INGVARSON, E. KOLLBERG, AND J. STAKE
Microwave Electronics Laboratory, Chalmers University of Technology
SE-412 96 Göteborg, Sweden

1. Introduction

The millimetre and submillimetre wavelength range (0.03-3 THz) has for a long time been used in spectroscopy since it is very rich in atomic, ionic and molecular lines. Important research applications include the study of the earth's atmosphere (e.g. to study the development of the ozone layer), radio astronomy and plasma diagnostics. Millimetre waves are now also used in commercial applications, e.g. radar for cars and radio transmission links, and it is likely that the upper frequency limit for mass applications will increase with improved future technology.

A major problem at these frequencies is the lack of convenient power sources. An oscillator converts DC-power directly to RF. Typical negative resistance devices for millimetre wave oscillators are impatt diodes, Gunn diodes and transistors. Shrinking dimensions and decreasing conversion efficiencies create heating problems at high frequencies. This results in the output power of oscillators decreasing rapidly with the frequency. Therefore, at frequencies above about 150 GHz, frequency multipliers are usually preferred to produce power.

A varactor is usually used as a non-linear element in multipliers at high frequencies since the conversion losses are much higher in varistors [1]. A commonly used varactor is a reverse biased Schottky diode, which produces both even and odd harmonics. For high conversion efficiency, currents at all harmonics up to the output frequency must be optimised. This makes it very difficult to realise a higher order multiplier using a Schottky diode. Nevertheless, high efficiencies and output powers have been obtained for Schottky frequency multipliers, especially for frequency doublers [2], [3].

In recent years, new varactor structures with C(V) characteristics of even symmetry have been proposed, the heterostructure barrier varactor (HBV) [4], the Schottky/ 2-DEG/ Schottky diode [5], and the back-to-back Barrier-N-layer-N+ diode (bbBNN) [6]. These varactors will only produce odd harmonics $(2n+1)\cdot f_p$ when pumped with power at the frequency f_p and require no bias circuit. The absence of even harmonics simplifies the realisation of higher order multiplier circuits. For the tripler case, it is possible to realise a multiplier circuit considering only the pump frequency, f_p, and the output frequency, $3\cdot f_p$. This simplification of the circuit will also make it possible to realise a frequency

R.E. Miles et al. (eds.), Terahertz Sources and Systems, 27–52.

multiplier able to operate over a wider frequency range [7]. Furthermore, the HBV has the advantageous ability of epitaxial stacking, which greatly increases its power-handling capacity [8], [9], [10].

Even in the first multiplier experiment, the HBV was shown to be capable of producing power in the upper millimetre wave range but the output power was relatively low, 2 mW at 221 GHz [11]. In recent years, the output powers from HBV multipliers have been improved considerably to 9 mW at 248 GHz [12].

In this paper, the basic principles of varactor frequency multipliers are described and different material systems for HBVs are investigated. Theoretical HBV models that can easily be used in large signal simulations are presented. A new fabrication process where an InP substrate is replaced with a copper substrate is described. This process improves the heat-sink and hence the power-handling capacity considerably (Chapter 2). Efficiency limitations of frequency triplers and frequency quintuplers using symmetric varactors are investigated (Chapter 3). Finally, results from HBV frequency multiplier measurements are presented (Chapter 4).

2. The Heterostructure Barrier Varactor

The heterostructure barrier varactor is a symmetric layered device consisting of a high bandgap semiconductor sandwiched between two low bandgap semiconductors, see Figure 1. The high bandgap material (barrier) prevents electron transport through the structure.

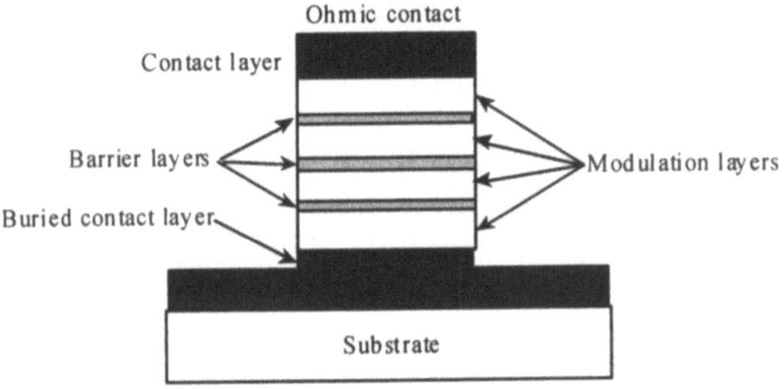

Figure 1. Epitaxial layer structure of a three-barrier HBV.

When the structure is biased, electrons are accumulated at one side of the barrier and depleted at the other side, Figure 2. Since the thickness of the depleted region is bias-dependent, the device capacitance changes with the bias, C(V). Electrons can, however, cross the barrier. Thermally excited electrons pass over the barrier or, when the barrier is thin enough, through the barrier by tunnelling. When this conduction current, I(V), becomes too large, it will degrade the multiplier efficiency. The structure can be grown with several barriers epitixially stacked in series which increases the break-down voltage and, therefore, also its power-handling capacity.

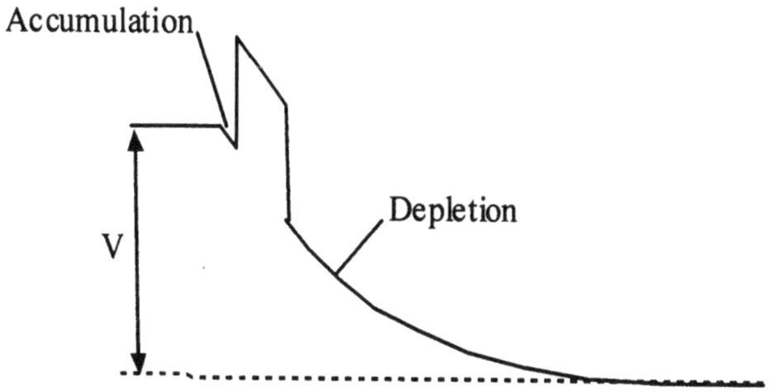

Figure 2. Conduction band of a biased single barrier HBV.

2.1 VARACTOR FREQUENCY MULTIPLIERS

The basic principle of frequency multipliers is as follows: a large AC signal is applied over a non-linear device, in this case an HBV diode, which generates harmonics. One of the harmonics is extracted to the output via a band-pass filter.

Passive frequency multipliers can use either a *varistor* (a variable resistance) or a *varactor* (a variable reactance) as the non-linear device. Varistor frequency multipliers can be broadband, but suffer from large conversion losses. It can be shown that the theoretical efficiency limit for a varistor frequency multiplier is $1/n^2$, where n is the multiplication factor [13]. In contrast, varactor frequency multipliers are generally less broadband but much more efficient. With no parasitic resistance and an ideal embedding circuit, they are can convert power without losses. This is borne out in practice where the losses are much smaller in the varactor case. Varactor frequency multipliers are therefore generally preferred at high frequencies where the input power and diode temperature are limiting factors.

A practical HBV diode is not an ideal varactor. The ohmic contacts and scattering in the bulk of the semiconductor produce losses which are represented by a series resistance, R_S. Moreover, there is current transport through the barrier due to thermionic emission and tunnelling, which is represented by a voltage dependent conductivity, $G(V)$. Figure 3 shows an equivalent circuit of an HBV diode where $S(V)=1/C(V)=dV/dQ$ is the differential elastance.

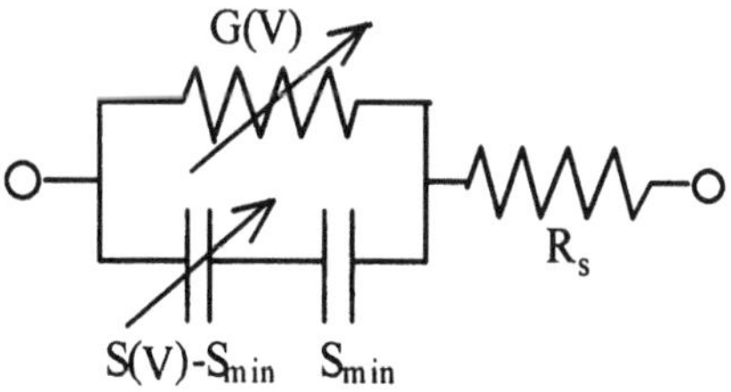

Figure 3. Equivalent circuit of a varactor diode.

The elastance can be visualised as consisting of a constant part, S_{min}, in series with a charge-dependent part increasing from zero to S_{max}-S_{min}, Figure 3. Since a constant elastance can be tuned out with a series inductance, the maximum varactor multiplier efficiency is dependent only on the non-linear part of the elastance and therefore independent of S_{min}. For a specific shape of the $S(V)$ characteristics, the maximum efficiency of a frequency multiplier depends on the ratio between the series resistance, R_s, and the impedance related to the non-linear part of the elastance, $(S_{max}-S_{min})/(2\pi f_p)$. S_{max} and S_{min} are the maximum and the minimum differential elastance during a pump cycle. This ratio can be written as

$$\frac{R_s}{(S_{\max}-S_{\min})/(2\pi f_p)}=\frac{f_p}{f_c} \tag{1}$$

where f_p is the pump frequency and f_c the dynamic cut-off frequency [1] defined as

$$f_c=\frac{S_{\max}-S_{\min}}{2\pi R_s}. \tag{2}$$

The dynamic cut-off frequency of a typical HBV is of the order of 1 THz.

2.2 HBV MATERIAL SYSTEMS

HBV materials are grown either by molecular beam epitaxy (MBE) or metalorganic vapour phase epitaxy (MOVPE). A general problem is that the performance of the HBV diode is strongly dependent on the material quality. Several different material systems have been proposed and fabricated since the invention of the HBV diode. Developments include improving the break-down voltage, decreasing the series resistance and decreasing the conductance current in order to improve the cut-off frequency and the power-handling capacity. If impact ionisation is the limiting break-down mechanism, the break-down voltage can roughly be estimated as

$$V_{d,\max}=60\left(\frac{E_g}{1.1}\right)^{3/2}\left(\frac{N_d}{10^{16}}\right)^{-3/4} \tag{3}$$

where $V_{d,max}$ is the avalanche voltage for a PN junction, E_g is the bandgap at room temperature, and N_d is the doping concentration [14]. However, if the barrier is not efficient, the conduction current will reach an unacceptable level at a much lower voltage than $V_{d,max}$.

The main constraint in the choice of materials is that the barrier material and the modulation layer material should have the same lattice constants to avoid dislocations. However, it is possible to grow dislocation-free layers with only a slight lattice mismatch if the layers are thinner than a critical thickness.

The doping concentration of the modulation layers is chosen to minimise the resistivity and to maximise the break-down voltage. Unfortunately, these demands are contradictory: low bandgap or high doping concentration gives a low resistivity but a low break-down voltage. A compromise has therefore to be made using a moderate doping concentration of about 10^{17} cm^{-3}. The thicknesses of the layers is chosen so that they are fully depleted at the maximum voltage over a pump cycle. Layers that are too thick add unnecessary series resistance and layers that are too thin decrease the maximum elastance S_{max} and therefore the cut-off frequency, Eq. (1).

A semiconductor with a large bandgap is chosen as the barrier material to achieve a large conduction band offset. This is of fundamental importance for minimising the conduction current. There are undoped spacer layers between the barrier and the modulation layers to avoid diffusion of dopands into the barrier. The contact layers are usually made of the same material as the modulation layer but with a higher doping concentration, ($\approx 10^{19}$ cm^{-3}), to minimise the resistivity and the ohmic contact resistance. The ohmic contact can be improved with a top layer of highly doped InAs closest to the contact metal [15]. For planar diodes, the buried contact layer should be thick, $\approx$4 μm, or else the resistance in this layer will greatly increase the total series resistance.

2.2.1 *GaAs/Al$_x$Ga$_{(1-x)}$As on GaAs*

The first HBV diode was fabricated in the GaAs/$Al_{0.7}Ga_{0.3}As$ material system grown on a GaAs substrate [11]. AlGaAs is used as a barrier material and GaAs is used in the modulation layers. GaAs/AlGaAs is well characterised and relatively simple to process. The drawback with this system is the large conduction current due to its low conduction band offset. Even when the aluminium content is high, the barrier remains low since for $Al_xGa_{(1-x)}As$ with $x > 0.45$, the X-valley is below the Γ-valley, i.e. the bandgap is indirect.

Table I: ITME 1596 HBV material structure

Layer	Material	Doping	Thickness
Contact	InAs	n^{++}	10 nm
Contact	$In_xGa_{1-x}As$ x=1->0.53	n^{++}	600 nm
Modulation	$In_{0.53}Ga_{0.47}As$	4.5×10^{16}cm^{-3}	400 nm
Barrier and spacer		i	
Modulation	$In_{0.53}Ga_{0.47}As$	4.5×10^{16}cm^{-3}	400 nm
Barrier and spacer		i	
Modulation	$In_{0.53}Ga_{0.47}As$	4.5×10^{16}cm^{-3}	400 nm
Barrier and spacer		i	
Modulation	$In_{0.53}Ga_{0.47}As$	4.5×10^{16}cm^{-3}	400 nm
Buried contact	$In_{0.53}Ga_{0.47}As$	n^{++}	2 μm
Substrate	InP	S.I.	

The barrier and spacer layers consist of 40 nm $In_{0.53}Ga_{0.47}As$, 5 nm $In_{0.52}Al_{0.48}As$, 3 nm AlAs, 5 nm $In_{0.52}Al_{0.48}As$ and 40 nm $In_{0.53}Ga_{0.47}As$.

Electrons scattered from the Γ-valley to the X-valley will therefore experience a bandgap lower than what is expected if only the Γ-valley is considered. Due to the low conduction band offset, the conduction current rapidly increases with the temperature. For high pump powers the diode temperature for GaAs/$Al_{0.7}Ga_{0.3}As$-HBVs goes up (self-heating), which results in an increased conduction current which leads to a lower efficiency. The current blocking can be improved if a thin AlAs layer is inserted in the middle of the barrier or by inserting InGaAs wells on each side of the barrier [16], [17].

2.2.2 *$In_{0.53}Ga_{0.47}As/Al_{0.48}In_{0.52}As$ on InP*

A very promising material system is $In_{0.53}Ga_{0.47}As/Al_{0.48}In_{0.52}As$ grown on an InP substrate. HBV diodes with very good current blocking characteristics have been fabricated in this material system [18], [19]. $In_{0.53}Ga_{0.47}As$ is a good compromise between high mobility and high break-down voltage and offers a large conduction band offset, especially if a layer of AlAs is inserted in the middle of the barrier. Drawbacks with this material system are that it is difficult to grow sufficiently thick epilayers with MBE and that InP substrates are more expensive than GaAs substrates. However, relatively thick, $\approx 4\mu m$, HBV material of good quality has recently been grown by MOVPE [20]. High multiplier efficiencies have been reported for HBVs in this material system [10], [12]. Table I shows an example of a state-of-the-art material grown by MOVPE.

2.2.3 *InAs/AlSb on InAs*

AlSb is slightly mismatched to InAs, but thin layers of AlSb, up to ≈ 140 Å, can be grown on InAs. This heterostructure offers a very large offset in the conduction band but a negative offset in the valance band and will therefore not block hole current. Therefore, a thin layer of AlAs should be inserted in the barrier. Very good current blocking characteristics together with the high mobility of InAs ($\approx 33\ 000$ cm^2/Vs) have attracted interest in this material system. However, experiments so far have been disappointing. The break-down voltage of InAs is low and experimental diodes have broken down before any noticeable change of capacitance. The absence of capacitance change indicates that there is a surface charge at the interface between InAs and AlSb.

2.2.4 *Metamorphic InGaAs/InAlAs on GaAs*

The metamorphic growth technique makes it possible to grow lattice mismatched dislocation free epilayers layers on a GaAs substrate. This is achieved by growing a buffer layer in InGaAs or InAlAs where the composition is gradually changed from GaAs to a material with the desired lattice constant. The thickness of this buffer layer is typically 1 μm. Using this technique, material systems with any indium content in the modulation layers are possible. This technology is relatively new and so far no successfully fabricated metamorphic HBVs have been reported.

2.2.5 *Si/SiO_2 on Si*

By bonding two thin silicon wafers with silicon dioxide on the bonding surfaces, a $Si/SiO_2/Si$ structure similar to HBV structures in III-V materials is achieved [21]. SiO_2 blocks the conduction current very efficiently but the mobility in silicon is low. The development of a method for stacking several barriers in series would make this material system interesting for lower frequencies where the series resistance is less critical.

2.2.6 *Summary*

$In_{0.53}Ga_{0.47}As/Al_{0.48}In_{0.52}As$ on InP seems to be the best material systems today. However, $GaAs/Al_xGa_{(1-x)}As$ on GaAs may be equally good if an efficient barrier can be developed. Further development could also make other material systems competitive, particularly for special applications such as very high frequencies or high power-handling capabilities. DC measurements show that GaAs-HBVs can handle much more current before break-down than InP-HBVs, which indicates that GaAs HBVs could be preferable for applications where reliability is more important than high efficiency.

2.3 HBV MODELS

Accurate models for the circuit elements in Figure 3 are needed in order to simulate the behaviour of a varactor diode accurately in a multiplier mount.

2.3.1 Capacitance-voltage models

If an abrupt space charge is assumed, the potential distribution across an HBV with homogeneously doped modulation layers can be calculated from Poisson's equation. The capacitance can be calculated as dQ/dV. The capacitance for zero bias, C_{max}, can be calculated using the theory for a surface space charge region in thermal equilibrium [14]

$$C_{\max} = \frac{A\varepsilon_b}{b + 2\frac{\varepsilon_b}{\varepsilon_d} L_d}, \quad \text{where} \quad L_d = \sqrt{\frac{kT\varepsilon_d}{N_d e^2}} \tag{4}$$

is the extrinsic Debye length for electrons. To model the screening effects for small bias, an empirical term is added to the solution for Poisson's equation, which gives the following voltage-charge model for an HBV diode:

$$V(Q) = N\left(\frac{bQ}{\varepsilon_b A} + 2\frac{sQ}{\varepsilon_d A} + \mathrm{sign}(Q)\left(\frac{Q^2}{2eN_d\varepsilon_d A^2} + \frac{4kT}{e}\left(1 - e^{-|Q|/(2L_d AeN_d)}\right)\right)\right) \tag{5}$$

where N is the number of barriers, b is the barrier thickness, s is the undoped spacer layer thickness, Q is the charge stored in the varactor, e is the elementary charge, A is the device area, N_d is the doping concentration in the modulation layer, ε_d is the dielectric constant in the modulation layer and ε_b is the dielectric constant in the barrier. Eq. (5) gives the C_{max} from Eq. (4) and that $dC/dV = 0$ for zero bias.

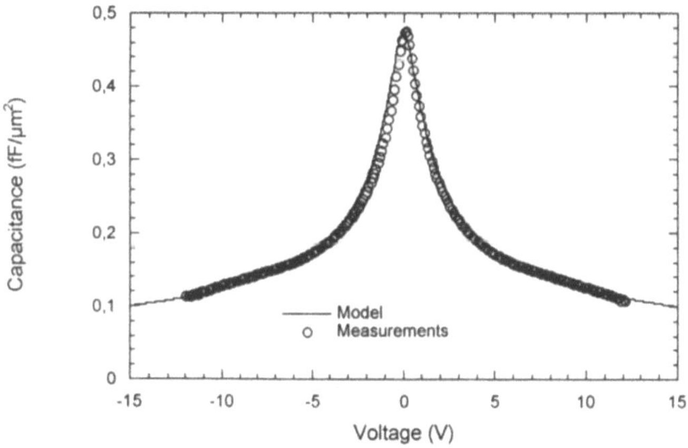

Figure 4. Capacitance-voltage characteristic generated from Eq. (5) compared with experimental capacitance-voltage characteristic measured at 1 MHz.

Figure 4 shows the capacitance-voltage characteristic for a 3-barrier HBV diode from the material ITME 1596. The epitaxial layer structure is shown in Table I. To fit the model to the measurements, the spacer layer thickness from the specifications (40 nm) was changed to 15 nm, which indicates that some silicon has diffused into the undoped spacer layers. This C(V) model is very convenient for multiplier simulations and can also be used for GaAs-HBVs [22].

A quantum mechanical screening effect makes the capacitance independent of the barrier thickness for thin barriers [23]. The maximum capacitance for very thin barriers is therefore smaller than predicted by Eq. (4).

Generally simple C(V) models are useful in a systematic analysis of varactor performances. The cubic varactor model introduced by Krishnamurthi [24] is based on a voltage-charge approach originally suggested by Tang [25]. In this model the voltage is given by a third order polynomial of the charge Q. The cubic model is often sufficient to analyse the performance limitations of frequency multipliers using symmetric varactors. The differential elastance in this model is uniquely defined from S_{min} to S_{max}. It is possible to modify the C(V) characteristic by modifying the doping profile, e.g. by introducing a δ-doping at the barrier or a position dependent doping profile in the modulation layers. In [26] this was investigated and it was found that a sharper C(V) yields a more efficient multiplier device.

2.3.2 *Current-voltage models*

A varactor is a much more efficient frequency multiplier than a varistor. Consequently, the conduction current in an HBV should be as low as possible. Since a considerable amount of power is dissipated in the diode, its temperature will increase. It is therefore not sufficient to characterise its conduction current at room temperature. The conduction current can be calculated by solving the Poisson equation and the Schrödinger equation [27], [28]. However, such calculations are time consuming and for multiplier simulations, it is convenient to use a simple analytical expression. The conduction current through an HBV from the material UVA-NRL-1174 is well described by the simple empirical model

$$J = aT^2 \sinh\left(\frac{E_b}{E_0}\right) e^{-\Phi_b / kT} \tag{6}$$

where E_b is the electric field in the barrier, T is the temperature in Kelvin, and a, E_0 and ϕ_b are constants [22]. The electric field in the barrier can be estimated from Poisson's equation to be

$$E_b = \frac{N_d e \varepsilon_d b}{\varepsilon_b^2}\left(\sqrt{1+\frac{2V\varepsilon_b^2}{N_d e \varepsilon_d b^2 N}} - 1\right) \tag{7}$$

where V is the voltage applied across the diode. Figure 5 shows measured and modelled conduction currents for GaAs/AlGaAs HBV diodes from the material UVA-NRL-1174 [29].

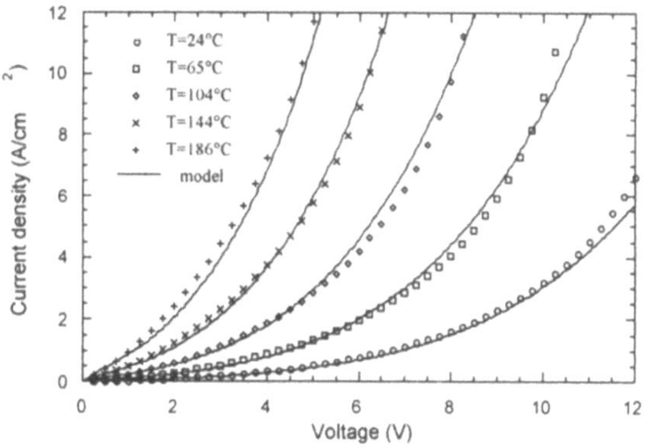

Figure 5. Measured and modelled conduction current density through a 4-barrier HBV diode (UVA-NRL-1174 with two diodes in series).

The conduction current is considerably smaller for the InGaAs/AlInAs material system, especially if a thin layer of AlAs is inserted in the middle of the barrier. In [27] it is shown that for the case with an AlAs layer, there is an optimum barrier thickness where the conduction current is a minimum. For a thin barrier, the conduction current decreases with increasing barrier thickness. This is because the tunnelling probability decreases with increasing barrier thickness. For a thick barrier with an AlAs layer, the conductance current increases with increasing barrier thickness, since for a biased diode, the barrier height ΔE decreases with increasing barrier thickness, see Figure 6.

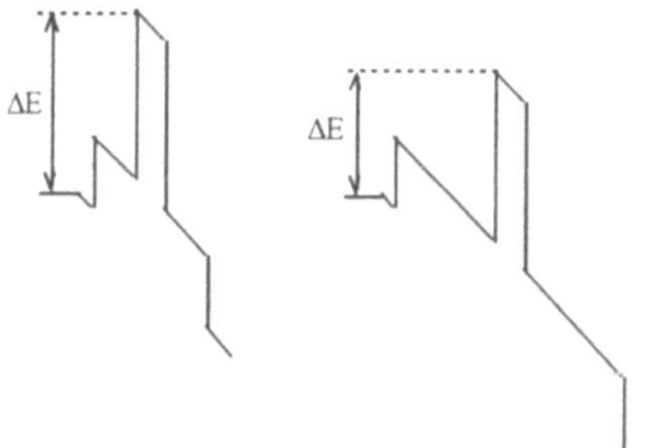

Figure 6. The conduction band for a biased barrier with an AlAs layer in the middle. The barrier height Δ E becomes smaller for a thicker barrier when the electric field is kept constant.

The conduction current through this kind of step-like HBV materials is not well described by Eq. (6). However, the conduction current through a step-like InGaAs/AlInAs HBV can be described by the simple empirical model

$$J = ae^{T/T_0}\sinh\left(\frac{V}{V_0}\right) \qquad (8)$$

where a, V_0 and T_0 are constants, T is the temperature in Kelvin, and V is the applied voltage. Figure 7 shows the measured and modelled conduction currents for HBV diodes from the material ITME 1596. The curves are calculated from Eq. (8) with a=2.2 A/(m^{2}K^2), T_0=74 K, and V_0=3.2 V.

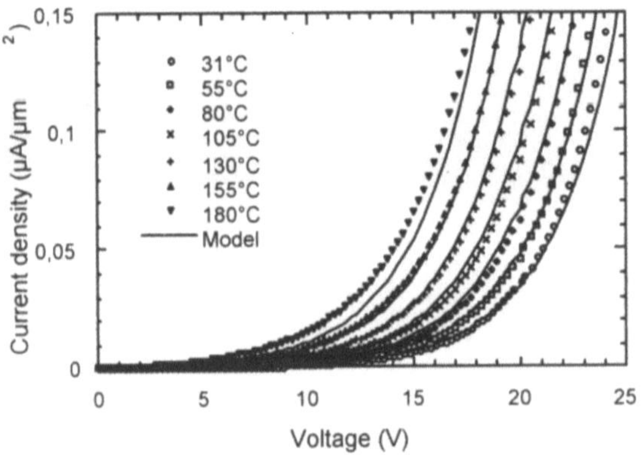

Figure 7. Measured and modelled conduction current density through a 3-barrier HBV diode (ITME 1596).

The modelling of the conduction current is less critical for this material system since the current density is small and will have much less influence on the multiplier efficiency. The conduction current for the two material systems are compared in Figure 8.

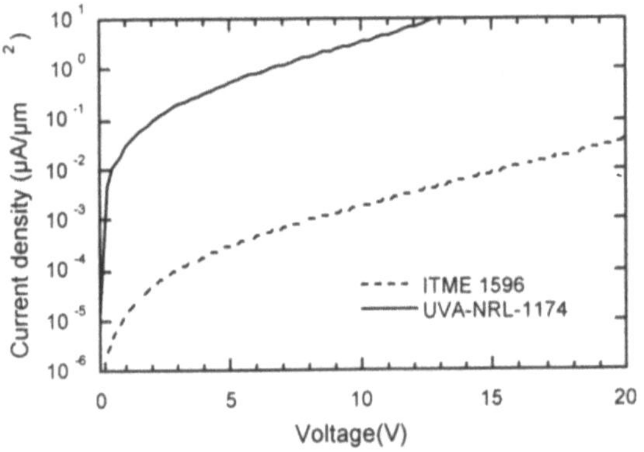

Figure 8. Measured conduction current density through a 3-barrier InGaAs/AlInAs HBV diode (ITME 1596) and a 4-barrier GaAs/AlGaAs HBV diode (UVA-NRL-1174 with two diodes in series) at room temperature.

2.3.3 The series resistance

For high cut-off frequency and high efficiency, the series resistance, R_S, should be as low as possible. Note that choosing a large diode area decreases the series resistance, but does not increase the cut-off frequency, since the elastance also decreases with the diode area. There are three main contributors to resistive losses in an HBV diode.

1. The resistance in the semiconductor layers in the mesa, especially in the modulation layers. This resistance can be decreased by avoiding unnecessarily thick semiconductor layers and by choosing a semiconductor with a high mobility or a high doping concentration. However, a high mobility or a high doping concentration results in a low break-down voltage and a low maximum elastance. The resistance of the modulation layers decreases when the thickness of the depleted regions increases.

However, this has only a small effect on the losses since the maximum of the displacement current occurs at zero bias. Both C(V) and dV/dt are a maximum at $V=0$.

2. The contact resistance. This resistance is related to the fabrication process, which makes the ohmic contact formation a critical fabrication step. It is important to choose proper contact layers; e.g. low contact resistance can be obtained with highly doped InAs closest to the contact metal. Note that the influence of the ohmic contact resistance is inversely proportional to the number of barriers.

3. The resistance in the buried contact layer. For planar diodes it is important that this layer is thick, preferably several microns. The diode geometry also has an impact on this resistance.

It is very difficult to measure R_S. For varactors that can be probed, R_S can be extracted from the reflection coefficients measured with a vector network analyser [30]. The series resistance can also be estimated from contact resistance measurements and resistivity data of the semiconductor. The resistivity of a semiconductor generally increases with temperature. For low doping concentrations the resistivity in GaAs is approximately proportional to the temperature [14].

The displacement current in a varactor diode must be matched with an electron current, i_e, through the undepleted region of the semiconductor:

$$i_e = AN_d e v_e \tag{9}$$

where v_e is the electron velocity. If the electrons are driven to high velocities, their kinetic energy can be sufficient to scatter a fraction of the electrons to a satellite valley. The electron mobility in the satellite valleys is considerably lower than in the lower (Γ) valley, which decreases the electron velocity. The electron velocity in a semiconductor can therefore not exceed a certain limit, v_{sat}, which is about $2 \cdot 10^5$ m/s for GaAs. This phenomenon, called current saturation, can for frequency multipliers be modelled with a current dependent series resistance:

$$R_s(i) = R_{s0}\left(1 + a(i_e)^6\right) = R_{s0}\left(1 + \left(\frac{v_e}{v_0}\right)^6\right) \tag{10}$$

where R_{s0} is the series resistance at small input powers and a and v_0 are constants [31]. A rapidly increasing series resistance results in a considerable decrease in the multiplier efficiency. The electron velocity can be calculated from the current waveform in Harmonic Balance simulations using Eq. (9). It can also be estimated from multiplier measurements as

$$P_{disp} = R_s \int_0^{\tau} i^2(t)dt \Big/ \tau = R_s i_{rms}^2 \Rightarrow v_{e,rms} = \sqrt{\frac{P_{disp}}{R_s}} \frac{1}{AN_d e} \tag{11}$$

where P_{disp} is the power dissipated in the series resistance of the diode. Eq. (11) can be used to evaluate R_s if the multiplier efficiency is degraded due to current saturation.

2.4 FABRICATION

The fabrication of an HBV diode depends on how the diode should be contacted. At long millimetre waves and lower frequencies, the diodes are large enough to be contacted with a bond wire. However, diodes for higher frequencies are generally too small to be contacted with a bond wire. One method for contacting a small varactor diode uses a planar topology involving air-bridges. An advantage with the planar topology is that it is mechanically robust.

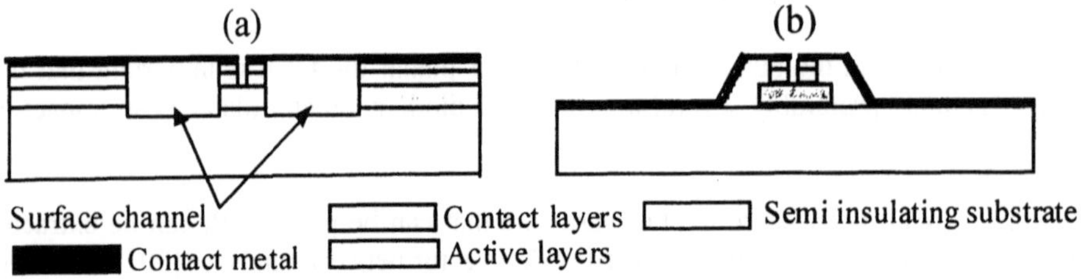

Figure 9. Two different planar topologies with HBV diodes in series configuration.

Figure 9 shows two different planar topologies. In Figure 9a the diodes are isolated by etching a surface channel. This topology, developed at the University of Virginia [32], [33], offers low parasitic capacitances due to the long air-bridge fingers. Figure 9b shows a topology where the contact pads sit directly on the semi-insulating substrate which may be more suitable for integrated circuits (MMIC). Drawbacks with a planar topology are that the buried contact layer must be thick to avoid large losses, problems with efficient heat transport and a complicated fabrication process.

The traditional way of contacting varactor diodes at submillimetre and short millimetre waves is with a whisker wire. In an array of hundreds of small circular diodes, one of the diodes is contacted with a sharp whisker wire. The other contact is on the back of the highly doped substrate and is soldered to the multiplier mount. Advantages with this method are that a very small diode area can be contacted, high cut-off frequencies can be obtained, and heat is efficiently transported away from the diode. Drawbacks are problems with mechanical reliability and that the method is not suitable for mass production or integration.

2.4.1 Fabrication of HBVs on a copper substrate

In a new HBV fabrication process, an InP substrate is replaced with a copper substrate [34], [35]. The benefits of a copper substrate are the excellent thermal and electrical conductivities. In this processes, the semiconductor substrate is removed before the device fabrication (post-processing). Advantages with the post-processing approach is a reduced risk of stress and alignment problems are avoided. One drawback is that some process parameters are influenced by the transfer of the epi-layers which makes the process development more complicated.

A Ti/Pt/Au ohmic contact is e-beam evaporated onto the top layer. Next, copper is electroplated to a thickness of about 50 μm and gold is electroplated on the copper for chemical protection. The InP substrate is etched away in a selective wet etch where the

InGaAs contact layer also serves as an etch stop layer. Note that the epitaxial layers are now stacked on copper in reversed order compared to the growth order on the original InP substrate.

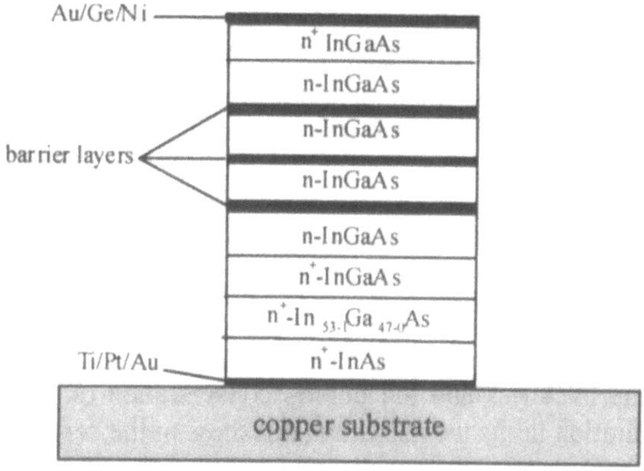

Figure 10. Schematic cross-section of an HBV diode fabricated on a copper substrate.

Diode mesas are then fabricated using standard photolithographic methods and e-beam evaporation of Ni/Ge/Au/Ti. The titanium serves as an etch mask. After annealing, diode mesas are etched using a reactive ion beam etch (RIBE) and a wet etch. To prevent the whisker from sliding off the mesa, some mechanical support is needed. We have tested two different methods. The first is to form a metal "cup" on each mesa. The cups were fabricated by photolithography, sputtering of titanium and gold, and lift-off. This method minimises the parasitic capacitance and dielectric losses. However, in the first attempt, the cups were too fragile. Moreover, it was difficult to contact the varactors correctly as the cups are small. The second method is to sputter a layer of SiO_2 onto the chip, and wet-etch openings for the whisker. This method resulted in very good mechanical support for the whisker but it is difficult to obtain a good electrical contact since after the annealing step, the top metal layer is an alloy and not pure gold. The electrical contact could be improved by depositing gold on top of the mesas after forming openings in the SiO_2.

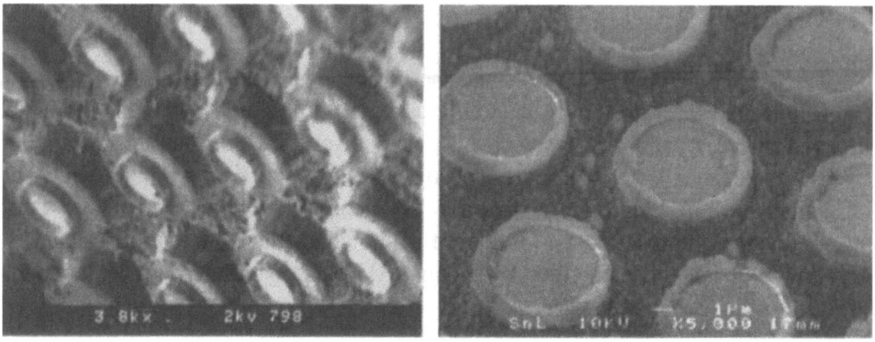

Figure 11. SEM pictures of 7 μm HBV diodes fabricated on copper substrates. Left: A gold 'cup' is formed on each mesa. Right: The chip is covered with SiO_2 with an opening on each mesa.

2.5 DESIGN OF HETEROSTRUCTURE BARRIER VARACTORS

When designing HBVs, the main goal is to maximise the efficiency for a certain input power, input frequency and multiplication factor [36]. Maximising the efficiency is closely related to maximising the cut-off frequency, Eq. (2), see section 3. For predictions of the efficiency, the relevant S_{max} in this expression is not the elastance at the break-down voltage, since the maximum elastance over a pump cycle can be considerably less due to current saturation or limited input power. The design procedure can be described as follows.

1. Choose material system. The material system $In_{0.53}Ga_{0.47}As/Al_{0.48}In_{0.52}As$ on InP with a thin layer of AlAs in the middle of the barrier is recommended, see section 2.2.

2. Calculate the thickness and the doping concentration of the modulation layers. Current saturation limits the maximum thickness of the depleted region, l_{max}, to

$$l_{max} = \frac{v_{sat}}{4kf_p} \qquad (12)$$

where v_{sat} is the saturation velocity, f_p is the pump frequency and k is the ratio between the saturation velocity and the average electron speed during a half pump cycle [36]. From Harmonic Balance calculations for HBV frequency triplers, k is typically between 1.5 and 2. A thicker modulation layer will increase the series resistance without increasing the maximum elastance. The doping concentration is usually chosento be as high as possible to minimise the series resistance but not so high that the modulation layer cannot be fully depleted without exceeding the break-down voltage. The voltage over the depleted region can be calculated from Poisson's equation

$$V_d = \frac{N_d q l^2}{2\varepsilon_d} \qquad (13)$$

where V_d is the voltage over the depleted region and l is the thickness of the depleted region, and the break-down voltage can be estimated from Eq. (3). Note that l_{max} is not always the optimum thickness. Especially at moderate frequencies, a considerably higher cut-off frequency can be obtained for a thickness thinner than l_{max}.

3. Choose the number of barriers and the device area. It is possible to optimise the impedance level and power-handling capacity through the choice of the number of barriers, N, and the device area, A. The power is proportional to A and N, and the impedance level is proportional to N and inversely proportional to A. A high impedance design (high N and low A) generally results in a higher cut-off frequency due to a reduced influence of the contact and spreading resistance. A low impedance design (low N and high A) generally results in a lower thermal resistance (better heat-sink).

4. Finally, the design can be optimised with a Harmonic Balance simulator using the capacitance-voltage model, Eq. (5), and a series resistance estimated from material data and contact resistance measurements.

3. Analysis of symmetrical varactor multipliers

We will now discuss how the cut-off frequency, Eq. 2, influences the efficiency of the multiplier. A systematic investigation was presented in [26]. It is shown that a C(V) characteristic like the one shown in Figure 4 yields an efficiency versus f_{out}/f_c that is almost the same for a tripler and a quintupler, see Figure 12.

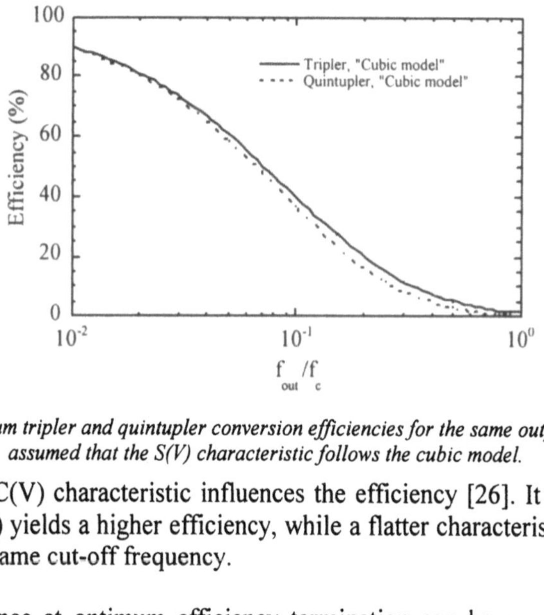

Figure 12. Maximum tripler and quintupler conversion efficiencies for the same output frequency. It is assumed that the S(V) characteristic follows the cubic model.

The shape of the C(V) characteristic influences the efficiency [26]. It was found that a more pointed C(V) yields a higher efficiency, while a flatter characteristic yields a lower efficiency for the same cut-off frequency.

The input impedance at optimum efficiency termination can be approximated by the following empirical expression

$$Z_{in} = \frac{(k_1 - jk_2)(S_{max} - S_{min}) - jS_{min}}{\omega_p} + (k_3 - jk_4)R_s \qquad (14)$$

and the optimum load impedances as

$$Z_n = \frac{(k_1 + jk_2)(S_{max} - S_{min}) + jS_{min}}{n\omega_p} + (k_3 + jk_4)R_s \qquad (15)$$

where the constants k_1 to k_4 are given in Table II. Note that for $f_p/f_c << 1$ the constants k_3 and k_4 can be neglected. For a tripler n=3 and for a quintupler n=5.

Table II: Coefficients for optimal impedances.

		k_1	k_2	k_3	k_4
Tripler	Z_{in}	0.046	0.21	0.78	0.25
Tripler	Z_3	0.28	0.35	0.55	0.08
Quintupler	Z_{in}	0.057	0.19	0.88	0.27
Quintupler	Z_3	0	0.35	0	0.29
Quintupler	Z_5	0.26	0.24	0.62	0.22

The embedding impedances at frequencies higher than the output frequency are optimally infinite.

For a quintupler, the idler impedance is important. If no current is allowed at the idler frequency (i.e. $Z_3=\infty$), the third harmonic will not cause any extra losses at all. However, it is very interesting to note that the efficiency is low, a fact which is due to the symmetric properties of the C(V). If the current at the idler frequency is maximised with an inductance in resonance with the diode capacitance at the idler frequency, $Z_3=jX_3$ where $S_{min}/(3\omega_p) < X_3 < S_{max}/(3\omega_p)$, maximum efficiency is obtained. Hence, the current at the third harmonic will, in combination with the input signal, increase the output power, although losses are produced in the series resistance.

It is interesting to compare quintuplers with triplers as power sources for the same output frequency. A varactor diode is usually dimensioned from the output impedance. Since the output impedances are approximately equal for triplers and quintuplers, triplers and quintuplers with the same diode capacitance can be compared. Figure 12 shows the maximum efficiency for triplers and quintuplers as a function of the output frequency. The maximum efficiencies are almost equal, except for output frequencies near the cut-off frequency. This indicates that quintuplers could be preferable since they use a lower input frequency where more powerful sources are available. The circuit is more complicated for quintuplers since the impedance at the idler frequency has to be optimised. A reasonable conclusion is that a quintupler should be able to reach a higher output power than a tripler if a close-to-optimal embedding circuit can be made.

In a practical varactor, conduction current is present. Conduction current can be modelled as a voltage dependent conductance in parallel with the elastance, Figure 3. The conductance increases with the bias, Figure 5. Therefore, the conductance reaches its maximum at maximum bias:

$$G_{\max} = \frac{I_{\max}}{V_{\max}} \tag{16}$$

Assuming the conductance to be constant, equal to its maximum I_{max}/V_{max}, an upper limit of the losses due to the conduction current can be estimated. Figure 13 shows the losses in a tripler including the conduction current for $f_p/f_c=0.1$ and for $f_p/f_c=0.01$. The S(V) characteristic is assumed to follow the cubic model with $S_{max}/S_{min}=3$. The calculations show that the loss due to the conduction current is less than 0.3 dB for $G/(\omega_p C_{min})<0.01$.

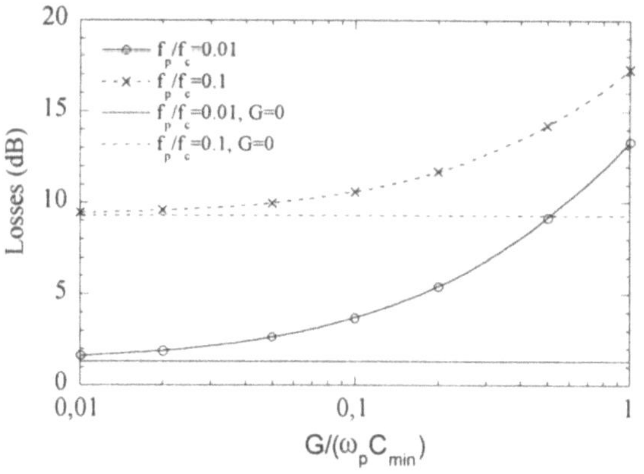

Figure 13. The tripler losses with f_p/f_c=0.1 and f_p/f_c=0.01, respectively, as a function of the conduction conductivity. The S(V) characteristic is assumed to follow the cubic model with S_{max}/S_{min}=3.

4. Frequency multiplier experiments

This section describes HBV multiplier measurements. In the efficiency simulations, a three-element quasi-static model is used, Figure 3. The capacitance voltage characteristic

Table III: Performance of some representative HBV frequency multipliers

Frequency (GHz)	Efficiency (%)	Output power (mW)	Reference	Comments
3x13.5	7.6	12	[37]	InP, 3 barriers
3x31	10	91	[10]	InP, 10 barriers
3x33	0.7	1250	[8]	GaAs,3 barriers Quasi optical
3x43.5	7	10	[38]	GaAs, NLTL
3x47	8	11.5	[39]	GaAs, 4 barriers, Quasi optical
5x34	0.78	0.029	[40]	InP, 1 barrier
3x62	2	0.80	[41]	GaAs, 1 barrier
3x70	2.8	1.4	[42]	GaAs,1 barriers
3x72	5.4	5	[12]	InP, 4 barriers
3x74	7.9	7.1	[35]	Cu, 3 barriers
3x74	5	2	[11]	GaAs, 1 barrier
3x82	4.8	4	[43]	GaAs, 4 barriers
3x82.5	12	9	[12]	InP,4 barriers
3x84	2.5	2	[33]	GaAs, 4 barriers

is calculated from Eq. (5) and the conduction current is calculated from Eq. (6) for GaAs/AlGaAs HBVs and Eq. (8) for InGaAs/InAlAs HBVs. The series resistance is assumed to be either constant or temperature-dependent as described in [29].

4.1 CROSSED WAVEGUIDE FREQUENCY MULTIPLIERS

The crossed waveguide configuration is a very common design for frequency multipliers in the millimetre and submillimetre wavelength range [44]. The widths of the crossed waveguides are chosen for the input and output frequency bands. The input signal frequency is below cut-off for the output waveguide which therefore acts as a high pass filter.

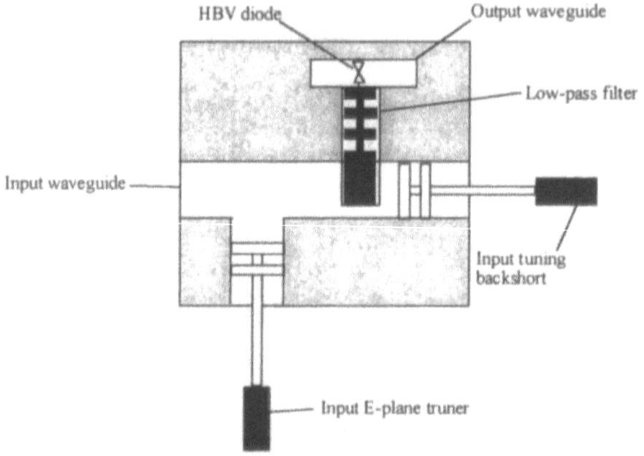

Figure 14. Cross section of an HBV crossed waveguide frequency tripler.

Figure 14 shows a cross section of a crossed waveguide frequency tripler with an HBV diode. The waveguides have movable short circuits to enable tuning of the embedding impedances, usually one or two at the output waveguide and two at the input waveguide. A drawback with tuneable short-circuits is that they add losses.

4.1.1 A planar four-barrier GaAs/AlGaAs HBV tripler

Four-barrier HBV diodes fabricated in the UVA-NRL-1174 material have been tested in a crossed waveguide frequency tripler [29]. The diodes were planar with a surface channel (Figure 9 a) and were fabricated by Jan Stake at the University of Virginia. The multiplier block was designed at the Rutherford Appleton Laboratory. Figure 15 shows the efficiency versus input power. It can be seen that the efficiencies saturate at about 50 mW input power. This is attributed to an increase in diode temperature when it is pumped (self-heating) and the conduction current increases rapidly with the temperature, see Figure 5. The temperature is assumed to increase by 2 K/mW of input power which is the temperature simulated by Robert Jones [45].

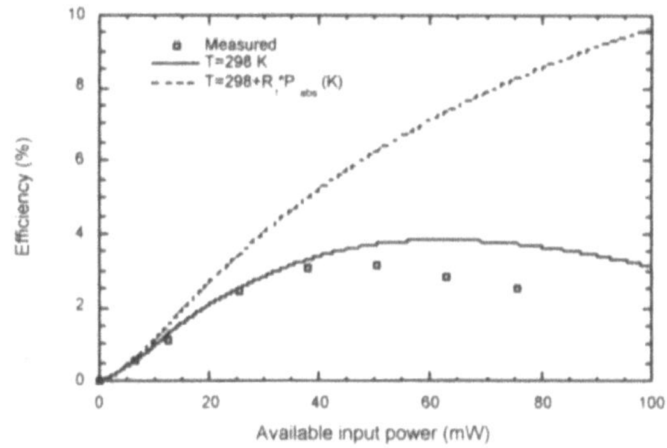

Figure 15. Measured and simulated efficiencies for a 3x87 GHz multiplier using a 57 μm^2 HBV diode. Measurements were performed at room temperature. For the simulations, an input loss of 1 dB and an output loss of 2 dB were assumed.

4.1.2 Cooled frequency tripler experiments

We have made cooled frequency multiplier measurements of HBV diodes from the UVA-NRL-1174 material to investigate how temperature affects the efficiency [29]. The HBV tripler was mounted in a cryostat and the input power was provided from a Gunn oscillator through a waveguide and a vacuum window. Output power was extracted through a horn antenna, a plastic window, and finally measured outside the chamber using a Thomas Keating power meter. The measured available input power into the tripler block at 85 GHz was $\approx$55 mW. The input tuners and the output tuner were optimised at room temperature for maximum tripler efficiency. After evacuating the chamber and during cooling, the only possibility for tuning the tripler was by changing the pump frequency.

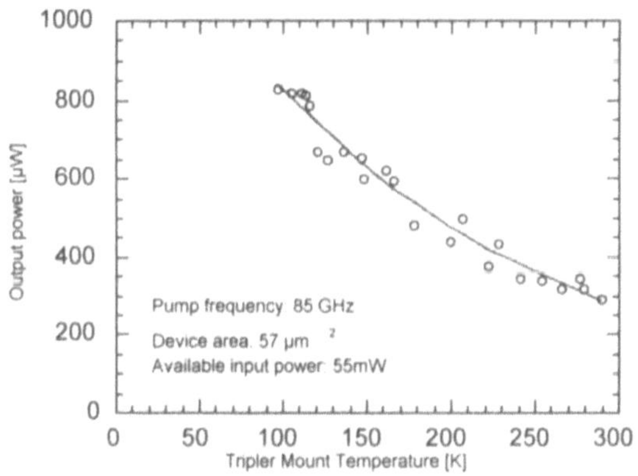

Figure 16. Measured output power at a fundamental of 85 GHz obtained for an 8 μm diameter device (UVA-NRL-1174-17) versus tripler mount temperature. The available input power is $\approx$ 55 mW.

The efficiency increases by almost 200% when the multiplier block is cooled from room temperature to 95 K which verifies the strong temperature dependence of the multiplier efficiency for this material system, see Figure 16. The lower third harmonic absolute power compared to the room temperature experiment, Figure 15, is due to larger losses at the output. Similar experiments with GaAs Schottky varactor doublers at 160 GHz resulted in an overall improvement of ≈25% [46]. Consequently, the strong temperature dependence of the low barrier height HBVs is mainly due to a reduced conduction current rather than the effect of improved electron mobility and lower circuit losses.

A computer controlled stepper motor system has recently been developed to allow for impedance matching at low temperatures [47].

4.1.3 A planar four-barrier GaAs/AlGaAs HBV tripler with improved diode geometry

To improve the heat-sink and to reduce the series resistance, HBV triplers with a new diode geometry have been tested [43]. The new planar HBV design incorporates a shorter finger with a larger cross section area to reduce the thermal resistance of the diode, see Figure 17. Furthermore, the distance between the anodes is reduced and their shape is changed from almost circular to rectangular to reduce the resistance between the two of them. In this experiment an efficiency of 4.8% was obtained at 246 GHz which is significantly more than the 3.1% obtained at 261 GHz with the earlier HBV design.

Figure 17. Planar HBV design incorporating a shorter finger with a large cross section area to reduce the thermal resistance of the diode.

4.1.4 Frequency triplers with whisker contacted HBVs on a copper substrate

The experiments for GaAs/AlGaAs HBVs showed that the efficiency is decreased due to the influence of self-heating, yielding an increased conduction current. By replacing the semiconductor substrate with a copper substrate, the heat-sink is improved and by using an improved material system the conduction current is decreased and less temperature dependent. We have tested HBVs from the material ITME 1622 fabricated on a copper substrate [35]. The material structure is similar to that shown in Table I but with a doping concentration of $7.2 \cdot 10^{16}$ cm^{-3} in the modulation layers. The chip was mounted in the output waveguide and two 7-μm HBVs (A=77 μm^2) were contacted in parallel using a planar whisker. A maximum output power of 7.1 mW was generated at 221 GHz as shown in Figure 18. The efficiency increases monotonically with the input power and reaches the top value, 7.9%, at the maximum output power, (90 mW,) from the Gunn oscillator. However, simulations predict that more than 150 mW of input power can be used without exceeding the break-down voltage (≈22 V). The measured efficiency agrees well with the simulations assuming room temperature operation. This indicates that there is no essential decrease in efficiency due to self-heating in this device.

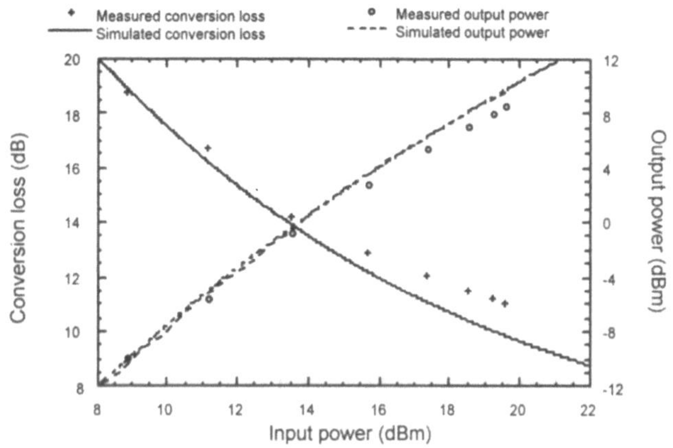

Figure 18. Measured and simulated tripler conversion efficiency versus available power for an HBV tripler at a fundamental frequency of 74 GHz. A maximum output power of 7.1 mW (8.5 dBm) was generated with a flange-to-flange efficiency of 7.9% (11.0-dB conversion loss). For the simulations, an input loss of 1 dB and an output loss of 3 dB were assumed.

In a recent experiment, we tested a single 7 μm diode from the same batch in a crossed waveguide frequency tripler. A maximum output power of 4.3 mW was generated at 267 GHz and the peak efficiency was 5.6%.

In order to investigate how the frequency multiplier behaves when the electrons are driven to high velocities, an HBV with a low doping concentration in the modulation layer has been tested in a frequency tripler. The electron velocity is inversely proportional to the doping concentration, see Eq. 11. The HBV tested was a single 7 μm diode fabricated on a copper substrate from the ITME 1557 material. The material structure is similar to that shown in Table I, but with a doping concentration of 2.8×10^{16} cm^{-3} in the modulation layers. The leakage current of the ITME 1557 material is less than that for the ITME 1622, making it even less sensitive to self-heating. Figure 19 shows the efficiency versus input power. It can be seen that the efficiency saturates at an input power of about 30 mW. The high series resistance, 45 Ω, partly depends on the low doping concentration. A poor electrical contact between the whisker and the HBV also added some resistance. The measurements agree well with simulations assuming an electron velocity dependent series resistance. The electron velocity at the maximum efficiency is estimated to be $2.1 \cdot \times 10^5$ m/s from the simulations. This velocity can be used to optimise the thickness of the modulation layers, see Eq. 12. However, more measurements with different HBVs at different frequencies are needed to verify the current saturation model and the velocity for optimum conversion efficiency.

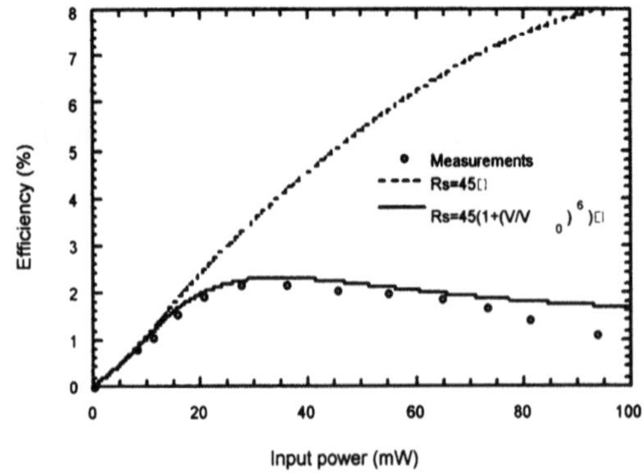

Figure 19. Measured and simulated tripler conversion efficiency versus available input power for an HBV tripler at a fundamental frequency of 90 GHz. For the dotted line a constant series resistance is assumed and for the solid line the series resistance is assumed to increase with the electron velocity as Eq. (10) with $v_0=2.2\cdot10^5$ *m/s. For the simulations, an input loss of 1 dB and an output loss of 2 dB were assumed.*

4.2 DISTRIBUTED FREQUENCY MULTIPLIERS

Non-linear transmission lines using HBVs as non-linear elements have recently been investigated by several authors [48], [49], [50], [12]. Wide bandwidth and high multiplication factors are advantages with distributed frequency multipliers. One drawback is that the peak efficiency is usually lower than for a single device multiplier. To simplify the analysis and achieve a realistic capacitance per unit length, a linear transmission line periodically loaded with non-linear devices is usually preferable to a fully distributed non-linear transmission line.

A multiplier consisting of a finline loaded with 15 HBVs has been tested [38]. A finline can be used because HBVs do not need any DC bias. The third harmonic is generated from propagation through the non-linear transmission line and radiated into free space from a tapered slot antenna at the output. A maximum output power of 10 mW was generated at 130.5 GHz with an efficiency of 7%. The 3-dB bandwidth was measured to 10% and the 6 dB bandwidth to 20%. To avoid reflection of the fundamental frequency at the output, which would degrade the multiplier performance, a radiating antenna was used. A future project could be to design a distributed frequency multiplier that works properly with a waveguide at the output.

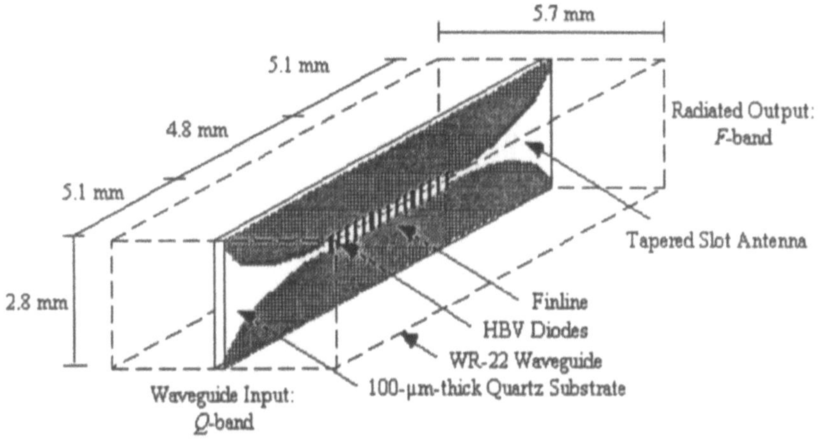

Figure 20. A frequency tripler consisting of a finline loaded with 15 HBVs.

4.3 QUASI-OPTICAL MULTIPLIERS

Losses in waveguide components increase with frequency and the machining becomes more difficult since the dimensions decrease. Quasi-optical circuits are therefore especially useful at high frequencies. By using quasi-optical spatial power combining of a grid of thousands of varactors, very high output powers can be produced [51]. 5 W of output power has been reported at 99 GHz by power combining 3100 varactors [8]. The varactor used was similar to HBVs but with Schottky contacts instead of ohmic contacts. One drawback with spatial power combining is that large input powers are needed and that the efficiencies are often low due to diffraction losses.

A new, simple quasi-optical frequency tripler [39] consists of two slot antennas loaded with HBVs located at the focal plane of a dielectric lens, see Figure 21. A maximum efficiency of 8% was generated at 141 GHz.

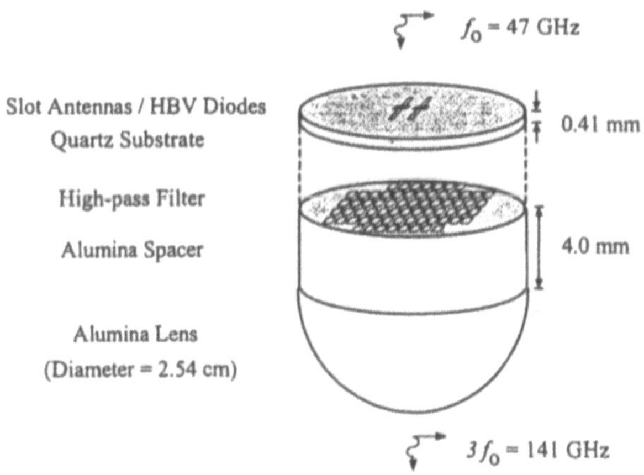

Figure 21. A quasi-optical tripler, consisting of two slot antennas loaded with HBV diodes located at the focal plane of a dielectric lens.

4.4 DISCUSSION

Experiments have shown that the HBV is capable of producing power in the upper millimetre wave range. In recent years, tripler performance in the frequency range 200-300 GHz has increased from 2 mW output power and 5% efficiency to almost 10 mW output power and 12% efficiency. This is close to the best reported results for Schottky multipliers. The most important improvement is the use of a material structure with a more efficient barrier, which decreases the conduction current. The device can be further improved by optimising the number of barriers, the doping concentration and the thickness of the modulation layers. Simulations show that the efficiency can be improved by at least a factor two. However, the low yield of the material growth and the difficulty in measuring the series resistance makes the development of HBV materials slow.

Over the last few years, new circuit topologies which take advantage of the symmetry of the HBV have been developed, and further development of HBV multiplier circuits can be expected in the near future. Frequency quintuplers, non-linear transmission lines and monolithic integrated multipliers (MMIC) are examples of interesting future work.

HBV multipliers have so far mainly been tested at frequencies around 250 GHz. However, the HBV could be even more competitive at higher frequencies since the gain in using a high multiplication factor increases with the frequency due to the fast drop in the available input power. HBV multipliers could also be an interesting choice for frequencies below 100 GHz. Using a copper heat-sink and a material with a high break-down voltage, it is realistic to handle more than 1 W of input power. Furthermore, the high Q-value of the HBV is advantageous for low noise applications. HBV multipliers can therefore be competitive as future power sources all the way from lower millimetre wave frequencies and up to over 1 THz.

References

1. P. Penfield and R.P. Rafuse (1962) *Varactor applications*, The M.I.T. Press, Massachusetts.
2. J. Thornton, C. Mann and P. Maagt (1997) A high power 270 GHz frequency tripler featuring a schottky diode parallel pair, *IEEE MTT-S Int. Microwave Symp. Dig.*, 957-958.
3. D. Porterfield, T. Crowe, R. Bradley and N. Erickson (1999) A high-power fixed-tuned millimeter-wave balanced frequency doubler, *IEEE Trans. Microwave Theory Tech.* **47**, 419-425.
4. E. Kollberg and A. Rydberg (1989) Quantum-barrier-varactor diodes for high-efficiency millimetre-wave multipliers, *Electronics Letters* **25**, 1696-1698.
5. W.C.B. Peatman, T.W. Crowe and M. Shur (1992) A novel schottky /2-DEG diode for millimeter- and submillimeter-wave multiplier applications, *IEEE Electron Device Lett.* **13**, 11-13.
6. U. Lieneweg, B.R. Hancock and J. Maserjian (1987) Barrier-intrinsic-n+ (BIN) diodes for near-millimeter wave generation, *12th Int. Conf. on Infrared and Millimeter Waves*, 6-7.
7. S. Mahieu, C.M. Mann, J. Stake, L. Dillner, S.H. Jones and J. Thornton (1998) A Broadband frequency tripler for SIS receiver, *Ninth Int. Symp. on Space Terahertz Tech., Pasadena,* 481-491.
8. H.-X.L. Liu, L.B. Sjogren, C.W. Domier, N.C. Luhmann, D.L. Sivco and A.Y. Cho (1993) Monolithic quasi-optical frequency tripler array with 5-W output power at 99 GHz, *IEEE Transaction on Electron Devices* **14**, 329-331.

9. K. Krishnamurthi, R.G. Harrison, C. Rogers, J. Ovey, S.M. Nilsen and M. Missous (1994) Stacked heterostructure barrier varactors on InP for millimeter wave triplers, *European Microwave Conf., Cannes*, 758-763.
10. A. Rahal, R.G. Bosisio, C. Rogers, J. Ovey and M. Missous (1995) A W-band medium power multi-stack quantum barrier varactor frequency tripler, *IEEE Microwave and Guided Wave Letters* **5**, 368-370.
11. A. Rydberg, H. Grönqvist and E.L. Kollberg (1990) Millimeter- and submillimeter-wave multipliers using quantum-barrier-varactor (QBV) diodes, *IEEE Transaction on Electron Devices* **11**, 373-375.
12. X. Mélique, A. Maestrini, E. Lheurette, P. Mounaix, M. Favreau, O. Vanbésien, M. Goutoule, G. Beaudin, T. Nähri and D. Lippens (1999) 12 % efficiency and 9.5 dBm output power from InP based heterostructure barrier varactor triplers at 250 GHz, *IEEE MTT-S Int. Microwave Symp. Dig., Anaheim*, 123-126.
13. C.H. Page (1956) Frequency conversion with positiv nonlinear resistors, *Journal of Research of the National Buroau of Standards* **56**, 179-181.
14. S.M. Sze (1981) *Physics of semiconductor devices, 2 ed*, John Wiley & Sons, New York.
15. Y. Shiraishi, N. Furuhata and A. Okamoto (1994) Influence of metal/n-InAs/interlayer/n-GaAs structure on nonalloyed ohmic contact resistance, *Journal of Applied Physics* **76**, 5099-5110.
16. K. Krishnamurthi, S.M. Nilsen and R.G. Harrison (1994) GaAs single-barrier varactors for millimeter-wave triplers: guidelines for enhanced performance, *IEEE Trans. Microwave Theory Tech.* **42**, 2512-2516.
17. V. Duez, X. Melique, O. Vanbésien, P. Mounaix, F. Mollot and D. Lippens (1998) High capacitance ratio with GaAs/InGaAs/AlAs heterostructure quantum well-barrier varactors, *Electronics Letters* **34**, 1860-1861.
18. V.K. Reddy and D.P. Neikirk (1993) High breakdown voltage AlAs/InGaAs quantum barrier varactor diodes, *Electronics Letters* **29**, 464-466.
19. E. Lheurette, P. Mounaix, P. Salzenstein, F. Mollot and D. Lippens (1996) High performance InP-based heterostructure barrier varactors in single and stack configuration, *Electronics Letters* **32**, 1417-1418.
20. W. Strupinski, J. Sass, L. Dillner, K. Kosiel, R. Jakiela, J. Stake and M. Ingvarson (2000) MOVPE strain layers - growth and application, *The Tenth International Conference on Metalorganic Vapor Phase Epitaxy (ICMOVPE-X)*.
21. Y. Fu, M. Willander, S. Bengtsson and L. Dillner (2000) Designing n-Si/SiO2/Si heterostructure barrier varactor diode, *to be published in Journal of Applied Physics*.
22. L. Dillner, J. Stake and E. Kollberg (1997) Modeling of the heterostructure barrier varactor diode, *Int. Semiconductor Device Research Symp, Charlottesville*, 181-182.
23. Y. Fu, L. Dillner, J. Stake, M. Willander and E.L. Kollberg (1998) Capacitance analyses for AlGaAs/GaAs and InAlAs/InGaAs heterostructure barrier varactor diodes, *Journal of Applied Physics* **83**, 1457-1462.
24. K. Krishnamurthi and R.G. Harrison (1993) Analysis of symmetric-varactor frequency triplers, *IEEE MTT-S Int. Microwave Symp. Dig.*, 649-652.
25. C.H. Tang (1966) An exact analysis of varactor frequency multipliers, *IEEE Trans. Microwave Theory Tech.*, 210-212.
26. L. Dillner, J. Stake, and E. Kollberg (1997) Analysis of symmetric varactor frequency multipliers, *Microwave and Optical Technology Letters* **15**, 26-29.
27. Y. Fu, J. Stake, L. Dillner, M. Willander and E.L. Kollberg (1997) AlGaAs/GaAs and InAlAs/InGaAs heterostructure barrier varactors, *Journal of Applied Physics* **82**, 5568-5572.
28. R. Havart, E. Lheurette, O. Vanbésien, P. Mounaix, F. Mollot and D. Lippens (1998) Step-like heterostructure barrier varactor, *IEEE Transaction on Electron Devices* **45**, 2291-2297.
29. J. Stake, L. Dillner, S. H. Jones, C. Mann, J. Thornton, J. R. Jones, W. L. Bishop, and E. Kollberg (1998) Effects of self-heating on planar heterostructure barrier varactor diodes, *IEEE Transaction on Electron Devices* **45**, 2298-2303.
30. J. Stake and H. Grönqvist (1995) An on-wafer method for C-V characterisation of heterostructure diodes, *Microwave and Optical Technology Letters* **9**, 63-66.
31. E.L. Kollberg, T.J. Tolmunen, M.A. Frerking and J.R. East (1992) Current saturation in submillimeter wave varactors, *IEEE Trans. Microwave Theory Tech.* **40**, 831-838.

32. W.L. Bishop, T.W. Crow, R.J. Mattauch and P.H. Ostdiek (1991) Planar schottky barrier mixer diodes for space applications at submillimeter wavelengths, *Microwave and Optical Technology Lett.* **4**, 44-49.

33. J.R. Jones, W.L. Bishop, S.H. Jones and G.B. Tait (1997) Planar multi-barrier 80/240-GHz heterostructure barrier varactor triplers, *IEEE Trans. Microwave Theory Tech.* **45**, 512-518.

34. L. Dillner, J. Stake, and E. Kollberg (1999) Heterostructure barrier varactors on copper substrate, *Electronics Letters* **35**, 339-341.

35. L. Dillner, W. Strupinski, S. Hollung, C. Mann, J. Stake, M. Beardsley and E. Kollberg (2000) Frequency multiplier measurements on heterostructure barrier varactors on a copper substrate, *IEEE Electron Device Letters* **21**, 206-208.

36. J. Stake, S. Jones, L. Dillner, S. Hollung, and E. Kollberg (2000) Heterostructure barrier varactor design, *IEEE Trans. Microwave Theory Tech. part II* **48**, 677-682.

37. K. Krishnamurthi, E. Boch and R.G. Harrison (1995) A Ka-band planar tripler based on a stacked symmetric InP heterostructure barrier varactor, *IEEE MTT-S Int. Microwave Symp. Dig., Orlando*, 549-552.

38. S. Hollung, J. Stake, L. Dillner, M. Ingvarson, and E. Kollberg (2000) A distributed heterostructure barrier varactor frequency tripler, *IEEE Microwave and Guided Wave Letters* **10**, 24-26.

39. S. Hollung, J. Stake, L. Dillner, and E. Kollberg (1999) A 141-GHz quasi-optical HBV diode frequency tripler, *Tenth Int. Symp. on Space Terahertz Tech.*, 492-500.

40. A.V. Räisänen, T.J. Tolmunen, M. Natzic, M.A. Frerking, E. Brown, H. Grönqvist and S.M. Nilsen (1995) A single barrier varactor quintupler at 170 GHz, *IEEE Trans. Microwave Theory Tech.* **43**, 685-688.

41. D. Choudhury, M. Ferking and P. Batalaan (1993) A 200 GHz tripler using a single barrier varactor, *IEEE Trans. Microwave Theory Tech.* **41**, 595-599.

42. R. Meola and J. Freyer (1998) 210 GHz tripler with monolithically integrated single barrier varactors, *Electronics Letters* **34**, 1756-1757.

43. J. Stake, C. Mann, L. Dillner, M.Ingvarson, S. Jones, and E. Kollberg (1999) Improved diode geometry for planar heterostructure barrier varactors, *Tenth Int. Symp. on Space Terahertz Tech.*, 485-491.

44. J.W. Archer (1984) An efficient 200-290 GHz frequency tripler incorporating a novel stripline structure, *IEEE Trans. Microwave Theory Tech.* **32**, 416-420.

45. J.R. Jones (1997) CAD of millimeter wave frequency multipliers: An experimental and theoretical investigation of the heterostructure barrier varactor, *Ph.D. Thesis, School of Engineering and Applied Science, University of Virginia.*

46. J.T. Louhi, A.V. Räisänen and N.R. Erickson (1993) Cooled schottky varactor frequency multipliers at submillimeter wavelength, *IEEE Trans. Microwave Theory Tech.* **41**, 565-571.

47. M. Ingvarson, L. Dillner, S. Hollung, E. Kollberg, C. Mann and J. Stake (2000) Computer controlled measurements on millimetre and submillimetre wave varactor multipliers, *GHz 2000, Symposium on GHz Electronics, Göteborg*, 297-300.

48. H. Shi, W.-M. Zhang, C.W. Domier, N.C. Luhmann, L.B. Sjogren and H.X. Liu (1995) Novel concepts for improved nonlinear transmission line performance, *IEEE Trans. Microwave Theory Tech.* **43**, 780-789.

49. J.R. Thorpe, P. Steenson and R.E. Miles (1997) Non-linear transmission lines for millimeter-wave frequency multiplier applications, *IEEE Sixth Int. Conf. on Terahertz Electronics, Leeds*, 54-57.

50. M. Li, K. Krishnamurthi and R.G. Harrison (1998) A fully distributed heterostructure barrier varactor nonlinear transmission line performance, *IEEE Trans. Microwave Theory Tech.* **46**, 2295-2301.

51. R.J. Hwu, L.P. Sandwick, N.C. Luhmann, D.B. Rutledge, M. Sokolich and B. Hancock (1989) Quasi-optical watt-level millimiter-wave monolithic solid-state diode-grid frequency multipliers, *IEEE MTT-S Int. Microwave Symp. Dig., Long Beach*, 1069-1072.

High Performance Heterostructure Barrier Varactors

D. LIPPENS, X. MELIQUE, S. ARSCOTT, T. DAVID, V. DUEZ, J. CARBONELL, P. MOUNAIX, O. VANBESIEN, F. MOLLOT
Institut d'Electronique et de Microélectronique du Nord
Université des Sciences de Lille 1
Avenue Poincaré BP 69, 59652 Villeneuve d'Ascq Cedex, France

We report on the physics, fabrication technology and characterisation of high performance Heterostructure Barrier Varactors diodes. State-of-the art results in terms of capacitance ratios of up to 10:1 and voltage handling have recently been published for InP-and GaAs-based devices with a typical breakdown voltage as high as 6 V in a single barrier scheme for the former. On this basis, vertical and lateral stacking of eight barriers was demonstrated yielding a voltage handling of 50 V with a zero-bias capacitance of 250 aF/μm^2. For tripler experiments integrated planar, devices were fabricated using e-beam patterning, RIE mesa etching and air-bridge techniques. Circular or finger-shaped devices, laterally integrated with a total of four barriers, demonstrate an output power of about 10 mW with a conversion efficiency of the order of 10% at 250 GHz. There are record results at the time of writing.

1. Introduction

At the end of 1980's, several research groups were involved in the nonlinear applications of Resonant Tunnelling Diodes. In order to use this device for harmonic generation, several attempts [1]-[3] were made to take advantage of the Negative Differential Conductance effects in an attempt to overcome the fundamental limitations of resistive multipliers [4]. It was soon recognised that the non linear feature of the diode capacitance characteristic also influences the overall efficiency. Subsequently, E. Kollberg and A. Rydberg [5] were the first to propose the concept of a single barrier varactor in 1989 with the acronym, at that time, of Quantum Barrier Varactor. However, for those structures consisting of a heterostructure barrier sandwiched between two cladding layers, the main quantum effects are on the leakage current characteristics due to tunnelling of carriers through the heterostructure barrier. It can be shown that Heterostructure Barrier Varactors (HBV's) have potential advantages over their Schottky counterparts and very promising results have been reported in the literature [6-15].

The key advantage is the symmetry of the capacitance-voltage characteristic about zero-bias. Such symmetry improves the functionality of devices dramatically since only odd harmonics are generated in the multiplication process. Symmetrical C-V relationship can also be achieved with back-to-back Schottky's but its was shown that self-biasing effects impede the variation of elastance when the devices are driven in the large signal regime [16]. Secondly, several degrees of freedom are afforded by the use of semiconductor heterojunctions and novel structures, notably those making use of a

R.E. Miles et al. (eds.), Terahertz Sources and Systems, 53–67.

quantum well/barrier configuration can be fabricated with tailored C-V characteristics. Another important advantage stems from the fact that the devices can be vertically stacked during epitaxy. This epitaxial integration is a welcome feature in terms of voltage handling and capacitance levels. In contrast, HBV's do exhibit some drawbacks. First of all, the epitaxial quality is of prime importance in order to avoid high leakage current through defect-assisted spurious tunnelling effects. Also, it is difficult to achieve a high barrier height, compared to the typical value of 0.7eV for the built-in potential of Schottky barriers. On the other hand, we have only ohmic contacts which complicate device integration especially for small-area devices.

In this overview, we report on the physics-based design, fabrication technology and Radio Frequence (RF) characterisation of high performance heterostructure devices. Section 2 deals with the design rules and is followed by a general discussion of material system and device scheme choice. Device technology and small signal characterisation are addressed in sections 3 and 4. Finally in Section 5, we report on record performances demonstrated at 250 GHz with an output power close to 10 mW and a conversion efficiency in the 10 % range. This is the highest performance reported to date for HBV devices at this operating frequency.

2. Device Physics

2.1 DESIGN RULES

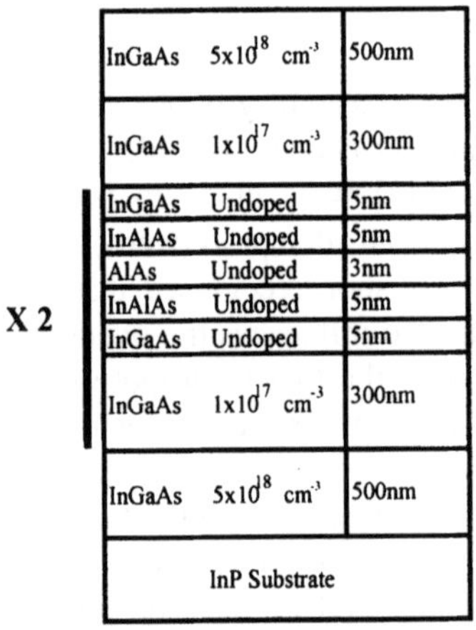

Figure 1: Epitaxial material for InP-based Heterostructure Barrier Varactor

In order to illustrate the design rules, let us consider the epitaxial material depicted in Figure 1 grown by gas source molecular beam epitaxy, starting from a semi-insulating Fe-doped InP substrate. Two basic $In_{0.53}Ga_{0.47}As$ /$In_{0.52}Al_{0.48}As$ /AlAs /$In_{0.52}Al_{0.48}As$ /$In_{0.53}Ga_{0.47}As$ layered structures were series integrated during the same run. It can be shown [17] that the device characteristics scale with the number of complete structures. Hence, the voltage breakdown is twice that of the single barrier device and the capacitance is half the value for one barrier. The design of epilayers was carried out by means of an in-house Schrödinger/Poisson equation solver.

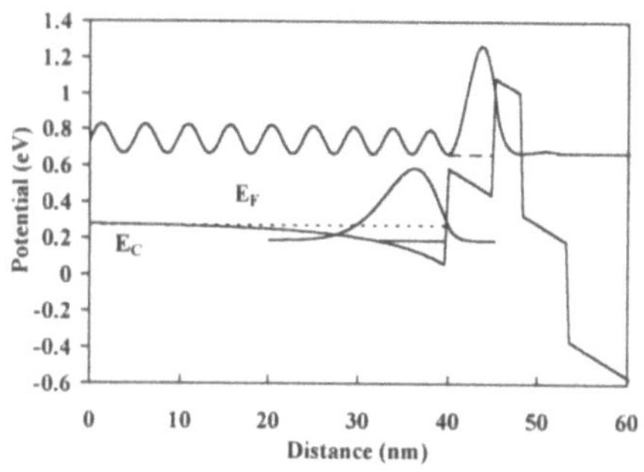

Figure 2: Illustration of design rules. The capacitance is due to the electron accumulation. The leakage current is due to tunneling.

Figure 2 illustrates this theoretical approach taken to calculate the I-V curve and C-V relationship. We plotted the conduction band edge as a function of distance for a one-barrier device (biased at 4 V) having the growth parameters listed in Figure 1. Charge accumulates in front of the heterostructure barrier in a quasi-triangular quantum well. A computation of the sheet carrier density, $n_{s,}$ from the electron wavefunction (as a function of voltage) is a direct calculation allowing for the variation of n_s with respect to voltage of the expected C-V characteristic.

In order to illustrate the conduction mechanisms, we have also plotted the wave-function corresponding to the leakage current via a tunneling mechanism [18]. This tunneling process explains the conduction mechanisms at moderate voltages. In contrast, at around 6V for the device under consideration, conduction near breakdown is due to impact ionization.

It is worth-noting that the characteristic tunneling energy is several hundred meV (>600 meV) and hence is much higher than the corresponding energy value involved in GaAs technology. For instance, the relevant characteristic energy would be about 180 meV for a GaAs/$Al_{0.7}Ga_{0.3}As$/GaAs device. This is the basic reason why the voltage handling of InP-based diodes is improved dramatically. In practice an improvement of a factor of three can be pointed out (6V instead of typically 2 V).

Figure 3 displays the nonlineraties achieved experimentally for two barriers in series i.e. a Dual Heterostructure Barrier Varactor (DHBV). The symmetry in the C-V characteristic is excellent. This is advantageous for efficient rejection of the even harmonics. The zero-bias capacitance is 1fF/μm^2 and capacitance ratio achieved is 6:1. Also plotted in Figure 3 is the voltage dependence of the conductance of the device. Note, the units of the conductance are nS/μm^2. The voltage handling of such a device is thus remarkable with a "safe" peak-to-peak operating voltage range of at least 20 V.

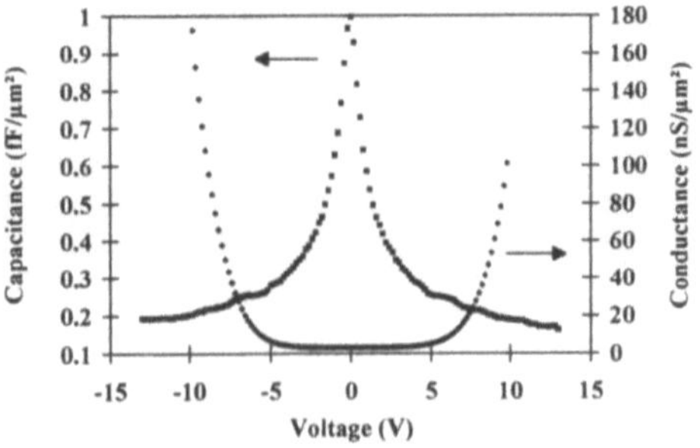

Figure 3: Conductance and capacitance versus voltage measured at 300 K for a dual barrier scheme The measurement frequency is 500 MHz for the C-V curve.

In addition to the careful design of the blocking barrier, special attention has also to be paid to the adjacent layer that greatly influences the trade-off between current saturation effects [19] and achievable capacitance ratio. The former can be overcome by increasing the doping concentration. The latter is related in a first approximation to the ratio between the barrier thickness (the screening length has also to be taken into account for a better estimate) and the depletion layer thickness as seen in the following section in GaAs technology. Also some band gap engineering has to be implemented to increase the conduction band offset.

2. 2 QUANTUM WELL BARRIER STRUCTURE

Figure 4 shows the growth sequence of the fabricated devices [20]. In comparison with the generic GaAs/AlAs structure, the AlAs barrier is now sandwiched between two quantum wells, formed by the pseudomorphic growth of InGaAs 5nm-thick layers. The indium concentration is 30%. The InGaAs and AlAs layers are undoped (u-d) while the adjacent 300nm-thick GaAs layers are doped to $1x10^{17}cm^{-3}$. This gradient in doping concentration results in a diffusion process. Electrons, in the vicinity of the GaAs/InGaAs heterojunction, experience real space transfer for occupying the two-dimensional (2D) density of states of the quantum wells. As a consequence under zero-bias conditions, the barrier is sandwiched, in very close proximity, by two highly populated regions while a medium doping level has been preserved in the adjacent layers.

The barrier thickness was chosen to be equal to 10 nm. This is a trade-off between the blocking capability of the barrier which depends on the ability of electrons to tunnel through it, as seen above, and the targeted capacitance value at zero-bias. For the latter, the thinner the barrier, the higher the capacitance.

GaAs	$2x10^{18}$ cm^{-3}	5000 Å
GaAs	$1x10^{17}$ cm^{-3}	3000 Å
InGaAs	u-d	50 Å
AlAs	u-d	100 Å
InGaAs	u-d	50 Å
GaAs	$1x10^{17}$ cm^{-3}	3000 Å
GaAs	$2x10^{18}$ cm^{-3}	1 μm
Substrate GaAs		

Figure 4: Epitaxial sequence grown by Gas Source Molecular Beam Epitaxiy (GSMBE), starting from a semi-insulating GaAs Substrate. InGaAs wells with an indium concentration of 30% layers are grown under pseudomorphic conditions.

Test structures, consisting of mesa-type diodes in a coaxial-like configuration with diameters ranging from 8 to 40 μm, were fabricated with a standard technology which included wet etching and ohmic contact material metal deposition by means of electron beam evaporation. Figure 5 shows the capacitance-voltage characteristic measured at room temperature and a frequency of 8 GHz. The diode diameter is 8 μm.

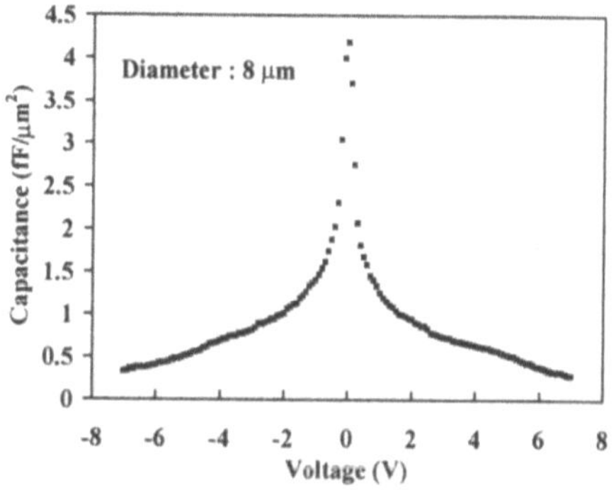

Figure 5: Capacitance voltage characteristic measured at 8 GHz. The capacitance at zero-bias is in excess of 4 fF/μm^2 as a consequence of the InGaAs pre-well and post-well which enable a significant charge accumulation at equilibrium

A high degree of symmetry was found between both bias directions with a zero-bias capacitance of ~ 4 fF/μm^2. Also because we characterise the samples at a relatively high frequency any defect related frequency deviations are avoided since the relevant time 1/f (f is the characterization frequency) is shorter than the corresponding time expected of traps. As a consequence of the high zero-bias value, the capacitance ratio is in excess of 10:1 an approximate improvement of a factor of 2 with respect to the best results published in the literature. Of greater interest is the steep decrease in the capacitance values over the voltage range [-2 ; +2 V]. It is believed that this narrowing of the voltage range over which the device is highly non linear is a welcome property for very high frequency application notably up to 1 THz

In order to analyse the reasons for these improvements in the capacitance ratio and subsequently in the non-linearity exhibited by the device, let us consider the schematic of the conduction profile plotted in figure 6(a) at a bias point very close to zero bias (V=50mV) and (b) under saturation conditions.

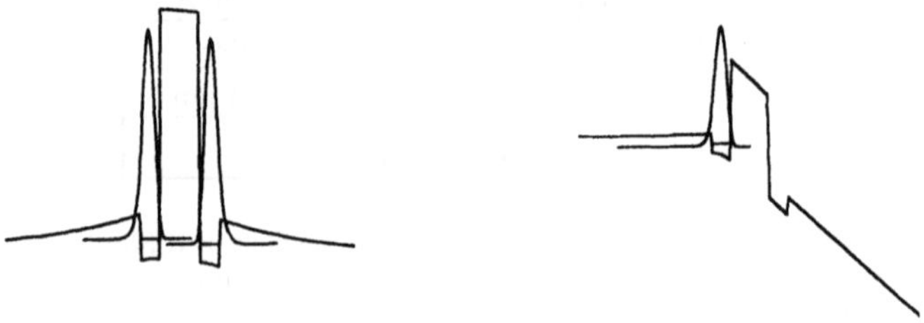

(a) at voltage 50mV (b) under full depletion

Figure 6 (a): Schematic of conduction profile. The screening lengths are substantially shortened with respect to conventional HBV structures.

In (a), the electrons are strongly localised within the quantum wells thus forming quasi-degenerate semiconductor regions. As a consequence, the length required for electric field screening (L_s) is expected to be very short. Under bias, the pre-well structure is progressively fed by the adjacent layer which plays the role of a reservoir. At the same time, carriers escape from the post-well structure until this quantum well is empty. At this stage, the mode of operation becomes classical with full depletion of cladding layer reached at approximately 8 V.

Having this physical interpretation in mind, a rough estimate of the ideal capacitance ratio can be deduced from the relevant lengths using a parallel plate capacitance approximation. At zero bias, the pertinent dimension between the two wave-functions maxima would be ~ 20nm. At saturation, the relevant dimension can be compared to the cladding layer thickness, namely 300nm. Under such a crude assumption, the limiting capacitance ratio would be 15:1. At the same time, the non linearity of the devices is enhanced drastically by a steep decrease in the capacitance at moderate bias (and hence pump power) when the quantum mode of operation dominates.

Let us now consider the choice of a material system for which this novel idea can be implemented. In fact, there are numerous trade-offs depending on the viewpoint, either technological or electrical. From the technological side, notably for the epitaxial growth, it is obvious that the use of GaAs-based devices is preferable. In practice, it was found that buried InGaAs pre-well and post-well structures with a 5nm width and an indium concentration of 0.3 can be grown by careful choice of growth temperature, in order to avoid any dislocations and to preserve a two-dimensional growth.

In contrast, the fabrication of InAs quantum well structures embedded in InGaAs layers lattice-matched to InP, is a formidable task at the required thickness of 5nm. This dimension is imperative for accommodating a quantum level within the wells and hence for trapping electrons. Recently [21], we succeeded in the growth of such epilayers provided the buried 5nm-thick InAs wells, over the 2.5nm critical thickness, were grown at a low temperature (~400°C) using the surfactant method.

There is also another drawback which results from the use of an Indium Phosphide technology. It is now well known that the growth of thick epilayers, typically several micrometers as is the case for stacked heterostructure barrier varactors, requires careful control of the growth conditions. The goal of a large number of devices integrated on the same wafer is still a challenging issue. In our group, in order to preserve the sample quality only two single barriers were stacked on the same epitaxy.

From the electrical side however, InP technology seems more suitable. We have shown that a step-like InAlAs/AlAs barrier scheme can be implemented with very good performance in terms of voltage handling. On the other hand, high quality ohmic contacts can be fabricated taking advantage of low gap material systems and the possibility to highly dope the structures. In contrast, to the GaAs-based system, the blocking capability of bare AlAs/GaAs heterostructures is relatively poor compared to InAlAs/InGaAs due to the well known Γ-X mixing effect. On this basis, InP seems more promising for ultra-high frequency whereas GaAs-based devices are likely to be more useful in the lower part of the Terahertz spectrum.

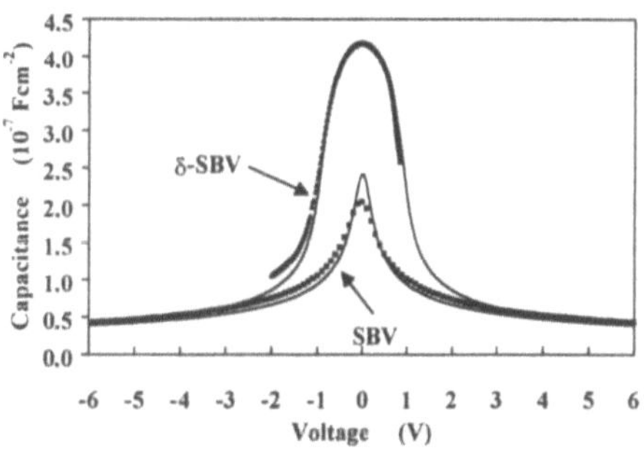

Figure 7: Comparison between the C-V of a planar doped HBV δ-SBV) and SBV [21]

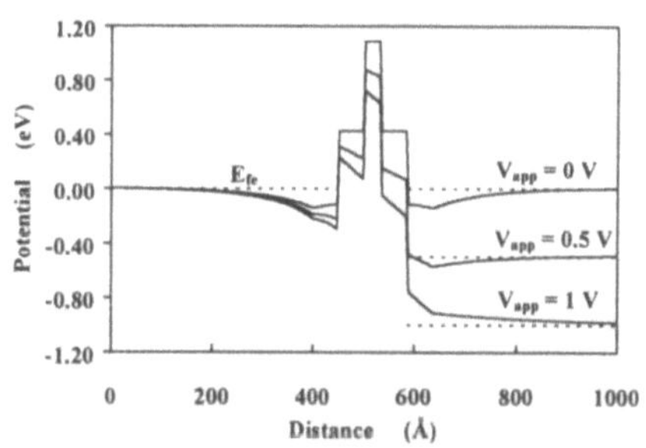

Figure 8: Conduction profile under various biases

Turning now to the choice of configuration, we demonstrated recently that the concept of a high capacitance ratio can also be implemented by means of planar doping concentrations, in close proximity to the barrier [21]. The underlying idea is the modulation of the electrron concentration in the cladding layer by means of a High-Low doping profile. Figure 7 shows a comparison between the capacitance-voltage characteristics measured and calculated for a planar doped structure and a conventional one. For the former, we have succeeded in improving the capacitance value at zero bias significantly. However, there is a large difference between the quantum well and the delta doping approaches. For the former, the buried wells are undoped and are formed by *heterojunctions*. In the delta doping case, there is a space charge effect which results in the *electrostatic* formation of adjacent quantum wells. The shape of these wells is highly non symmetric as seen in Figure 8 which illustrates the conduction band profile calculated self-consistently by taking into account the space-charge effect and the electron trapping process. The limiting case would be a sheet doping density located exactly at the hetero-interface. It can be demonstrated that such an effect lowers the barrier height with a resulting degradation of the blocking properties of the crystalline barrier.

3. Device Technology

Turning now to the device technology challenges, firstly we have to mention that several stages can be distinguished aimed at the characterisation of devices at the wafer level or in a multiplier block. It seems preferable to fabricate a set of devices of varying topology. Towards this goal, we included in the mask set a large variety of configurations in the device area and in the level of planar integration. Figure 9 is an optical view of the layout prior to the implementation of the air bridges. Three kinds of device have been patterned on the mask set. The simplest is of the coaxial-type previously discussed. It consists of a mesa-etched circular-shaped device with a concentric ring shaped side contact. It is used for rapid assessments of current-voltage and capacitance-voltage characteristics without the requirement for an air-bridge contacting technology.

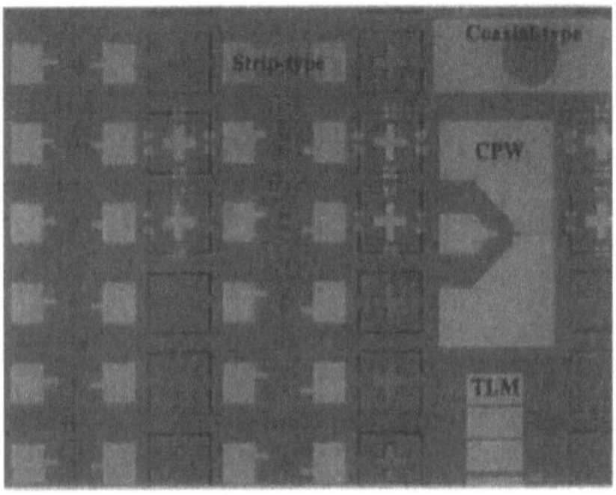

Figure 9: Plan view of the mask set before air bridge

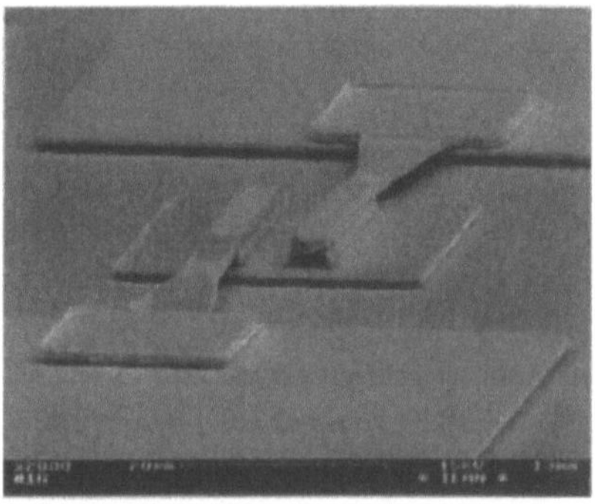

Figure 10: Scanning Electron Micrograph of implementation. The layout includes coaxial-coplanar- planar integrated device and microstrip-type configuarations

The second test configuration is of coplanar-type (CPW). This microwave compatible technology permits us to characterize small-area devices connected to the-CPW line by means of an air bridge.The third test scheme, labeled strip-type, consists of two pads facing each other (Figure 10). In this configuration, planar integration of two diodes with a possible extension to four was also carried out. Several devices, with areas ranging between 10 μm^2 and 100 μm^2, have been written on the mask set in order to test the devices at different frequencies. Finally, the layout includes a TLM pattern for assessing the quality of the ohmic contact. In our recent developments, we systematically implemented finger-shaped top ohmic contacts which can be connected to the pads or between them by means of freestanding metallic structure. Figure 11 illustrates this technology of integration which permits us to planar integrate several diodes in series. For the present example, the number of equivalent barrier is eight (two barriers stacked vertically and four laterally).

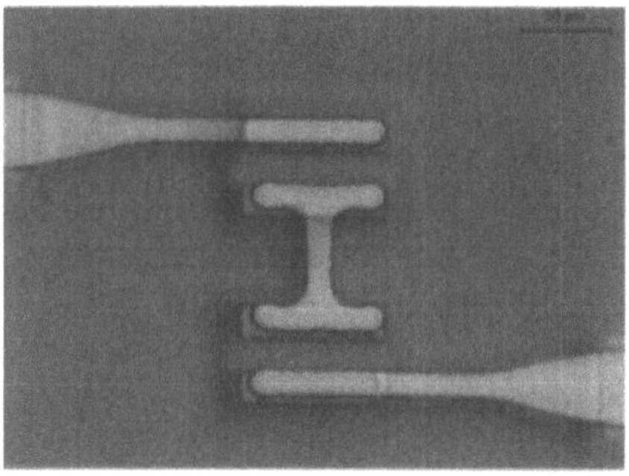

Figure 11: Close-up view of a typical device

Figure 12: SEM of a height barrier device

Figure 12 is a Scanning Electron Microphotograph (SEM) of the final device the technique involved opening the way for tree-dimensional integration. From the fabrication side, the technological effort was concentrated on the mesa etching by means of Reactive Ion Etching (RIE) in order to avoid some undercutting effects and on the fabrication of self-sustaining metallic bridges for interconnecting the top ohmic contacts to the pads. Also, for good definition, the writing steps were performed using an electron beam (LEICA High Resolution Electron Beam Pattern Generator). Conventional fabrication technologies were employed with sequential Ni/Ge/Au/Ti/Au overlay for the ohmic contact fabrication and $H_3PO_4/H_2O_2/H_2O$ solution for the wet etching of isolation mesas taking advantage of the high selectivity between InGaAs and InP. We employed either electroplating or evaporation for the metallization of the air-bridges. For the latter option, we developed a photo-resist pyrolyzation technique which results in a convex shaped temporary mold for thin metal film evaporation.

The current voltage characteristic of an eight barrier device measured using probes is displayed in Figure 13. The symmetry of current values under reverse and forward bias conditions is excellent attesting to the high quality of epilayers. The breakdown voltage is of the order of 50 V.

4. Small signal measurements

The main purpose of the small signal measurements was to derive equivalent circuits of devices for the subsequent harmonic balance simulations. As outlined in section 2, the intrinsic nonlinear capacitance-voltage characteristics can be measured without de-embedding techniques by means of coaxial type configuration not only at low frequency but also at a frequency corresponding to the pump frequency (83.3 GHz for the application reported in the present work). On the other hand it is worth mentioning that these preliminary experiments give some estimates of the series resistance on the basis of the analysis reported in [22].

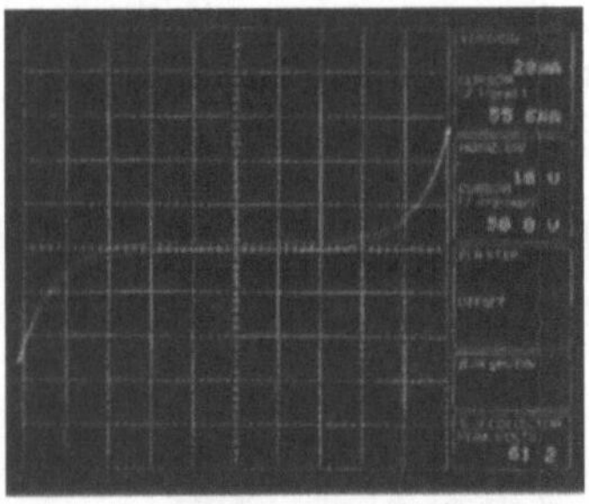

Figure 13: I-V characteristic of an eight barrier device

Experimentally, we measured a few ohms, the exact values depending on the area, despite the fact that large anode contacts were used. In addition, TLM measurements by means of the resistance ladder (Figure 9) were performed yielding a resisivity of typically $1x10^{-7}$ Ω cm^2 . This clearly shows the benefits of using a low band gap InGaAs contacting layer and high doping concentration (~ $5x10^{18}$ cm^{-3}). This means that the contact resistance is less than 1Ω for an area greater than 10μm^2. Subsequently the CPW test pattern was used to deduce the equivalent circuit for airbridge-contacted devices. In that case, we must include the extrinsic elements which are, to a first approximation, the self inductance due to the air bridge and the pad-to-pad capacitance.

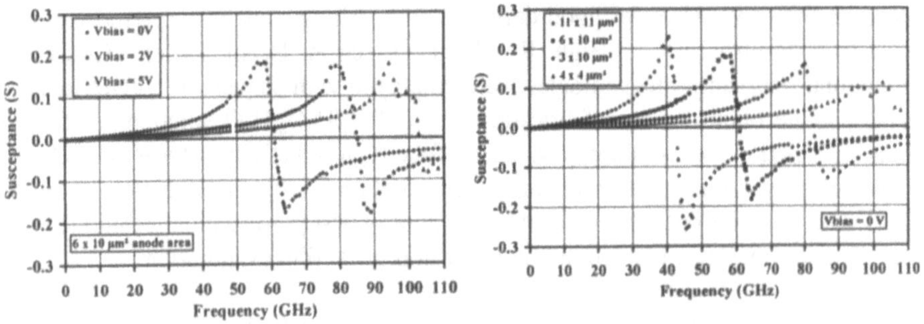

(a) (b)

Figure 14: Small signal impedance as a function of frequency for various area devices

The self-inductance resonates with the capacitance elements, notably the diode capacitance, which can be tuned either by a change in the bias voltage or in the diode area. This well defined resonance which can be carefully recorded, provided the measurements are performed in a broad band, permits us to derive a complete equivalent circuit without ambiguity.

In order to illustrate the de-embedding technique we have plotted in Figure 14 (a) and (b), the frequency dependence of the susceptance of HBV's measured from dc up to 110 GHz. As expected the self-inductance due to interconnecting elements gives rise to a resonance effect, here at ~ 60 GHz for a 60 μm^2 device. This resonance is particularly useful for fitting a lumped-element equivalent circuit which can then be used in the subsequent harmonic balance analysis [23].

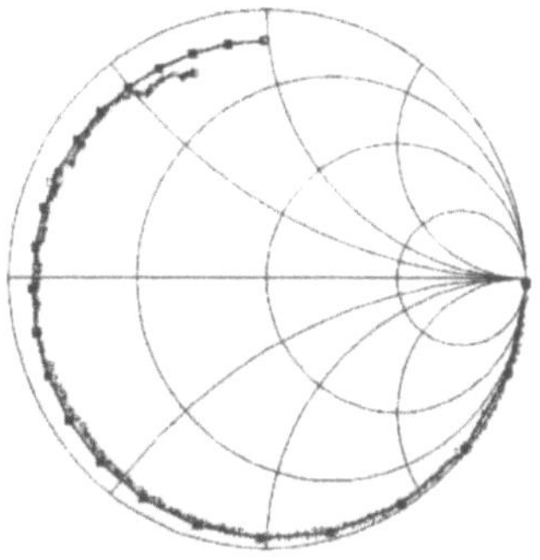

Figure 15: S_{11} versus frequency at zero bias measured (+) and calculated (■).

Confidence in the accuracy of the measurements can also be gained by full electromagnetic analysis of the CPW circuit and connecting elements [24]. In Figure 15 we have thus measured and calculated the reflection coefficient of a DHBV on a Smith Chart up to 110 GHz. The layout used for the Electromagnetic simulation using High Frequency Structure Simulator, HFSS, is reported in Figure 16.

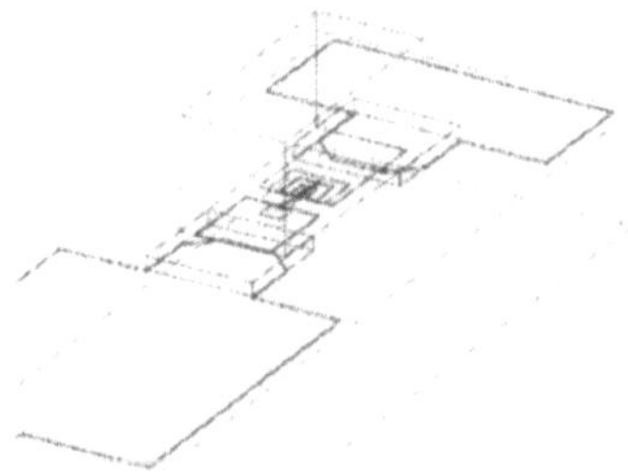

Figure 16: layout of the air bridged device structure simulated by means of HFSS [23]

Using this procedure we have measured the series resistance which is the key parameter for achieving a high cut-off frequency and hence high device performance: R_s =3.8 Ω for a 30μm^2, 2.5 Ω for 60 μm^2 and 2.2 Ω for a 121 μm^2 device. One can note that the series resistance does not scale with area as expected when the spreading resistance plays the major role. From the data obtained from both types of measurements (coaxial- and CPW-type) an estimate of the cut-off frequency of the diodes can be obtained. We thus calculate f_c = 1.5 THz and 2.1 THz respectively for DBH's of 60 and 30 μm^2 area.

5. Tripler experiment

The multiplier block used for the tripler measurements at 250 GHz is a crossed wave guide type mount with a design similar to Archer's [25] with wave-guides coupled through a low-pass stripline filter. The pump power incident in the full-height WR-8 waveguide is fed to the planar integrated diode through a stripline E-plane transition and through the low pass filter implemented on a 75 µm-thick quartz substrate. Impedance matching at the pump power is achieved using two sliding non contacting backshorts. The output was also equipped with two backshorts. On the other hand, the multiplier block differs from Archer's tripler design by the full-height output waveguide where the diode chip is mounted in a flip-chip technology using a conductive epoxy. Figure 17 shows an optical view of the tripler block.

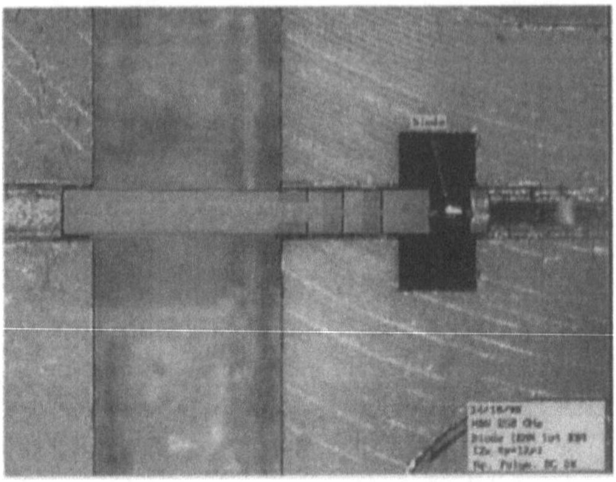

Figure 17: Optical view of the multiplier block. Courtesy of Matra Marconi Space France [27]

For mounting purpose, the diode samples were first thinned to a thickness of about 100µm by a HCl wet etching and then diced into discrete devices. The overall chip dimensions are thus 100x220x100 μm^3.

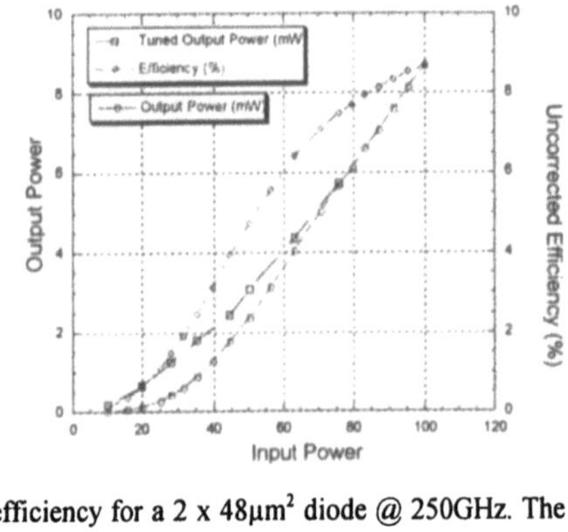

Figure 18: Output power and efficiency for a 2 x 48µm² diode @ 250GHz. The measurements were carried out at RAL.

The first experiments were carried out at the Rutherford Appleton Laboratories (RAL) [9]-[26]. A new waveguide block was equipped with a single backshort at the output and both backshort and E-plane tuners at the input. Input power was delivered by a Carlstrom Gunn which can be electronically tuned over the frequency range 72 GHz to around 89 GHz and gives around 100 mW. Both input and output powers were measured using separate Anritsu power heads which had previously been compared with a Thomas Keating power meter. To date, the maximum output power P_{out} = 8.7 mW was obtained at a pump power of 100mW at 246 GHz.

Figure 18 shows the variation of output power and efficiency versus input power level. These measurements were carried out with a DBH with 4x12μm^2 finger shaped diodes in a double configuration (the equivalent of 4 barriers). There is no saturation in the increase of output power and the maximum efficiency under the present conditions is close to 9 %. In a second stage some experiments were performed at Ecole Normale Supérieure and Observatoire de

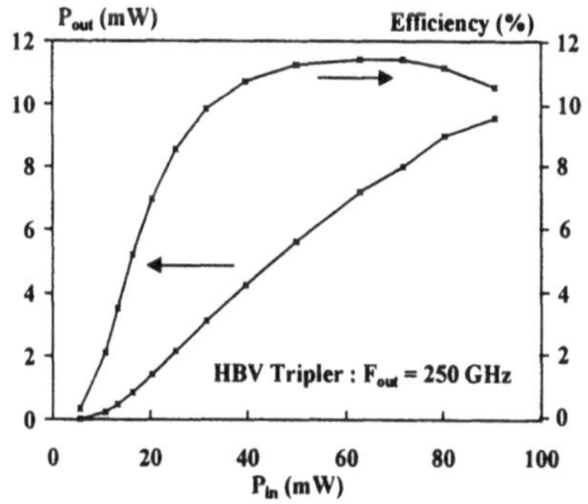

Figure 19: Output power and conversion efficiency as a function of input power. Measurements were perfomed at Observatoire de Paris and Ecole Normale Supérieure in Paris

Paris ENS/OBS[27] with a 28 μm^2-area device mounted in the Matra Marconi Space tripler (Figure 10). The pump source was a carcinotron (Thomson, serial number C040) which can deliver high power in the 77-82.25 GHz range but has power jumps in the frequency response. Figure 19 shows the output power and conversion efficiency as a function of input power for an output frequency of 247.5 GHz. The maximum efficiency was obtained with about 60 mW input power. The efficiency variations around this maximum are relatively flat. The maximum output power was 9.8 dBm (9.55mW) and was obtained with an efficiency of 10.7 %.In comparison to our previous experiment with 2x12μm^2 samples reported in [9] with 5% efficiency and 5mW output power at 216 GHz a significant increase in efficiency and frequency capability has been achieved. The new low-loss block multiplier is responsible for this improvement. On the other hand, the influence of device impedance which is varied by changing the device areas is now clear. In the RAL experiment, large area devices have been used (48μm^2). No saturation effects in the output power variation versus input power can be seen and it is believed that much higher power could be delivered at higher pump level. In the ENS/OBS experiments smaller area devices have been tested (28 μm^2). Under these conditions the devices can be driven at reduced power in the non linear regime. Finally, let us mention that in this later case the output waveguide was WR 4 and this permitted us to measure the second harmonic content which was found to be -25 dB below the third harmonic. This very good rejection of the even harmonic is a direct consequence of the excellent symmetry in the capacitance-voltage characteristics. In the short term, it is believed that this very good performance achieved simultaneously with high output power and efficiency can be further improved with the rf testing of eight barrier samples.

6. Conclusions

Record performance with an efficiency in excess of 10 % and an output power around 10 dBm has been demonstrated in a 250 GHz tripler. To realize this, high quality InP-based Heterostructure Barrier Varactors have been designed and fabricated in stacked and planar integrated configurations. Small signal measurements show high capacitance ratio (in excess of 5:1), low series resistance of a few ohms, and a potential voltage handling capability of the order of 50 V. Further improvements in the frequency and power capabilities of the devices should be achieved in the short term by the fabrication of lower capacitance devices by means of lift-off and transfer techniques on quartz and/or monolithic integration on semiconductor membranes.

7. References

1. Sollner, T.C.L.G., Brown,E.R., Goodhue, W.D. and Correa, C.A. (1988) Harmonic multiplication using resonant tunneling, *J. Appl. Phys.* Vol. **64**, 42-48

2. Rydberg, A., and Grönqvist, H. (1989) Quantum well high efficiency millimeter-wave frequency multiplier, *Electronics Letters*, **25**, 348-349

3. Bouregba, R. , Lippens, D. , Palmateer, L , Bockenhoff, E. , Bogey, M., Destombes, J. L. ,. Lecluse, A (1990) Frequency multiplication using resonant tunneling diodes with output at submillimeter wavelengths, *Electronics Letters*, **26**, 1804-1805

4. Page C. H. (1958) Harmonic multiplication with ideal rectifiers, Proceedings of the IRE, **46**, 1738-1740

5. Kollberg, E. and Rydberg, A. (1989) Quantum-barrier-varactor diodes for high-efficiency millimeter wave multipliers, *Electronics Letters*, **25**, 1696-1697

Rydberg, A., Grönqvist, H. and Kollberg, E. (1990) Millimeter - and Submillimeter-Wave Multipliers using Quantum Barrier-Varactor (QBV) Diodes, *IEEE Electron Device Letters*, **11**, 373-375

7. Choudhury, D. , Frerking, M. A. and Batelaan, P. D. (1993) A 200 GHz Tripler using a single barrier varactor, *IEEE Transactions on microwave Theory and Techniques*, **41**, 595-599

8. Stake, J., Dillner, L., Jones, S. H.,. Mann, C. M. , Thornton, J., Jones, J. R., Bishop, W. L. and Kollberg, E. L. (1998) Effects of self heating on planar heterostructure Barrier varactor diodes, *IEEE Transactions on Electron Devices*, **45**, 179-182

9. Mélique, X. , Mann, C. M. , Mounaix, P. , Thornton J., Vanbésien, O, Mollot, F. , and Lippens, D. (1998) 5mW and 5 % efficiency 216 GHz InP-based Heterostructure barrier varactor tripler, *IEEE Microwave and Guided Wave Letters*, **8**, 384-386

10.Rahal, A. ,. Bosisio, R. G., Rogers, C., Ovey, J. , Sawan, M., and Missous, M.(1995) A W-band medium power multistack quantum barrier varactor frequency multiplier, *IEEE Microwave and Guided Wave Letters*, **5**, 368-370

11. Jones, J.R. , Bishop, W. L. , Jones, S. H., and Tait, G. B. (1997) Planar multibarrier 80/240 GHz Heterostructure Barrier Varactor triplers, *IEEE Transactions on Microwave Theory and Techniques*, **45**, 512-518

12. Krisnamurthi, K. , Nilsen, S. M. , and Harrison, R. G. (1994) GaAs Single-barrier varactors for millimeter-wave triplers: guidelines for enhanced performances, *IEEE Transactions on Microwave Theory and Techniques*, **42**, 2512-2516

13. Rahal, A., Boch, E., Rogers, C. , Ovey, J., and Bosisio, R. G. (1995) Planar multistack quantum barrier varactor tripler evaluation at W-band, *Electronics letters*, **31**, 2022-1995

14. Krisnamurthi, K. and Harrison, R. G. (1996), Millimeter-wave frequency tripling using stacked heterostructure Barrier Varactor on InP, *IEEE Proc. Microwave Antennas Propagation*, **143**, 272-276
15. Räisäinen, A. V. , Tolmunen, T. J. , Natzic, M. , Frerking, M. A. , Brown, E. , Grönqvist, H. and. Nilsen, S. M. (1995), A single barrier quintupler at 170 GHz, *IEEE Transactions on Microwave Theory and Techniques*, **43**, 685-688
16. Bradley, R. F., Kerr, A. R. , and Erikson, N. (1997) Why don't back-to-back abrupt junction frequency triplers work, *Proceedings of the eight International Symposium on Space Terahertz Technology*, Havard Smithsonian, Cambridge,
17. Lheurette, E. , Mounaix, P., Salzenstein, P. , Mollot, F., and Lippens, D. (1996) High performance InP-based Heterostructure barrier varactors in single and stack configuration, *Electronics Letters*, **32**, 1417-1418
18. Havart, R., Lheurette, E., Vanbésien, O., Mounaix, P., Mollot, F. and Lippens D., (1998) Step-like heterostructure barrier varactor", *IEEE Transactions on Electron Devices*, **45**,
19. Kollberg, E. , Tolmunen, T. J. Frerking, M.A., and East, J. (1995) Current saturation in submillimeter wave varactors, *IEEE Transactions on Microwave Theory and Techniques*,. **40**,. 831-838
20. Duez, V., Mélique, X., Vanbésien, O. , Mounaix, P. , Mollot, F. and Lippens, D. (1998), High capacitance ratio with GaAs/InGaAS/AlAs heterostructure quantum well-barrier varactors, *Electronics letters*, **34**, 1860-1861
21 Lheurette, E. , Mélique, X., Mounaix, P., Mollot, F. , Vanbésien, O. and Lippens, D. (1998) Capacitance engineering for InP-based Heterostructure Barrier Varactor, *IEEE Electron Device Letters*, **19**, 338-340
22. Bhaumik, M. , Gelmont, B. , Mattauch, R. J., and Shur (1992) M. Series impedance of planar Schottky diodes operated to 500 GHz *IEEE Transactions on Microwave Theory and Techniques*,. **40**,. 880-885
23 Carbonell, J. , Havart, R., Mélique, X. , Mounaix, P. , Vanbésien, O., Goutoule, J. M. , and Lippens, D. (1998) Reverse engineering through electromagnetic and harmonic balance simulations Proceedings of the European Microwave Conference, 123-128, Amsterdam
24. Mélique, X., Carbonell, J. , Havart, R. , Mounaix, P. , Vanbésien, O. , and Lippens, D. (1998) InGaAs/InAlAs/AlAs heterostructure barrier varactors for harmonic multiplication, *IEEE Microwave and Guided Wave Letters*, **8**, 254-256,
25 Archer J. (1984) An efficient 200-290 GHz frequency tripler incorporating a novel stripline structure, *IEEE Transactions on Microwave Theory and Techniques*, **32**
26 Mann C. M., Rutherford Appleton Laboratory Chilton Oxon, UK, Private communication
27. Mélique, X. , Maestrini, A., Farre, R., Mounaix, P., Favreau, M. , Vanbésien, O. , Goutoule, J. M. , Beaudin, G. , Närhi, T. , and Lippens, T. , (2000) Fabrication and performance of InP based heterostructure barrier varactor in a 250 GHz waveguide tripler, *IEEE Transactions on Microwave Theory and Techniques*, **48**, 1000-1005

ACTIVE TWO-TERMINAL DEVICES FOR TERAHERTZ POWER GENERATION BY MULTIPLICATION

Heribert Eisele

Solid-State Electronics Laboratory
Department of Electrical Engineering & Computer Science
University of Michigan
Ann Arbor, Michigan 48109-2122

Abstract

Devices from two major groups of active two-terminal devices, *i.e.*, transit-time diodes and transferred-electron devices, have been employed successfully to generate RF power levels of more than 1 mW at frequencies above 200 GHz. The extraction of higher harmonics from the internal frequency multiplication in these inherently nonlinear devices has often been the means for reaching submillimeter-wave frequencies. The RF power levels at high millimeter-wave and even submillimeter-wave frequencies are also high enough to reach higher terahertz frequencies with external frequency multiplication. At frequencies up to 300 GHz, Si IMPATT diodes yielded the highest RF power levels from any fundamental solid-state source, *e.g.*, 50 mW at 245 GHz and 7.5 mW at 285 GHz. However, the RF power of more than 1 mW from InP Gunn devices around 315 GHz is the highest from any such fundamental RF source above 300 GHz. GaAs TUNNETT diodes yielded 100 ± 5 mW in the fundamental mode at 100–107 GHz. Operating as efficient self-oscillating frequency multipliers, they generated RF power levels of more than 10 mW at 202 GHz and more than 4 mW at 235 GHz. GaAs IMPATT diodes, *e.g.*, yielded 2 mW at 232 GHz and 1 mW at 242 GHz. This paper reviews the power generation capabilities and basic properties of these two-terminal devices as fundamental RF sources. It compares the above-mentioned devices, but also other two-terminal devices such as RTDs and SLEDs directly at the RF power level, but also in terms of RF power per unit area as a figure of merit. The paper also compares up-conversion efficiencies and dc power consumption of these two-terminal devices.

R.E. Miles et al. (eds.), Terahertz Sources and Systems, 69–86.

1. Introduction

Systems for rapidly emerging applications at submillimeter-wave frequencies such as upper atmospheric imagery, remote sensing, array receivers in radio astronomy, high-resolution near-object analysis, and ultra-wide bandwidth communications require reliable and compact radio frequency (RF) sources with low direct current (dc) power consumption as one of their key components. RF power generation at high millimeter-wave frequencies was already demonstrated with various types of solid-state devices [1] thus making them good candidates either for sources at submillimeter-wave frequencies or for driving frequency multipliers. The practical application of "classical," but also more recent principles in device physics [2]–[4] resulted in major advances in device performance, which have also been enabled by epitaxial materials of superior properties or quality as well as much improved processing technologies. Amplifiers with three-terminal devices reach higher and higher frequencies [5]–[9], and improvements in their RF output power [10]–[12] as well as their increasing use in systems applications are being reported up to millimeter-wave frequencies. Nonetheless, there have been a few practical demonstrations of fundamental oscillators with three-terminal devices above 100 GHz, and, in addition, only low RF power levels of typically less than 2 mW were reported [13]–[17]. This paper gives an overview of the basic properties of different two-terminal devices for continuous-wave (CW) RF power generation at submillimeter-wave frequencies and mainly focuses on devices that are either well suited for driving frequency multipliers or, in the laboratory, already yielded RF power at frequencies above 200 GHz.

2. RTDs and Other Quantum-Well Devices

Resonant tunneling through discrete energy levels of a so-called quantum well was first demonstrated in 1974 [4]. However, only after major advances in growth techniques, such as molecular beam epitaxy (MBE) and metalorganic chemical vapor deposition (MOCVD) in the 1980s, were device structures suitable for oscillators grown and evaluated. As can be seen from Figure 1, the highest oscillation frequency of any fundamental solid-state RF source were reported from InAs/AlSb resonant tunneling diodes (RTDs) [18], [19]. The dc-to-RF conversion efficiencies, however, are below 1% at millimeter-wave frequencies and the RF power levels of, *e.g.*, 3 μW at 360 GHz and 0.3 μW at 712 GHz [19] are low, which severely limits the use of RTDs in systems applications.

More recently, superlattice electronic devices (SLEDs) were demonstrated as millimeter-wave oscillators [20]–[23], and the results are included in Figure 1. The devices were typically not mounted in a resonant cavity, but they were in quasi-planar structures and were contacted with different ground-signal or ground-signal-ground (GSG) probes for *V*-band (50–75 GHz) or *W*-band (75–110 GHz) [21]–[22]. Exceptions are the quasi-planar circuits in a WR-15/WR-6 waveguide combination [20] and in a WR-6 waveguide as shown in Figure 2 [23]. As an example, Figure 2 also illustrates the setup with a GSG probe for *W*-band [22]. RF power levels (and corresponding dc-to-RF conversion efficiencies) of 0.4 mW (1%) were reported at the fundamental frequency of 56 GHz and 40 μW (0.1%), at the second-harmonic frequency of 112 GHz from the quasi-planar circuit configuration in a waveguide [20]. With the GSG probe and a spectrum analyzer, an RF power of 0.5 mW (0.3%) was determined at 103 GHz [22].

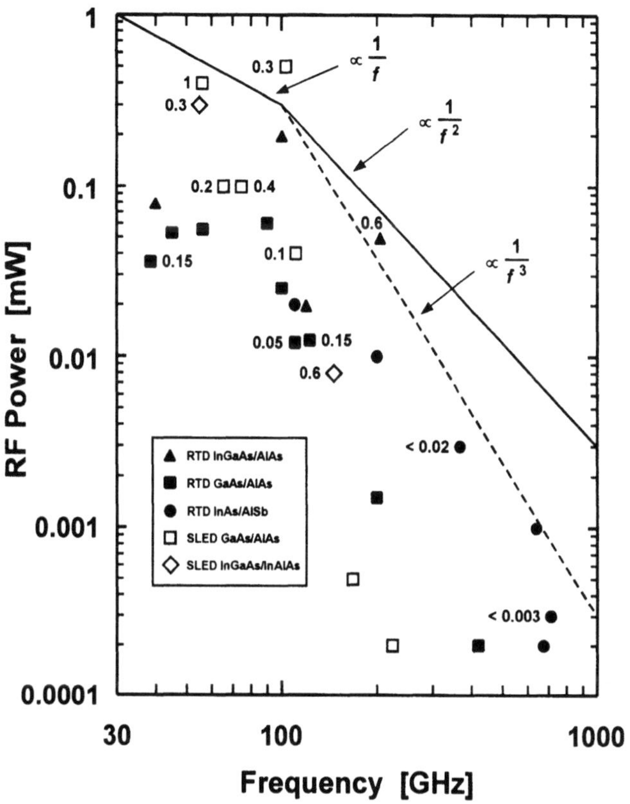

Figure 1. Published state-of-the art results from RTDs and SLEDs in the 30–1000 GHz frequency range. Numbers next to the symbols denote dc-to-RF conversion efficiencies in percent, where applicable.

The length of the active region ranges approximately from 0.44 to 0.64 μm in these GaAs/AlAs SLEDs. Therefore, the higher oscillation frequency of 103 GHz is attributed to much narrower barriers of the quantum wells, which are only two monolayers of AlAs thick and cause a larger miniband width of 120 meV [22]. At a similar length of the active region, an InGaAs/InAlAs SLED with a miniband width of 190 meV generated 80 μW (0.6%) at the higher oscillation frequency of 147 GHz [23]. Recorded spectra of these SLEDs in free-running oscillators, however, are not as clean as, *e.g.*, those of Gunn devices in free-running oscillators, but are quite narrow-band even without a resonant circuit of a moderate-to-high quality factor *Q*.

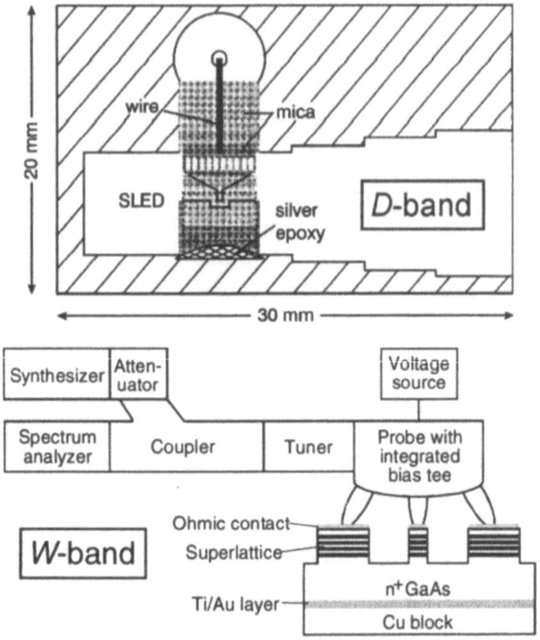

Figure 2. Experimental setup for evaluation of SLEDs at *W*-band and *D*-band frequencies, respectively. After E. Schomburg *et al.* [22], [23].

RF power levels from SLEDs are higher than those from RTDs of the same material system up to *D*-band (110–170 GHz) frequencies, whereas dc-to-RF conversion efficiencies are comparable. They are still much lower than the RF power levels from impact avalanche transit-time (IMPATT) and tunnel injection transit-time (TUNNETT) diodes as well as Gunn devices at the same frequencies and not yet sufficient to drive frequency multipliers.

The device area *A* varies considerably among different two-terminal devices. Therefore, the RF power density Φ_{RF} ($= P_{RF}/A$) provides a better comparison of the power generating capabilities of various two-terminal devices than the RF power P_{RF} alone. Therefore, a comparison on the basis of Φ_{RF} should be considered analogous to the performance comparisons quoted as RF power per gate length for three-terminal devices, such as field-effect transistors (FETs) or high electron mobility transistors

(HEMTs). Figure 3 shows Φ_{RF} as derived from published state-of-the-art results and device areas in the 30–1000 GHz frequency range for device types that, so far, have yielded measurable RF power levels above 200 GHz.

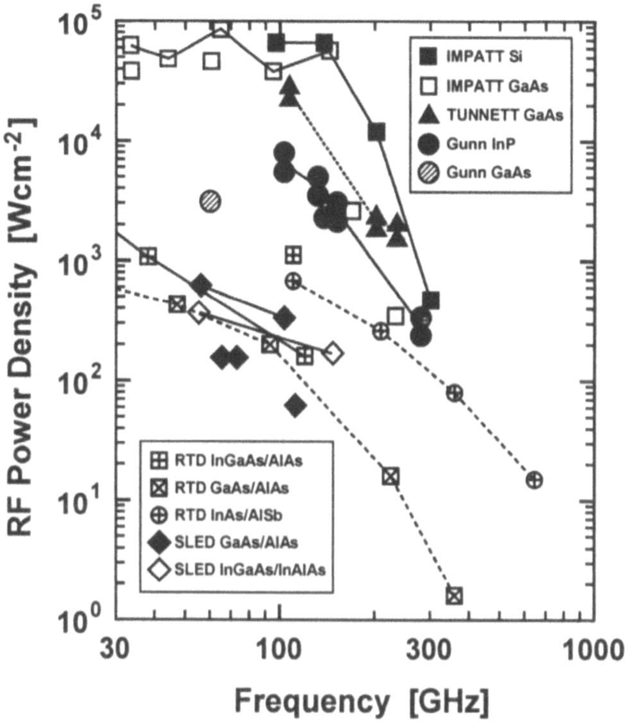

Figure 3. RF power density Φ_{RF} in two-terminal devices as derived from published state-of-the art results. Ranges are given for Φ_{RF} of InP Gunn devices and GaAs TUNNETT diodes at particular frequencies if device areas are not known exactly. Lines between data points at different frequencies are only as a guide to the eye.

3. Transit-Time Diodes

Si IMPATT diodes were the first semiconductor devices to generate RF power above 300 GHz. As shown in Figure 4 (and in conjunction with Figures 1 and 5), they yielded the highest RF power levels from any solid-state fundamental RF source up to 300 GHz. Exemplary RF power levels (and corresponding dc-to-RF conversion efficiencies) of 50 mW (1.3%) at 202 GHz [25]; 44 mW (1.2%) at 214 GHz [25]; 50 mW (< 2%) at 217 GHz [26], [27]; 50 mW (< 1%) at 245 GHz [27]; 12 mW (< 0.5%) at 255 GHz [27]; 7.5 mW (0.35%) at 285 GHz [28]; 1.2 mW (< 0.05%) at 301 GHz [25]; and 0.2 mW at 361 GHz [28] were measured at very high operating junction temperatures of typically more than 300 °C with waveguide circuits at room temperature [26]. Higher RF power levels of, *e.g.*, 4.5 mW (0.13%) at 295 GHz and 2.2 mW (0.047%) at 412 GHz, but also higher oscillation frequencies of up to 430 GHz were attained by cooling the heat sink of the diode and the waveguide circuit to 77 K (liquid nitrogen) [29].

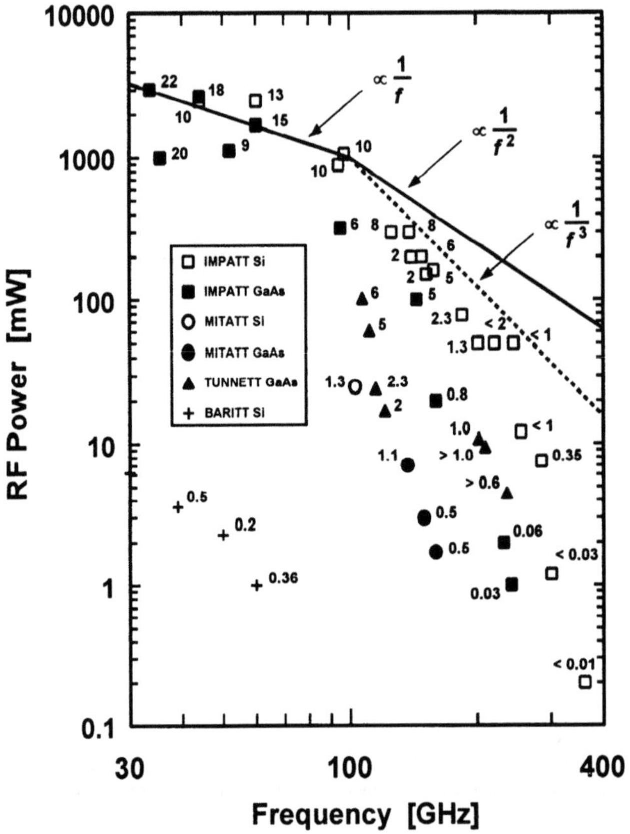

Figure 4. Published state-of-the-art results from Si and GaAs transit-time diodes under CW operation at room temperature in the frequency range of 30–400 GHz. Numbers next to the symbols denote dc-to-RF conversion efficiencies in percent.

Impact ionization as the carrier generation mechanism in IMPATT diodes is a major contributor to the noise of free-running oscillators compared to, *e.g.*, the smaller contribution from mainly thermal noise in the domain formation process of Gunn devices [1], [2]. Therefore, IMPATT diodes are generally considered quite noisy, although techniques are known on how to reduce noise contributions from impact ionization at the price of typically lower dc-to-RF conversion efficiencies [1], [2]. The noise properties of IMPATT diodes restrict their use as local oscillator (LO) sources to either receiver applications where the highest sensitivity is not of primary concern or sensitive low-noise receivers in which well-balanced Schottky diodes as subharmonically pumped mixers cancel out the noise from the LO [30].

TUNNETT diodes, on the other hand, are based on a fast and quiet primary carrier injection mechanism [1], [2], which makes them another prime candidate for RF power generation at high millimeter-wave and submillimeter-wave frequencies. RF power levels of 100 ± 5 mW (and corresponding dc-to-RF conversion efficiencies of around 6%) were measured with GaAs TUNNETT diodes on diamond heat sinks in the fundamental mode at oscillation frequencies of 100–107 GHz [24].

Power extraction at higher harmonic frequencies was investigated with Si [25], [31] and GaAs [32], [33] IMPATT diodes as well as GaAs TUNNETT diodes [24], [34], [35], and the best results are also included in Figure 4. Diode structures and properties suggest operation in a second- or even third-harmonic mode for some of the aforementioned state-of-the-art results from Si IMPATT diodes above 200 GHz. As an example, the RF power of 1.2 mW at 301 GHz was thought to be generated at the third-harmonic frequency [25], but for many other results at these high millimeter-wave frequencies, no attempts to determine the exact mode of operation were reported. GaAs TUNNETT diodes, operated in a second-harmonic mode up to 237 GHz, yielded RF power levels exceeding 10 mW at 202 GHz, 9 mW around 210 GHz, and 4 mW around 235 GHz [34], [35]. These power levels correspond to dc-to-RF conversion efficiencies around 1% at 202 GHz and 210 GHz as well as above 0.6% around 235 GHz [34], [35]. The diodes operate as self-oscillating multipliers with a large modulation of the depletion region, which is akin to that of high-performance varactor diodes and is responsible for the high up-conversion efficiencies of more than 20% [34]. In addition, very clean spectra were recorded from these GaAs TUNNETT diodes in the fundamental as well as in a second-harmonic mode [24], [34], [35]. TUNNETT diodes also yielded the lowest small-signal frequency-modulation (FM) noise measure reported from any oscillator with a two-terminal device [1], [24].

These GaAs TUNNETT diodes were not designed for optimum operation as self-oscillating multipliers, and the experimental results indicate that they reach the performance peak below the maximum permissible bias current density [34]. Therefore, higher RF power levels and higher operating frequencies are expected from TUNNETT diodes that are designed for this mode of operation and have shorter active regions as well as doping concentrations appropriate for higher current densities. Simulations also indicated that RF power levels of the order of 10 mW can be generated in the 240–280 GHz frequency range with GaAs single-drift TUNNETT diodes operating in the fundamental mode [36].

The measured RF power levels and, particularly, conversion efficiencies of these TUNNETT diodes in a second-harmonic mode are comparable to those obtained from all-solid-state RF sources with frequency multipliers where, *e.g.*, Schottky-barrier varactor diodes are driven by Gunn devices [37], [38] and/or transistor amplifiers [39], [40]. Figure 6 compares results from various frequency multipliers in the 100–1000 GHz frequency range with select results from GaAs TUNNETT diodes, InP Gunn devices, and RTDs.

GaAs IMPATT diodes generated RF power levels up to 2 mW at *Y*-band (170–260 GHz) frequencies [32], [33] and RF power was detected up to 290 GHz [33]. However, contrary to preliminary results from GaAs TUNNETT diodes [24], a very preliminary study of the RF and noise properties of single-drift GaAs IMPATT diodes in a second-harmonic mode found that the measured spectra of free-running oscillators with these diodes [33] were clearly not as clean as those of free-running oscillators with TUNNETT diodes (or Gunn devices). This result was indeed expected from approximately 7–8 dB higher values of the small-signal FM noise measure and at least 6 dB higher values of the large-signal FM noise measure of GaAs single-drift IMPATT diodes in the fundamental mode at *W*-band frequencies [1], [24]. Furthermore, dc-to-RF conversion efficiencies of, *e.g.*, 0.04% at 194 GHz from a single-drift structure [33] or 0.06% at 232 GHz from a double-drift structure [32] were inferior to those of the TUNNETT diodes in a similar circuit configuration [24], [34].

4. Transferred-electron devices

GaAs or InP transferred-electron (or Gunn) devices are quite frequently employed in all-solid-state RF sources for submillimeter-wave frequencies, where they generate RF power at medium millimeter-wave frequencies of 60–150 GHz and then drive one or more stages of frequency multipliers with Schottky-barrier varactor diodes [30], [37], [38] or, more recently, heterojunction-barrier varactor diodes [41], [42]. More than ten semiconductor materials in the III-V and II-VI groups are known to exhibit the transferred-electron effect, [2], [43], however, so far, only GaAs and InP have been exploited commercially. GaAs Gunn devices are not a topic of this paper since only InP Gunn devices have so far yielded RF power above 200 GHz.

The highest dc-to-RF conversion efficiencies at low to medium millimeter-wave frequencies and excellent performance up to *D*-band frequencies were reported from InP TEDs with n^-n^+ layer structures and current-limiting contacts at the cathode side of the active region. Examples of RF power levels (and corresponding conversion efficiencies) are 380 mW at 57 GHz (10.6%) in the fundamental mode [44] as well as 175 mW (7%) at 94 GHz, 85 mW (3.8%) at 125 GHz, and 65 mW (2.6%) at 138 GHz in a second-harmonic mode [45], [46]. As shown in Figure 5, the highest RF power levels from any Gunn device at any frequency above 100 GHz were achieved with devices that had an $n^+n^-n^+$ layer structure and a graded doping profile in the active region as shown schematically in the inset of Figure 7. With a few exceptions, they were all mounted on diamond heat sinks [24], [43], [47] for efficient heat removal and low operating temperatures in the active region. Operation in the fundamental mode was observed up to 165 GHz. As examples, RF power levels (and corresponding dc-to-RF conversion efficiencies) exceeded 200 mW (2.6%) at 103 GHz [43]; 130 mW (2.2%) at 132 GHz; 80 mW (1.4%) at 152 GHz; and 30 mW at 162 GHz [24], [43]. Such high RF power levels and the excellent noise performance [1], [24], [35], [43], [47] make these InP Gunn devices ideally suited for driving high-performance Schottky-diode frequency multipliers in sensitive terahertz receivers.

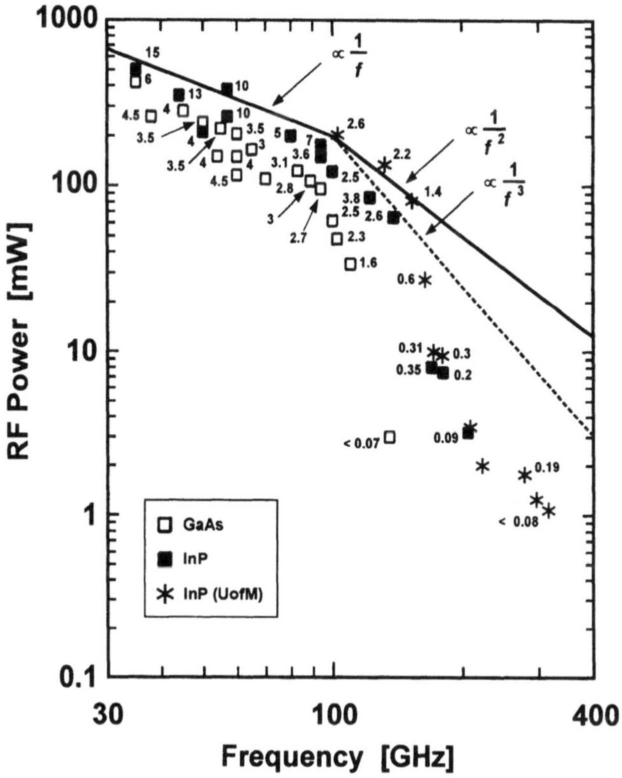

Figure 5. Published state-of-the-art results from GaAs and InP Gunn devices under CW operation at room temperature in the frequency range of 30–400 GHz. Numbers next to the symbols denote dc-to-RF conversion efficiencies in percent.

As a characteristic of operation in the fundamental mode, oscillators with these InP Gunn devices on diamond heat sinks also exhibit wide mechanical tuning ranges of typically several GHz [47]. In addition, they can be electronically tuned through the bias voltage over at least several hundred GHz at *D*-band frequencies [33]. As one example, the RF power of 20–30 mW is available over an electronic tuning range of 152.5–153.5 GHz. As another example, the oscillator with a mechanical tuning range of 131–135.5 GHz [47] can be tuned electronically from 133.4 GHz to 134.0 GHz and generates RF power levels of 70–90 mW if a back short position different from that of the maximum RF power [47] is chosen.

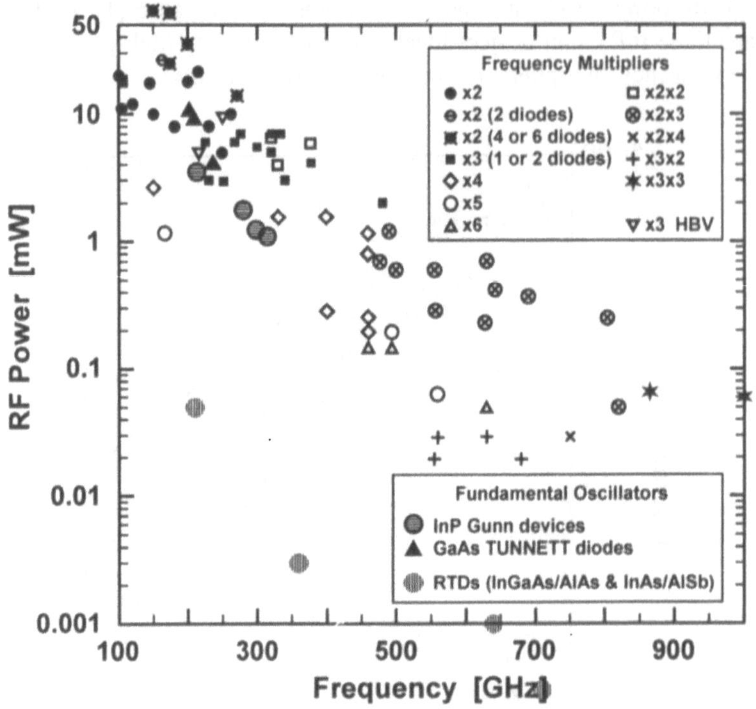

Figure 6. Published state-of-the-art results from frequency multipliers with GaAs Schottky-barrier varactor diodes or InP-based heterojunction-barrier varactor (HBV) diodes in the 100–1000 GHz frequency range in comparison with published state-of-the-art results from GaAs TUNNETT diodes, InP Gunn devices, and RTDs above 200 GHz. Frequency multipliers and fundamental oscillators are all operated at room temperature.

Operation in a second-harmonic mode helps overcome the inherent fundamental frequency limit in Gunn devices. RF power levels of 10 mW at 171.5 GHz and 9.5 mW at 179.8 GHz were measured with devices on integral heat sinks in the circuit configuration of Reference [48]. More than 3.5 mW at 214 GHz, more than 2 mW around 220 GHz, as well as more than 1 mW up to 315 GHz were measured [35], [49] with devices on diamond heat sinks. Impedance matching at the second-harmonic frequency requires device areas that are smaller than those of devices generating RF power in the fundamental mode in the same WR-6 waveguide cavity [35]. Figure 7 summarizes the best results from several devices with diameters in the range of 30–45 μm. The RF power levels above 300 GHz are the highest reported to date from any solid-state fundamental RF source and, as can be seen from Figure 6, they approach those from frequency multipliers [37]. These RF power levels even exceed early performance predictions [50]–[53], but the results of more recent simulations indicate that significant increases in RF power levels can be expected from more optimized device structures and circuits up to at least 320 GHz [54], [55], [56]. The trend in the observed up-conversion efficiencies up to 300 GHz is another indication of the strong RF power generating potential of InP Gunn devices in a second-harmonic mode at submillimeter-wave frequencies. At medium millimeter-wave frequencies, typical up-conversion efficiencies are in the 10–20% range for GaAs [59] and InP [44]–[46], [51] Gunn devices. Around 180 GHz, they are in the 10–13% range for InP Gunn devices [33], [48] and still more than 9% for the best result at 300 GHz in Figure 7.

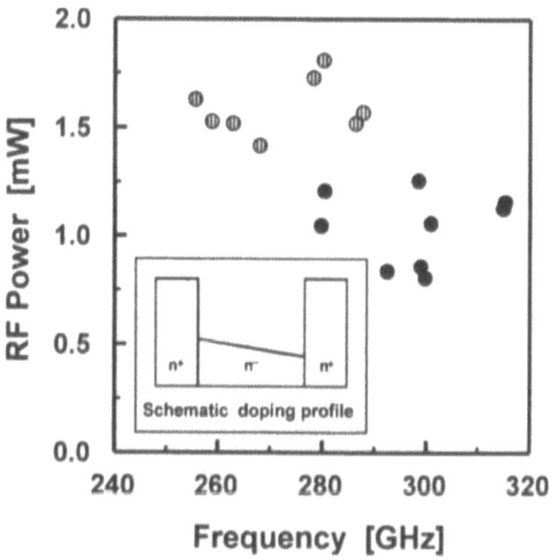

Figure 7. Measured CW RF power levels from InP Gunn devices with an $n^+n^-n^+$ structure and a doping gradient in the active region for the 240–320 GHz frequency range. Numbers next to the symbols denote dc-to-RF conversion efficiencies in percent.

Figure 8 compares the experimental results in the fundamental mode, as well as those in a second-harmonic mode with the recent predictions from detailed simulations [55], [56] that employed a Monte Carlo-based harmonic balance technique [54]. Monte Carlo simulations were originally used in the design of the above-mentioned *D*-band structures [57]. Monte Carlo simulations also showed good agreement between performance predictions and measured results for different device structures at frequencies from 75 GHz to 150 GHz [33], [54], [58], [60]. The predicted results in Figure 8 are for device structures, which were designed for operation in a second-harmonic mode in the 200–500 GHz frequency range and typically had a doping gradient in the active region as shown schematically in the inset of Figure 7 [55], [56].

The measured RF power levels in the fundamental mode at *D*-band frequencies and the predicted RF power levels for operation in a second-harmonic mode at *J*-band (220–325 GHz) frequencies follow a clear $1/f^3$ roll-off up to about 300 GHz. This trend evidently shows that substantial performance improvements can be expected from optimized Gunn device structures and waveguide circuits at *J*-band frequencies. The trend above 300 GHz may also indicate that the transferred-electron effect in InP could be utilized even above 320 GHz and possibly up to 500 GHz [56] with devices operating in a second- or third-harmonic mode. In addition, the experimental results from InP Gunn devices above 220 GHz imply that the derivations of fundamental physical frequency limits for the transferred-electron effect in InP should be revisited and re-investigated in greater detail.

5. Conclusions

RF power generating capabilities of several two-terminal devices at millimeter- and submillimeter-wave frequencies were reviewed and compared. Compared to GaAs, InP is the better semiconductor material system to reach submillimeter-wave frequencies with Gunn devices. Experimental results from both GaAs and InP Gunn devices indicate that the fundamental frequency limit of the transferred-electron effect may be higher than previously thought. More detailed experimental studies are necessary to determine the ultimate fundamental frequency limit, *i.e.*, if InP Gunn devices are capable of generating significant RF power levels in the 350–500 GHz frequency range. Nonetheless, simulations clearly indicate that substantial performance improvements are feasible at *J*-band frequencies, which may help eliminate one or two stages in a multiplier chain to reach terahertz frequencies. Predictions from different simulation tools also clearly show that, in any of the two-terminal devices presented in this paper, the contact resistances of ohmic contacts may cause severe performance degradation. Further improvements beyond current state-of-the-art technologies may be necessary to exploit the full potential of these two-terminal devices at submillimeter-wave frequencies.

As can be seen from the published results in Figures 1, 4, and 5, the RF power levels from InP Gunn devices on diamond heat sinks are the highest reported to date from any solid-state fundamental RF source operated at room temperature and for frequencies above 300 GHz. Likewise, GaAs TUNNETT diodes are considered the second-most powerful solid-state fundamental RF source operated at room temperature and for frequencies above 200 GHz. Reliable long-term operation requires the heat dissipation in the device to be minimized. Both Gunn devices and TUNNETT diodes on diamond heat sinks meet these requirements since the above-mentioned state-of-the-art-results were obtained at "cool" operating active-layer temperatures estimated to be generally below 200 °C, but most often well below 150 °C [1], [34], [35],[43], [47], [49], [58].

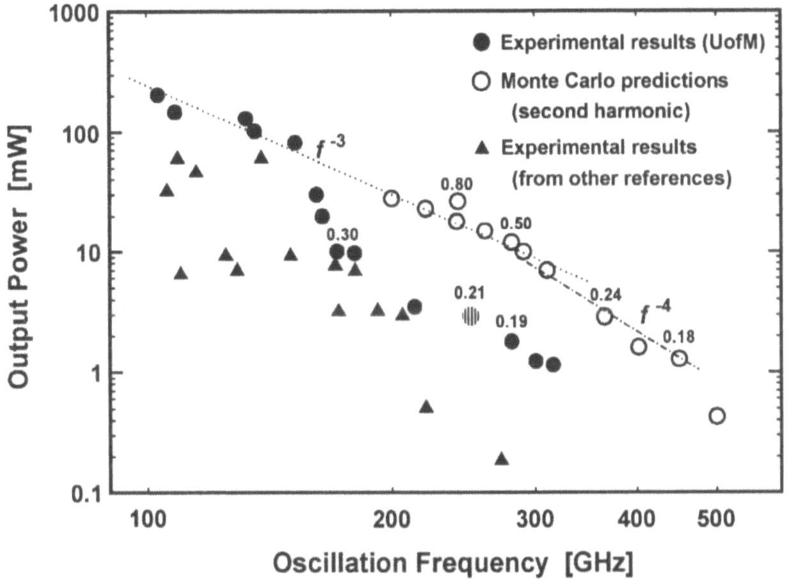

Figure 8. Comparison of predicted [54], [55], [56] and measured [24], [49] CW RF power levels from InP Gunn devices with an $n^+n^-n^+$ structure and a doping gradient in the active region for the 100–500 GHz frequency range. Numbers next to the symbols denote corresponding predicted or measured dc-to-RF conversion efficiencies in percent. Published results from other InP Gunn devices are also shown as a reference.

Other semiconductor material systems, such as GaN or InN, are known to exhibit a transferred-electron effect [43] that could be utilized for millimeter-wave Gunn devices. However, various Monte Carlo simulations predict its onset to occur in these semiconductor materials at five to more than ten times higher threshold electric fields than in InP [61]–[64]. Furthermore, much smaller values for the negative differential mobility at electric fields above twice the threshold field, as well as lower values for the low-field electron mobility, require much higher doping × length products for rapid domain formation than in GaAs or InP. Higher electric fields and, as a result, bias voltages, will enhance the RF power generating capabilities, but, together with higher doping × length products, may cause dc input power densities that are at least one order of magnitude higher than in InP and, even on diamond heat sinks, result in extreme overheating of the device. As a consequence, major advances in material growth, material characterization, and experimental verification of material properties and parameters, as well as appropriate device fabrication technologies, are required before the potential and capabilities of any of these material systems can be assessed fully and eventually utilized.

Acknowledgments

Funding from JPL under contracts 961299 and 961527 as well as from NSF under grant ECS 98-03781 is kindly acknowledged.

References

[1] H. Eisele and G. I. Haddad, "Active Microwave Diodes," in *Modern Semiconductor Devices*, S. M. Sze, Ed., Ch. 6, John Wiley & Sons, New York, 1997, pp. 343–407.

[2] S. M. Sze, *Physics of Semiconductor Devices*, 2nd Ed., John Wiley & Sons, New York, 1981.

[3] L. Esaki and R. Tsu, "Superlattice and negative differential conductivity in semiconductors," *IBM J. Res. Develop.*, **14**(1), 1970, pp. 61–65.

[4] L. L. Chang, L. Esaki, and R. Tsu, "Resonant tunneling in semiconductor double-barriers," *Appl. Phys. Lett.*, **24**(12), 1974, pp. 593–595.

[5] B. Agarwal, A. E. Schmitz, J. J. Brown, M. Matloubian, M. G. Case, M. Le, M. Lui, and M. J. W. Rodwell, "112-GHz, 157-GHz, and 180-GHz InP HEMT traveling-wave amplifiers," *IEEE Trans. Microwave Theory Tech.*, **46**(12), 1998, pp. 2553–2559.

[6] S. Weinreb, T. Gaier, R. Lai, M. Barsky, Y. C. Leong, and L. Samoska, "High-gain 150–215-GHz MMIC amplifier with integral waveguide transitions," *IEEE Microwave Guided Wave Lett.*, **9**(7), 1999, pp. 282–284.

[7] R. Lai, M. Barsky, T. Huang, M. Sholley, H. Wang, Y. L. Kok, D. C. Streit, T. Block, P. H. Liu, T. Gaier, and L. Samoska, "An InP HEMT MMIC LNA with 7.2-dB gain at 190 GHz," *IEEE Microwave Guided Wave Lett.*, **8**(11), 1998, pp. 393–395.

[8] Q. Lee, B. Agarwal, D. Mensa, R. Pullela, J. Guthrie, L. Samoska, and M. J. W. Rodwell, "A > 400 GHz f_{max} transferred-substrate heterojunction bipolar transistor IC technology," *IEEE Electron Device Lett.*, **19**(3), 1998, pp. 77–79.

[9] D. Mensa, Q. Lee, J. Guthrie, S. Jaganathan, and M. J. W. Rodwell, "Transferred-substrate HBTs with 254 GHz f_τ," *Electron. Lett.*, **35**(7), 1999, pp. 288–290.

[10] Y. C. Chen, D. L. Ingram, R. Lai, M. Barsky, R. Grunbacher, T. Block, H. C. Yen, and D. C. Streit, "A 95-GHz InP HEMT MMIC amplifier with 427-mW power output," *IEEE Microwave Guided Wave Lett.*, **8**(11), 1998, pp. 399–401.

[11] C. Pobanz, M. Matloubian, V. Radisic, G. Raghavan, M. Case, M. Micovic, M. Hu, C. Nguyen, S. Weinreb, and L. Samoska, "High performance MMICs with submillimeter wave InP-based HEMTs," *Proc. 12th Int. Conf. Indium Phosphide Related Mat.*, Williamsburg, VA, May 14–18, 2000.

[12] H. Wang, L. Samoska, T. Gaier, A. Peralta, H. H. Liao, Y. C. Chen, M. Nishimoto, and R. Lai, "Monolithic Power Amplifiers Covering 70–113 GHz," *IEEE MTT-S Int. Microwave Symp. Dig.*, Boston, MA, June 11–16, 2000.

[13] Y. Kwon, D. Pavlidis, T. L. Brock, and D.C. Streit, "A *D*-band monolithic fundamental oscillator using InP-based HEMT's," *IEEE Trans. Microwave Theory Tech.*, **41**(12), 1993, pp. 2336-2344.

[14] S. E. Rosenbaum, B. K. Kormanyos, L. M. Jelloian, M. Matloubian, A. S. Brown, L. E. Larson, L. D. Nguyen, M. A. Thompson, L. P. B. Katehi, and G. M. Rebeiz, "155- and 213-GHz AlInAs/GaInAs/InP HEMT MMIC oscillators," *IEEE Trans. Microwave Theory Tech.*, **43**(4), 1995, pp. 927–933.

[15] K. Uchida, I. Aoki, H. Matsuura, T. Yakihara, S. Kobayashi, S. Oka, T. Fujita, and A. Miura, "104 and 134 GHz InGaP/InGaAs HBT oscillators," *Proc. 1999 GaAs IC Symposium*, Monterey, CA, 1999, pp. 237–240.

[16] K. W. Kobayashi, A. K. Oki, L. T. Tran, J. C. Cowles, A. Gutierrez-Aitken, F. Yamada, T. R. Block, and D. C. Streit, "A 108-GHz InP-HBT monolithic push-push VCO with low phase noise and wide tuning bandwidth," *IEEE Solid-State Circuits*, **34**(9), 1999, pp. 1225–1232.

[17] K. Uchida, H. Matsuura, T. Yakihara, S. Kobayashi, S. Oka, T. Fujita, and A. Miura, "A power-combined W-band HBT oscillator," *IEEE MTT-S Int. Microwave Symp. Dig.*, Boston, MA, June 11–16, 2000, pp. 51-54.

[18] E. R. Brown, C. D. Parker, A. R. Calawa, M. J. Manfra, C. L. Chen, L. J. Mahoney, and W. D. Goodhue, "High-frequency resonant-tunneling oscillators," *Microwave Optical Technol. Lett.*, **4**(1), 1991, pp. 19-23.

[19] E. R. Brown, J. R. Söderström, C. D. Parker, L. J. Mahoney, K. M. Molvar, and T. C. McGill, "Oscillations up to 712 GHz in InAs/AlSb resonant tunneling diodes," *Appl. Phys. Lett.*, **58**(20), 1991, pp. 2291–2293.

[20] E. Schomburg, J. Grenzer, K. Hofbeck, T. Blomeier, S. Winnerl, S. Brandl, A. A. Ignatov, K. F. Renk, D. G. Pavel'ev, Yu. Koschurinov, V. Ustinov, A. Zhukov, A. Kovsch, S. Ivanov, and P. S. Kop'ev, "Millimeter wave generation with a quasi planar superlattice electronic device," *Solid-State Electron.*, **42**(7–8), 1998, pp. 1495–1498.

[21] S. Brandl, E. Schomburg, R. Scheuerer, K. Hofbeck, J. Grenzer, K. F. Renk, D. G. Pavel'ev, Yu. Koschurinov, A. Zhukov, A. Kovsch, V. Ustinov, S. Ivanov, and P. S. Kop'ev, "Millimeter wave generation by a self-sustained current oscillation in an InGaAs/InAlAs superlattice," *Appl. Phys. Lett.*, **73**(21), 1998, pp. 3117–3119.

[22] E. Schomburg, M. Henini, J. M. Chamberlain, D. P. Steenson, S. Brandl, K. Hofbeck, K. F. Renk, and W. Wegscheider, "Self-sustained current oscillation above 100 GHz in a GaAs/AlAs superlattice," *Appl. Phys. Lett.*, **74**(15), 1999, pp. 2179–2181.

[23] E. Schomburg, R. Scheuerer, S. Brandl, K. F. Renk, D. G. Pavel'ev, Yu. Koschurinov, V. Ustinov, A. Zhukov, A. Kovsch, and P. S. Kop'ev, "InGaAs/InAlAs superlattice oscillator at 147 GHz," *Electron. Lett.*, **35**(17), 1999, pp. 1491–1492.

[24] H. Eisele and G. I. Haddad, "Two-terminal millimeter-wave sources," *IEEE Trans. Microwave Theory Tech.*, **46**(6), 1998, pp. 739–746.

[25] T. Ishibashi and M. Ohmori, "200-GHz 50-mW CW oscillation with silicon SDR IMPATT diodes," *IEEE Trans. Microwave Theory Tech.*, **24**(11), 1976, pp. 858–859.

[26] N. B. Kramer and R.A. Johnson, "Generating power at mm-wave frequencies," *Microwaves & RF*, **23**(5), 1984, pp. 243–249.

[27] K. Chang, W. F. Thrower, and G. M. Hayashibara, "Millimeter-wave silicon IMPATT sources and combiners for the 110-260-GHz range," *IEEE Trans. Microwave Theory Tech.*, **29**(12), 1981, pp. 1278–1284.

[28] M. Ino, T Ishibashi, and M. Ohmori, "C.W. oscillation with p^+-p-n^+ silicon IMPATT diodes in 200 GHz and 300 GHz bands," *Electron. Lett.*, **12**(6), 1976, pp. 148–149.

[29] T. Ishibashi, M. Ino, T. Makimura, and M. Ohmori, "Liquid-nitrogen-cooled submillimeter-wave silicon IMPATT diodes," *Electron. Lett.*, **13**(10), 1977, pp. 299–300.

[30] I. Mehdi, P. H. Siegel, D. A. Humphrey, T. H. Lee, R. J. Dengler, J. E. Oswald, A. Pease, R. Lin, H. Eisele, R. Zimmermann, and N. Erickson, "An all solid-state 640 GHz subharmonic mixer," *IEEE MTT-S Int. Microwave Symp. Dig.*, Baltimore, MD, June 7–12, 1998, pp. 403–406.

[31] M. Ohmori, T. Ishibashi, and S. Ono, "Dependency of the highest harmonic oscillation frequency on junction diameter of IMPATT diodes," *IEEE Trans. Electron Devices*, **24**(12), 1977, pp. 1323–1329.

[32] H. Böhm, J. Freyer, and M. Claassen, "CW harmonic power generation of GaAs IMPATT diodes above 200 GHz," *Proc. Terahertz Spectroscopy Applications II*, Munich, Germany, SPIE **3828**, 1999, pp. 81–88.

[33] H. Eisele, unpublished results.

[34] H. Eisele, "Efficient second-harmonic power extraction from GaAs TUNNETT diodes above 200 GHz," *Electron. Lett.*, **34**(13), 1998, pp. 1324–1326 **and** *Electron. Lett.*, **34**(15), 1998, pp. 1531.

[35] H. Eisele, A. Rydberg, and G. I. Haddad: "Recent advances in the performance of InP Gunn devices and GaAs TUNNETT diodes for the 100–300-GHz frequency range and above," *IEEE Trans. Microwave Theory Tech.*, **48**(4), 2000, pp. 626–631.

[36] C-C. Chen, R. K. Mains, G. I. Haddad, and H. Eisele, "Numerical simulation of TUNNETT and MITATT devices in the millimeter and submillimeter range," *Proc. 4th Int. Symp. Space Terahertz Technol.*, Los Angeles, CA, March 30–April 1, 1993, pp. 362–376.

[37] A. V. Räisänen, "Frequency multipliers for millimeter and submillimeter wavelengths," *Proc. IEEE*, **80**(11), 1992, pp. 1842–1852.

[38] T. W. Crowe, T. C. Grein, R. Zimmermann, and P. Zimmermann, "Progress toward solid-state local oscillators at 1 THz," *IEEE Microwave Guided Wave Lett.*, **6**(5), 1996, pp. 207–208.

[39] N. R. Erickson, G. Naryanan, R. P. Smith, S. C. Martin, T. W. Crowe, and W. L. Bishop, "Planar frequency doublers and triplers for FIRST," *Proc. 11th Int. Symp. Space Terahertz Technol.*, Ann Arbor, MI, May 1–3, 2000.

[40] E. Schlecht, J. Bruston, R. P. Smith, S. Martin, A. Fung, R. Tsang, I Mehdi, "200 and 400 GHz Schottky diode Multipliers Fabricated with Integrated Air-Dielectric (Substrateless) Circuitry," ," *Proc. 11th Int. Symp. Space Terahertz Technol.*, Ann Arbor, MI, May 1–3, 2000.

[41] X. Mélique, A. Maestrini, P. Mounaix, M. Favreau, O. Vanbésien, J.M. Goutoule, G. Beaudin, T. Nähri, and D. Lippens, "Record performance of a 250 GHz InP-based heterostructure barrier varactor tripler," *Electron. Lett.*, **35**(11), 1999, pp. 938–939.

[42] X. Mélique, A. Maestrini, R. Farré, P. Mounaix, M. Favreau, O. Vanbésien, J.-M. Goutoule, F. Mollot, G. Beaudin, T. Nähri, and D. Lippens, "Fabrication and performance of InP-based heterostructure barrier varactors in a 250-GHz waveguide tripler," *IEEE Trans. Microwave Theory Tech.*, **48**(6), 2000, pp. 1000–1006.

[43] H. Eisele, "Gunn or transferred-electron devices," in *Encyclopedia of Electrical and Electronics Engineering*, J. G. Webster, Ed., vol. 8, John Wiley & Sons, New York, 1999, pp. 523–537.

[44] Fank, B., Crowley, J., Tringali, D., Wandinger, L.: "Basics and Recent Applications of High-Efficiency Millimeter Wave InP Gunn Diodes", *Proc. First Int. Conf. Indium Phosphide Related Mat.*, Norman, OK, SPIE **1144**, 1989, pp. 534–546.

[45] J. D. Crowley, C. Hang, R. E. Dalrymple, D. R. Tringali, F. B. Fank, and L. Wandinger, "140 GHz indium phosphide Gunn diode," *Electron. Lett.*, **30**(6), 1994, pp. 499–500.

[46] J. D. Crowley, R. E. Dalrymple, C. Hang, D. R. Tringali, F. B. Fank, and L. Wandinger, "InP Gunn diodes serve millimeter-wave applications," *Microwaves & RF*, **33**(3), 1994, pp. 143–146.

[47] H. Eisele and G. I. Haddad, "High-performance InP Gunn devices for fundamental-mode operation in *D*-band (110–170 GHz)," *IEEE Microwave Guided Wave Lett.*, **5**(11), 1995, pp. 385–387.

[48] A. Rydberg, "High efficiency and output power from second- and third-harmonic millimeter-wave InP-TED oscillators at frequencies above 170 GHz," *IEEE Electron Device Lett.*, vol. 11, pp. 439–441, Oct. 1990.

[49] H. Eisele, "Second-harmonic power extraction from InP Gunn devices with more than 1 mW in the 260–320 GHz frequency range," *Electron. Lett.*, **34**(25), 1998, pp. 2412–2413.

[50] L. Wandinger, "mm-wave InP Gunn devices: status and trends," *Microwave J.*, **24**(3), 1981, pp. 71–78.

[51] I. G. Eddison, "Indium phosphide and gallium arsenide transferred-electron devices," in *Infrared and Millimeter Waves*, Vol. 11, *Millimeter Components and Techniques, Part III*, Academic Press, Orlando, 1984, pp. 1–59.

[52] P. A. Rolland, G. Salmer, E. Constant, and R. Fauqembergue, "Comparative frequency behavior of GaAs, InP, and GaInAs transferred-electron devices—derivation of a simple comparative criterion," *IEEE Trans. Electron Devices*, **28**(3), 1981, pp. 341–343.

[53] M. R. Friscourt and P.A. Rolland, "Optimum design of n^+-n-n^+ InP devices in the millimeter-range frequency limitation—RF performances," *IEEE Electron Device Lett.*, **4**(5), 1983, pp. 135–137.

[54] R. Kamoua, "Monte Carlo-based harmonic balance technique for the simulation of high-frequency TED oscillators," *IEEE Trans. Microwave Theory Tech.*, **46**(10), 1998, pp. 1376–1381.

[55] R. Kamoua, "Potential of second-harmonic power generation in InP Gunn oscillators above 200 GHz," *Proc. 4th Int. Conf. Millimeter Submillimeter Waves Applications*, San Diego, California, July 20–24, 1998, pp. 32–37.

[56] R. Kamoua and H. Eisele, "Power generation with fundamental and second-harmonic mode InP Gunn oscillators—performance above 200 GHz and upper frequency limits," *Proc. 11th Int. Symp. Space Terahertz Technol.*, Ann Arbor, MI, May 1–3, 2000.

[57] R. Kamoua, H. Eisele, and G. I. Haddad, "*D*-band (110–170 GHz) InP Gunn devices," *Solid-State Electron.*, **36**(11), 1993, pp. 1547–1555.

[58] H. Eisele and G. I. Haddad, "Recent advances in the performance of GaAs- and InP-based two-terminal devices as high-power millimeter-wave sources," *Proc. 4th Int. Conf. Millimeter Wave Far Infrared Science Technol.*, Beijing, China, August 12–15, 1996, pp. 2–5.

[59] J. E. Carlstrom, R. L. Plambeck, and D. D. Thornton, "A continuously tunable 65–115-GHz Gunn oscillator," *IEEE Trans. Microwave Theory Tech.*, **33**(7), 1985, pp. 610–619.

[60] H. Eisele and R. Kamoua, unpublished results.

[61] J. Kolnik, I. H. Oğuzman, K. F. Brennan, R. Wang, P. P. Ruden, and Y. Wang, "Electronic transport studies of bulk zincblende and wurtzite phases of GaN based on an ensemble Monte Carlo calculation including a full zone band structure," *J. Appl. Phys.*, **78**(2), 1995, pp. 1033–1038.

[62] U. V. Bhapkar and M. S. Shur, "Monte Carlo calculation of velocity-field characteristics of wurtzite GaN," *J. Appl. Phys.*, **82**(4), 1997, pp. 1649–1655.

[63] S. Krishnamurthy, M. van Schilfgaarde, A. Sher, and A.-B. Chen, "Bandstructure effect on high-field transport in GaN and GaAlN," *Appl. Phys. Lett.*, **71**(14), 1997, pp. 1999–2001.

[64] E. Bellotti, B. K. Doshi, K. F. Brennan, J. D. Albrecht, and P. P. Ruden, "Ensemble Monte Carlo study of electron transport in wurtzite InN," *J. Appl. Phys.*, **85**(2), 1999, pp. 916–923.

Theme 3

Intersubband Sources

Far-infrared intersubband luminescence from quantum cascade sources

Jérôme Faist, Michel Rochat, Mattias Beck
University of Neuchâtel, CH-2000 Neuchâtel, Switzerland
Jerome.Faist@iph.unine.ch
Ursula Oesterle
EPFL Ecublens 1015 Lausanne

Abstract

A systematic study of different active regions for the realization of a far-infrared quantum cascade source based on intersubband transitions is presented. In particular, emission from structures based on a vertical transition and fabricated from GaAs/AlGaAs and InGaAs/AlInAs are reported and their temperature dependence systematically compared. The guiding loss of a vertical transition emitter embedded in a double plasmon waveguide is studied as a function of injected current density.

I INTRODUCTION

The first theoretical discussions of intersubband-based emitters and lasers were focused on the far-infrared portion of the electromagnetic spectrum[1-4]. The first report of intersubband electroluminesence was also performed in a structure emitting in the far-infrared[5]. The general understanding at that time was that the intersubband lasers would be easier to manufacture for photon energies below the optical phonon energy because the latter was known to lead to very short intersubband lifetimes (~1ps). These early investigations assumed that the non-radiative intersubband transitions would be limited by the emission of acoustic phonons, yielding to low-temperature lifetimes of 100ps or more[2-4]. These long lifetimes were actually confirmed by experiments under low excitation conditions[6,7].

Nevertheless, the first quantum cascade lasers were realized in the mid-infrared[8]. These devices have now demonstrated very high level of performances with high power operation at temperatures much above 300K [9].

Intersubband electroluminescence experiments have been performed in the far-infrared since the work of M. Helm and coworkers[5] in quantum cascade structures based on

R.E. Miles et al. (eds.), Terahertz Sources and Systems, 89–100.

diagonal transitions[10], vertical transitions[11] and in chirped superlattices [12,13] but so far no laser action has been reported to our knowledge. Contrary to original expectations, an intersubband quantum cascade laser turns out to be much more difficult to realize at far-infrared wavelengths compared to the mid-infrared, for two main reasons: an underestimate of the intersubband scattering rate and the difficulty in manufacturing good optical waveguides.

II INTERSUBBAND SCATTERING RATES

First, as pointed out by many authors, Auger-like electron-electron intersubband scattering will be more efficient than acoustic phonons as soon as the sheet carrier density per well is larger than $\sim 10^9 cm^{-2}$. The first evaluations of the Auger electron-electron scattering were reported by Borenstein and Katz[14]. Although the latter group made a technical mistake in the calculation which was corrected by subsequent authors[15-18], the general estimate of the order of magnitude of the effect was correct. The lifetime was now predicted to be between 10 and 100ps for low densities, falling to ps levels for high ($>10^{11} cm^{-2}$) electron densities. These short lifetimes were measured experimentally using picosecond luminescence experiments[19].

The general picture that now emerges is illustrated in Fig. 1. At high temperatures (T>~60K), the intersubband scattering is dominated by the emission of optical phonons by thermally activated electrons. The activation barrier for this process is $\hbar\omega_{LO} - E_{21}$ where $\hbar\omega_{LO}$ is the optical phonon energy ($\hbar\omega_{LO} = 36meV$ in GaAs) and E_{21} is the transition energy. This process leads to lifetimes of the order of hundreds of femtoseconds at room temperature. For completeness, one should also mention the absorption of optical phonons (although it will be significant only at high temperature because of its significantly larger activation energy).

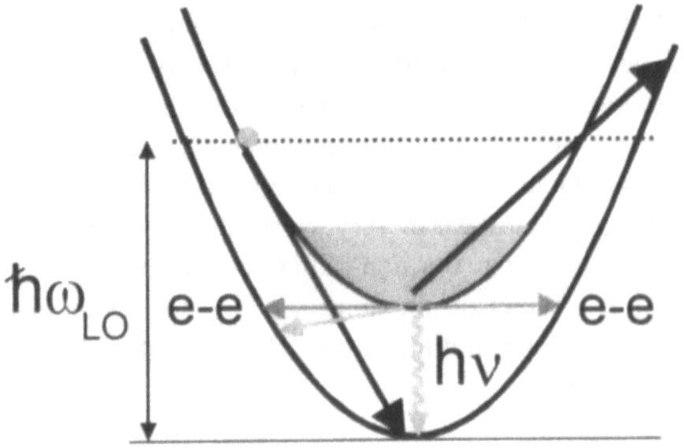

Fig. 1 Schematic diagram of the various intersubband scattering mechanisms.

For low temperatures but non-negligeable electron densities, as mentioned above, electron-electron scattering is going to be the dominating scattering mechanism. Another important elastic process is interface roughness scattering. Because this process is inherently dependent on the interface morphology and therefore on the growth conditions, the strength of this scattering mechanism is very difficult to assess. Some numerical estimates have been attempted for periodic superlattices by Fereirra and Bastard[19]. Finally, for extremely low carrier densities, the only significant scattering mechanism is the emission of acoustic phonons.

This overall picture is very well supported by our experimental evidence. We have demonstrated that the luminescence intensity in our vertical transition structure has a square-root dependence on the injected current[5]. A study of the temperature dependence of the electroluminescence intensity gives additional information on the scattering mechanisms.

I) VERTICAL TRANSITIONS GAAS SAMPLE

Our structure was grown by molecular beam epitaxy on a n-doped GaAs substrate and consists of 35 periods. One period of our structure consists of four GaAs quantum wells separated by thin $Al_{0.15}Ga_{0.85}As$ tunnel barriers. The emission occurs in the 28nm well through a vertical transition, i.e. the optical transition proceeds between two states with a strong spatial overlap.

For the experiments, the samples were processed into 450μm x 450μm mesas and the light coupled using a metal grating with a 15μm periodicity. The samples were mounted on the cold finger of a variable temperature He flow cryostat, and the current driven in pulsed mode. The light was then collected by an off-axis parabolic mirror with an f/1.5 aperture and sent through a Fourier transform infrared spectrometer (FTIR) on a liquid-helium cooled Si Bolometer detector The entire optical path was purged by dry air to minimize water vapour absorption.

Special care was taken to minimize the effect of blackbody emission from the sample. The overall duty cycle was kept at 50% and the phase of the emission was adjusted at low temperature and kept constant as the temperature was raised. Lattice heating was therefore efficiently quenched using this procedure because its phase is at Pi/2 compared to the intersubband luminescence.

As shown in Fig. 2, intersubband electroluminescence is clearly visible in this structure up to temperatures of 100K. The narrow peak observed occurs at an energy (14meV) which corresponds very well with the calculated transition energy of the structure (14.1meV). The linewidth broadens slightly with increasing temperature, from 1meV at low temperature, to 1.4meV at 100K. Above this temperature, the peak broadens significantly and the emission disappears altogether at 120K.

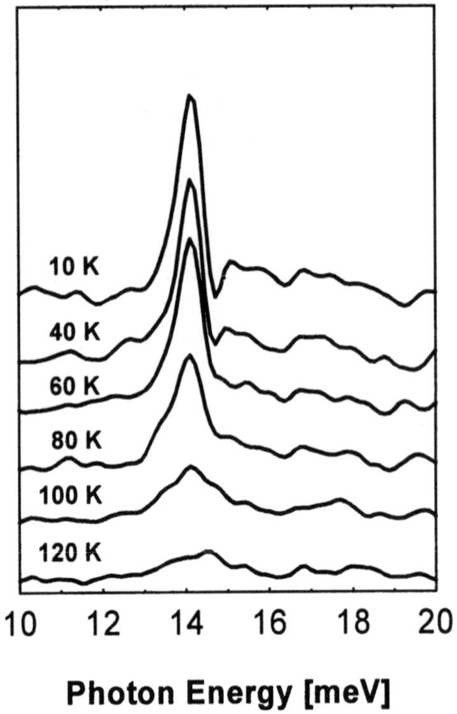

Fig. 2. Luminescence spectra as a function of temperature for a fixed current (68mA) The electroluminescence intensity decreases with increasing temperature and vanishes altogether at T = 120K

The luminescence intensity dependence on temperature confirms our picture of the non-radiative processes, because the intersubband luminescence intensity at fixed injected current is simply proportional to the non-radiative lifetime.

In Fig. 3, the intensity of the peak plotted in Fig.2 is displayed as a function of the inverse temperature. The intersubband scattering rate at finite temperature for our square well is given by the sum of a contribution from optical phonon emission from electrons thermally excited above the optical phonon energy and electron-electron scattering [16]:

$$\tau^{-1} = \frac{\hbar n_s}{\pi m^*}\frac{Ry}{E_{21}}|U(q_\Delta)|^2 I_p + w_{isb}\exp\left[-\frac{\hbar\omega_{LO} - E_{21}}{kT}\right] \qquad (1).$$

In Eqn. 1, Ry = 5.5meV is the effective Rydberg energy for GaAs, $|U(q_\Delta)|^2 \sim 1$ the square matrix element and I_s = 0.78 the phase space factor for electron-electron scattering, evaluated for the momentum q_Δ corresponding to the intersubband transition [11], w_{isb} is the optical phonon intersubband scattering rate for electrons thermally excited above the optical phonon energy. In Fig. 3, the first term representing the electron-electron as well as the collection efficiency are fitted from the data, using the calculated value of the intersubband scattering rate w_{isb}.

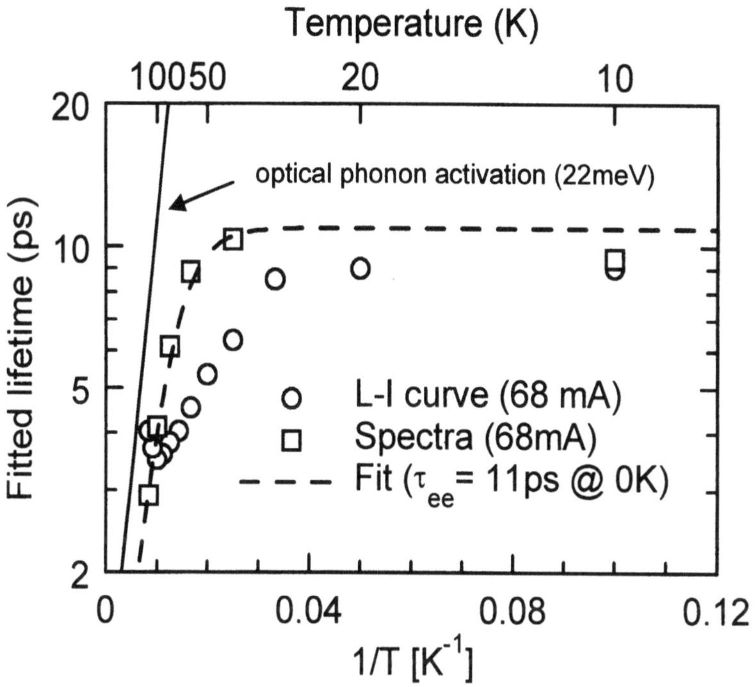

Fig. 3. Luminescence intensity as a function of the inverse temperature for a fixed current (68mA) as measured from the spectra (squares) or from the L-I curves (circles). The dashed line is the fit from equation 1, the full line displays the slope of the optical phonon activation.

As shown in Fig. 3, this model fits the experimental data very well and allows us to derive the electron-electron scattering time at this current. The fit yields the intersubband lifetime $\tau = 11$ps for a current density of J=36A/cm^2, corresponding to an upper state density of $n_s = 2.5 10^9$cm^{-2}. For comparison, Fig. 3 also displays the intensity derived from the L-I curve (squares). The difference from the data derived from the L-I curve and the spectra is attributed to residual blackbody emission.

II) VERTICAL TRANSITIONS INGAAS/ALINAS SAMPLE

It is expected that the electron-electron lifetime should not be dependent on the effective mass of the electron. To verify this fact, a structure with a design very similar to the preceding one was grown in the InGaAs/AlInAs material and also based on a vertical transition. The intersubband luminescence spectra as a function of temperature are displayed in Fig. 4. In this case, the low temperature emission peak is slightly broader (1.2meV) but persists up to a higher temperature of 140K.

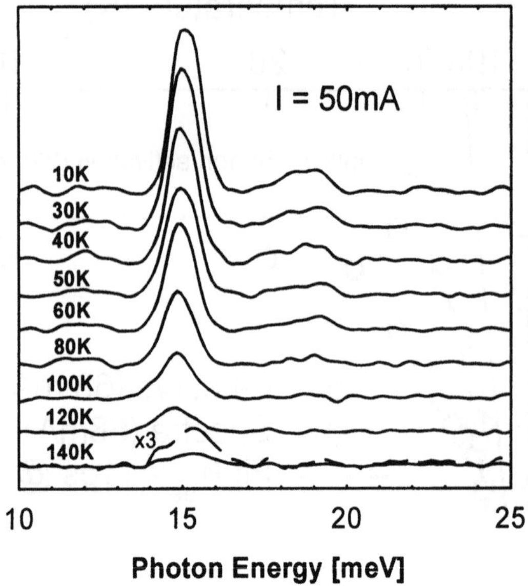

Fig. 4. Luminescence spectra as a function of temperature for a fixed current (50mA) for the InGaAs sample. The electroluminescence intensity decreases with increasing temperature and vanishes altogether at T = 140K

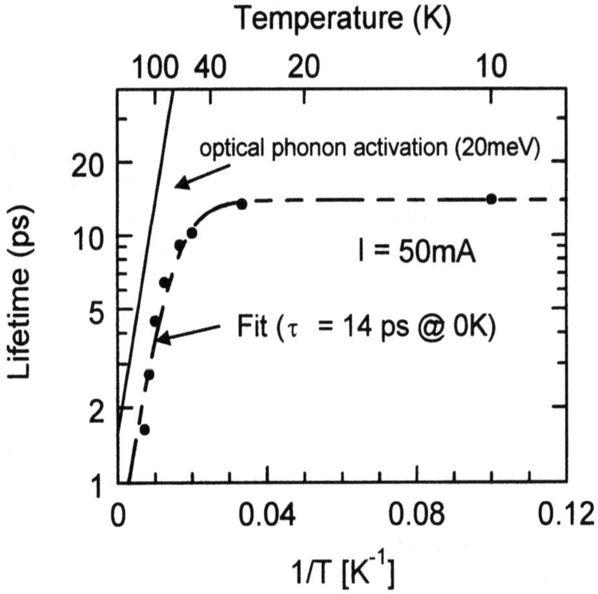

Fig. 5. Luminescence intensity as a function of temperature for a fixed current (50mA) as measured from the spectra. The dashed line is the fit from equation 1, the full line displays the slope of the optical phonon activation.

The slightly higher temperatures obtained in this sample are attributed to the smaller optical phonon scattering rates of the InGaAs-based material compared to the GaAs-based material. Similarly, the integrated intensity of the peak is plotted as a function of inverse temperature and fitted with the same expression as the one used for the GaAs-based sample, as shown in Fig. 5. The fitted value for the e-e scattering time (14ps) is well within the range of the one measured in the GaAs sample.

III WAVEGUIDES

The second roadblock for the fabrication of very long wavelength intersubband quantum cascade lasers stems from the difficulty in the manufacture of suitable waveguides. In the far-infrared, i.e beyond the *reststrahlen* region (~30μm for GaAs), the design of an optical waveguide is hampered by several problems. At these wavelengths, free carrier absorption, proportional to the square of the wavelength in the Drude model becomes prohibitive. This can be overcome by having high gain active regions and very low doping levels. In the mid-infrared, good resonators are obtained by growing low refractive index cladding below and over the high refractive index active region, which can be done with by molecular beam epitaxy as the required cladding thickness is of the order of 3μm [4]. In the far-infrared however, waveguides based on purely dielectric confinement would require cladding thicknesses be of the order of 20μm on both sides of the active region. Recently, *Sirtori* et al. have demonstrated QC laser waveguides based on surface plasmons [20]. In this case, the metal is directly deposited above the active region of the device and used as a guiding interface. Compared to traditional cladding, this techniques allows a drastic reduction of the cladding growth, but at the expense of higher absorption as the mode penetrates partly into the metal. However, this technique is promising for the far-infrared as these losses become less effective as the wavelength is increased. However this method still requires the growth of a thick (about 20μm at FIR wavelengths) lower cladding layer between the active region and the substrate. In order to overcome this problem, we have chosen to use a double surface plasmon mode confinement waveguide. Evaporating metal on both sides of the active region leads to serious technological difficulties, so we have chosen to replace the metal by highly doped GaAs grown on both sides of the active region, allowing us to have metal-like semiconductor interfaces. This approach leads to similar confinement at the expense of slightly higher waveguide losses (calculated to be $51 cm^{-1}$ instead of $38\ cm^{-1}$ for metal confinement). A parabolically graded $Al_xGa_{1-x}As$ transition region has been introduced between the doped and undoped regions to give an abrupt carrier distribution profile. The calculated mode intensity profile is displayed in Fig. 6.

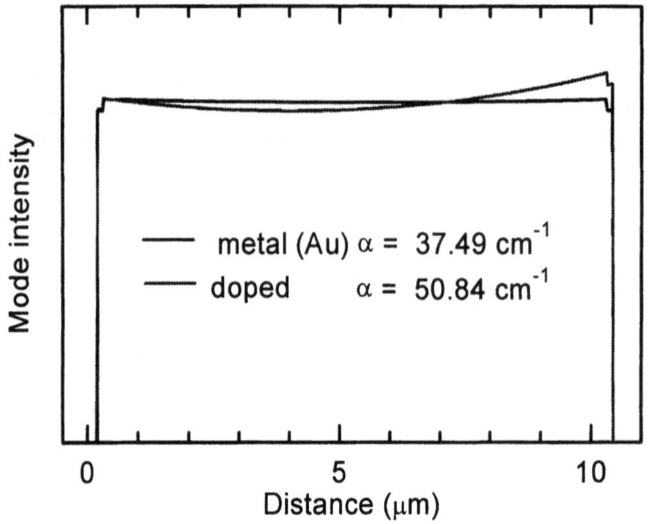

Fig. 6. Intensity of the mode as a function of distance for a double-plasmon waveguide.

The samples were processed in 500μm large ridges by wet chemical etching. 10/400nm Ti/Au ohmic contacts were deposited on top of the ridges sample by e-gun evaporation. The samples where then cleaved at different lengths ranging from 0.1mm to 1.5 mm and mounted on the cold finger of a liquid He flow cryostat. The injected current was provided by a HP-8114A pulse generator programmed to deliver bursts of 1.93μs long pulses with a duty-cycle of 80% at a repetition rate of 413 Hz, yielding an overall duty-cycle of 40%. The luminescence signal was collected by a wide numerical aperture gold coated parabolic off-axis mirror and sent through a Fourier Transform Infrared Spectrometer (FTIR). The resulting signal was then detected with a liquid helium cooled Si bolometer.

The waveguide geometry, in contrast to measurements done through a grating coupler, allowed us to verify the polarization selection rules of intersubband transitions. Fig. 7 shows the electroluminescence spectrum from a device measured through a polarizor. As expected, light emission at a photon energy hν=16meV, corresponding to the intersubband transition, vanishes completely when the polarizor is set to select the light polarized parallel to the layers. Note however that a certain degree of polarization is induced by the waveguide itself because the transverse electric mode is cut-off at such long wavelengths.

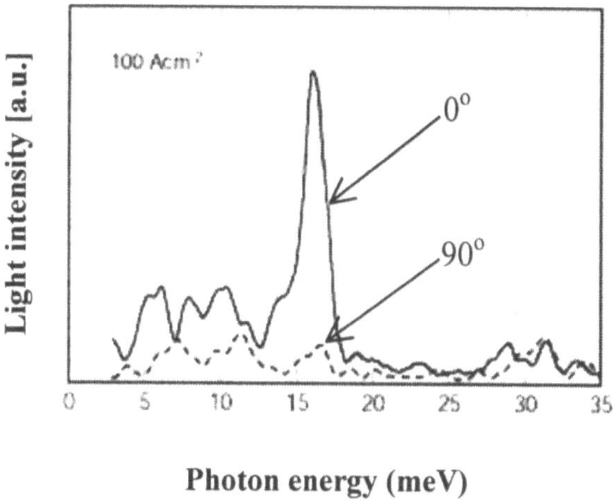

Fig. 7 Waveguide emission observed through a polarizor. The results show that the luminescence has a TM polarisation.

The extremely low signal intensities emitted by our samples preclude the use of conventional techniques to measure the gain in semiconductor lasers. We decided in contrast to use a technique based on the analysis of the luminescence intensity as a function of the waveguide length. At fixed injected current density J, the luminescence intensity as a function of waveguide length L and waveguide loss α_w is written as

$$I(L) = \int_0^L I_0 \exp((g-\alpha_w)x)dx = \frac{I_0}{g-\alpha_w}\left[\exp((g-\alpha_w)L)-1\right] \qquad (2)$$

The luminescence intensity per unit length I_0 and the gain (or loss) of the transition g are functions of the injected current density J. Performing measurements for different cavity lengths and at various injection current densities allows us to extract both g and $\alpha\omega$ separately. A special photolithographic mask has been designed in order to process waveguides of different length side by side on the same chip, and their light- and voltage- versus current characteristics are measured. Light intensity for a fixed current density as a function of waveguide length, extracted from these curves are reported in Fig. 8. As shown in the latter figure, this measurement was repeated five times for each length. As expected, the luminescence intensity increases with waveguide length and eventually saturates, and the dependence can be fitted using equation (2) (see dotted lines).

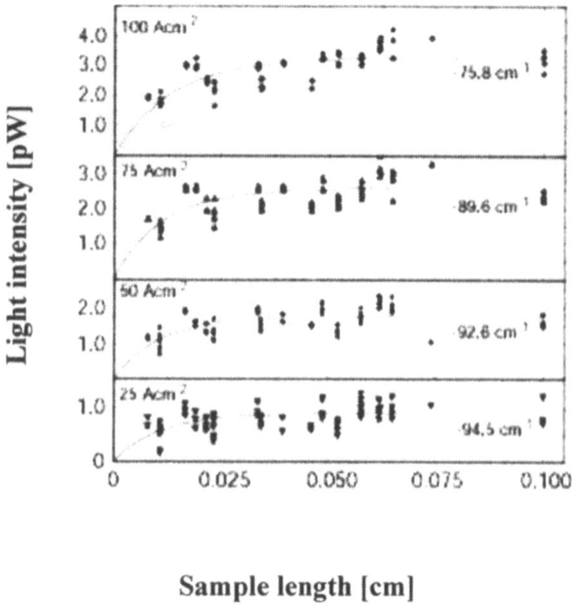

Fig. 8 Light intensity versus waveguide length is plotted for various injection current densities. Full lines are fits to data obtained by using the relation (Eq. 2) linking the emission intensity with the sample length.

At the injection current density of $25A/cm^2$ the waveguide net absorption g-α was found to be $94.5cm^{-1}$. At the maximum injection current density of $100A/cm^2$, we obtained a net absorption of $76cm^{-1}$.

These results show that a waveguide with reasonable losses can be manufactured using MBE growth techniques and that the device shows differential gain (i.e. the cavity loss decreases with the injected current). However, the path to a true laser might still be quite some distance away, as shown in Fig. 9 where the calculated gain for our structure is shown as a function of injected current for various lower state lifetimes. From this figure, it is clear that a lower state lifetime shorter than 0.5ps is necessary for the laser threshold to be reached. The reason is that the increase in injected current density is obtained at the cost of a shorter upper state lifetime due to electron-electron scattering. It is therefore important that future devices

i) have even further reduced waveguide losses

ii) display a shorter lower state and/or a longer upper state lifetime.

The latter conditions will likely lead to structures based on superlattices[22] or photon-assisted tunneling transitions[23] which have better lifetime ratios between the upper and lower states than structures based on vertical transitions.

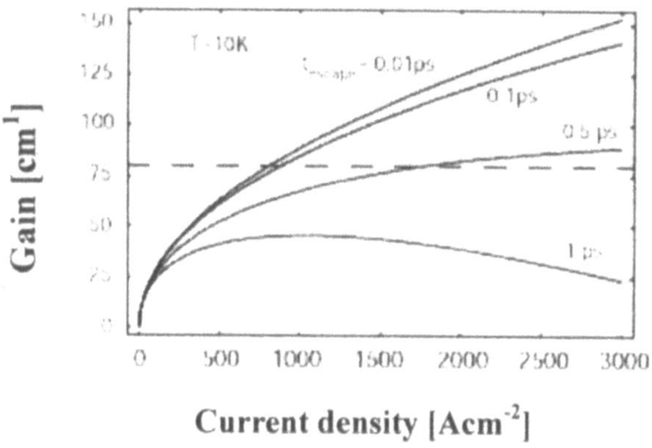

Fig. 9. Computed gain as a function of injected current density for a structure based on a vertical transition and for various lower state lifetimes.

Acknowledgments

This work is supported by the Swiss national foundation for science.

References

1. R. F. Kazarinov and R. A. Suris, " Possibility of amplification of electromagnetic waves in a semiconductor with a superlattice" Fiz. Tekh. Poluprov., 1971, **5**, pp. 797-800; transl. in Sov. Phys. Semicond. , 1971, **5**, pp 707-709
2. H.C. Liu, Jour. of Appl. Phys. **63**, 2856 (1988)
3. S.I. Borenstain, J. Katz, Appl. Phys. Lett. **55**, 654 (1989)
4. Q. Hu, S. Feng, Appl Phys. Lett. **59**, 2923 (1991)
5. M. Helm, P. England, E. Colas, F. DeRosa, and S. J. Allen, Jr. Phys. Rev. Lett. **63**, 74 (1988)
6. J. N. Heyman, K. Unterrainer, K. Craig, J. Williams, M.S. Sherwin, K. Campman, P.F. Hopkins, A.C. Gossard, B.N. Murdin, and C.J.G.M. Langerak, Appl. Phys. Lett. **68**, 3019, 1996
7. J. Faist, F. Capasso, C. Sirtori, D. L. Sivco, A. L. Hutchinson, S. N. G. Chu, and A. Y. Cho, Appl. Phys. Lett. **64**, 1144 (1994)
8. J. Faist, F. Capasso, D. L. Sivco, C. Sirtori, A. L. Hutchinson, and A. Y. Cho, Science **264**, 553 (1994)
9. J. Faist, F. Capasso and C. Sirtori, "Quantum Cascade Lasers", In semiconductors and semimetal, Landold-Berstein series, Edited by F. Capasso, Vol 53 New York, 2000
10. B. Xu, Q.Hu, M.R.Melloch, Appl. Phys. Lett. **71**, 440 (1997)
11. M. Rochat, J. Faist, M. Beck, U. Oesterle, M. Ilegems, Applied Physics Letters **73**, 3724, (1998)
12. M. Rochat, J. Faist, M. Beck, U. Oesterle, Physica E 7, 44 (2000)
13. A. Tredicucci, M. Wanke, F. Capasso, A.Y. Cho, this proceedings.
14. Borenstein and Katz, Phys. Rev. B. 39, 10852 (1989)

15. J. H. Smet, C. G. Fonstad, Q. Hu, J. Appl. Phys. **79**, 9305 (1996)
16. P. Hyldgaard, J. W. Wilkins, Phys. Rev. B **53**, 6889 (1996)
17. S.-C. Lee and I. Galbraith, Physical Review B **59,** 15796 (1999)
18. K. Donovan, P. Harrison, and R. W. Kelsall, Appl. Phys. Lett. **75**, 1999 (1999)
19. M. Hartig, S. Haake, P.E. Selbmann, B. Deveaud, R.A. Taylor and L. Rota, , Phys. Rev. Lett. **80**, 1940 (1998)
20. R. Ferreira and G. Bastard, Phys. Rev. B **40**, 1074 (1989)
21. C. Sirtori, C. Gmachl, F. Capasso, J. Faist, D. L. Sivco, A. L. Hutchinson, and A. Y. Cho, Optics Lett. **23**, 1366 (1998)
22. G. Scamarcio, F. Capasso, C. Sirtori, J. Faist, D. L. Sivco, A. L. Hutchinson, and A. Y. Cho, Science **276**, 773 (1997)
23. J. Faist, F. Capasso, C. Sirtori, D. L. Sivco, A. L. Hutchinson, and A. Y. Cho, Nature **387**, 777 (1997)

Superlattice QC lasers towards the far-infrared

ALESSANDRO TREDICUCCI,[1,2] CLAIRE GMACHL,[1] MICHAEL C. WANKE,[1] FEDERICO CAPASSO,[1] ALBERT L. HUTCHINSON,[1] DEBORAH L. SIVCO,[1] SUNG-NEE G. CHU,[1] and ALFRED Y. CHO [1]
[1] *Bell Laboratories, Lucent Technologies, 600 Mountain Avenue, Murray Hill, NJ 07974*
[2] *Scuola Normale Superiore and INFM, Piazza dei Cavalieri 7, I-56126 Pisa (Italy)*

Key words: Superlattices, quantum cascade lasers, unipolar semiconductor lasers, mid-infrared, far-infrared, electroluminescence, inter-miniband transitions

Abstract: At present efficient, compact, solid-state sources for the spectral range 1-10 THz, also known as the "terahertz gap", are still lacking. Quantum cascade (QC) lasers possess the potential for the efficient generation of far-infrared light, although many important physical questions have to be addressed when the transition energy is reduced below the optical phonon frequency. Significant progress towards far-infrared QC lasers can be obtained using inter-miniband transitions in semiconductor superlattices and employing non-conventional laser waveguide designs. Surface plasmon modes at the interface between a metal and the semiconductor are exploited in the design of a $\lambda \sim 19$ μm (15.8 THz) superlattice QC laser, which represents the longest wavelength III-V semiconductor laser to date. Single-mode operation is achieved at $\lambda \sim 16$ μm with the adoption of a dual-metal Bragg grating which modulates the skin depth of the surface plasmon. Preliminary electroluminescence results at ~ 3.5 THz are also discussed.

1. INTRODUCTION

The recent flourishing of activity in the field of quantum cascade (QC) lasers [1] has stimulated great interest in the development of such lasers in new material systems [2] or for new frequency ranges [3,4]. Being based on electronic transitions between quantum confined states in a semiconductor heterostructure, the operation of QC lasers is limited on the short wavelength side by the magnitude of the conduction band offset. On the other hand, at very long wavelengths (far infrared) other problems arise. Non-radiative relaxation processes forbid operation in the proximity of the materials *reststrahlenband.* In addition, free carrier absorption in doped semiconductors, which scales in the first approximation as λ^2 [5], leads to increased optical losses at longer wavelengths, making QC-laser action in the far-infrared a challenging proposition. As a result, QC lasers have been limited to the mid-infrared spectral region, ranging from 3.4 μm [3] to 17 μm [6].

R.E. Miles et al. (eds.), Terahertz Sources and Systems, 101–114.

Active regions relying on inter-miniband transitions in so-called "chirped", i.e. appropriately graded, superlattices [7] have, however, recently shown very encouraging possibilities [6,8]. They guarantee, in fact, a large material gain even for low transition energies, compounding all the advantages of superlattice QC lasers with the low doping requirement for the injectors [9], which is particularly advantageous for long wavelength devices [6,8,10].

In parallel, new types of waveguide designs have recently been proposed in order to improve the optical confinement and reduce the absorption losses. The main target of some of these investigations [11,12] has also been the reduction of the total thickness of the semiconductor material required to efficiently confine the light; in fact, for $\lambda > 15$ μm, conventional dielectric waveguides [6] are already at the practical limits of molecular beam epitaxy (MBE) growth techniques. A possible alternative to thick semiconductor waveguides is to employ air as cladding. In this way a larger optical confinement factor is achieved, due to the higher refractive index contrast with the active core, together with lower absorption losses. This requires, however, special device geometries, like microdisks, in which the active region is, at least partially, freestanding, to allow air to act as the upper and lower cladding [12]. Another possibility is offered by surface plasmons, which are TM electromagnetic modes confined at the interface between materials with dielectric constants of opposite sign [13]. Negative dielectric constants are typical of metals below the plasma frequency, thus guided surface plasmon modes at a metal-semiconductor interface appear as an excellent waveguiding solution for QC lasers [11], owing to the intrinsic TM polarization of intersubband transitions. At the wavelengths of the visible, near - and mid-infrared spectrum (up to $\lambda \sim 10$ μm), surface plasmon waveguides cannot compete with conventional dielectric waveguides because of the high optical absorption in the metal. For wavelengths approaching the far infrared ($\lambda \geq 15$ μm), however, the penetration depth (skin depth) in the metal is largely reduced, and the large mode confinement Γ with the accompanying reduced thickness of the waveguide layers can be advantageously used to demonstrate QC semiconductor lasers in a spectral region difficult to access with conventional dielectric waveguides. Here we present surface plasmon superlattice QC lasers operating at wavelengths ranging from 16 μm to 19 μm. In addition we extend the concept of DFB to such surface plasmon lasers. A two metal grating deposited on top of the semiconductor produces a strong index contrast due to the large spatial modulation of the skin depth. This allows selection of a single mode for laser action with a side mode suppression ratio comparable to those found in lasers used in fiber optic communications. Preliminary spontaneous emission data at far infrared wavelengths are reported as well.

2. DESIGN CONSIDERATIONS

The relationship $\mathrm{Re}[\varepsilon] = n^2 - k^2$ between the dielectric constant ε, the refractive index n, and the extinction coefficient k implies that materials with negative $\mathrm{Re}[\varepsilon]$ are usually very absorping. For this reason the attenuation coefficient α of a surface plasmon guided mode depends strongly on the penetration (or skin) depth δ in the metallic layer, decreasing as the latter is decreased. With ε_1 being the dielectric constant of the metal and ε_2 that of the semiconductor, δ is given by [13]:

$$\delta = \frac{1}{2}\left|\mathrm{Re}\left[\frac{\omega\varepsilon_1}{c}\sqrt{\frac{-1}{\varepsilon_1+\varepsilon_2}}\right]\right|^{-1}, \tag{1}$$

where $\omega/2\pi$ is the frequency of the electromagnetic wave and c the speed of light in vacuum. This means that a large negative $\mathrm{Re}[\varepsilon_1]$ ($k^2 >> n^2$) implies a small δ which entails a lower optical loss. With some straightforward algebra, α can easily be derived for the case of a real positive ε_2 [13]:

$$\alpha = \frac{\varepsilon_2^{3/2}\,\mathrm{Im}[\varepsilon_1]}{\left\{\mathrm{Re}[\varepsilon_1]\left(\varepsilon_2+\mathrm{Re}[\varepsilon_1]\right)^3\right\}^{1/2}}\frac{\omega}{c}. \tag{2}$$

In a first crude approximation, one can represent the frequency dependent dielectric constant of a metal with the simple Drude free-carrier expression:

$$\mathrm{Re}[\varepsilon_1(\omega)] = \varepsilon_\infty\left(1-\frac{\omega_p^2}{\omega^2+\gamma^2}\right),\quad \mathrm{Im}[\varepsilon_1(\omega)] = \frac{\varepsilon_\infty\omega_p^2\gamma}{\omega\left(\omega^2+\gamma^2\right)}, \tag{3}$$

where ε_∞ is the background dielectric constant and γ^{-1} is a phenomenological scattering time [14]. Evidently, surface plasmon waveguides become less lossy as the radiation wavelength is increased, owing to the Lorentzian dependence of ε. Metallic waveguides are in fact commonly used in microwave applications, and, although they present more complicated three-dimensional geometries allowing also for TE and TEM mode propagation, in the simplest designs ("microstrips") the fundamental TM mode is again of a surface plasmon nature [15].

Our devices employ QC active materials with variable period superlattices of the type described in Refs. 6 and 7. The potential of the artificial crystal, obtained by periodically alternating nanometer-thick layers of two different semiconductors, splits the conduction band in narrower energy bands ("minibands") separated by energy gaps ("minigaps") [16]. Laser action is achieved between the states at the edge of the first minigap through unipolar electron injection via miniband transport. The applied electric field is compensated in the superlattices with an "effective" quasi-electric field [17] generated by "chirping" the SL period, so that a flat miniband condition can be reached without doping the SL, which would broaden the optical transition and negatively affect population inversion at high electronic temperatures. The large oscillator strengths of inter-miniband transitions and the "built-in" population inversion resulting from the difference between intra-miniband and inter-miniband scattering times are, however, preserved. According to the QC scheme [3], multiple active regions can then be stacked together, bridged by specially designed injectors which, under the appropriate bias conditions, extract electrons from the lower miniband of one superlattice to inject them into the second miniband of the following one.

The samples were grown by solid-source MBE in the InGaAs/AlInAs material system lattice matched to InP substrates low doped to $n_e \sim 2\times10^{17}$ cm^{-3} (see Fig. 1 for the band diagram of a $\lambda \sim 17$ μm one, whose complete layer sequence is given in Ref. 8). The wafers were processed into ridges by wet chemical etching through the epilayers and silicon dioxide was then deposited on the sides for insulation. Top and back contacts of the type described in [11] were then electron-beam evaporated, however leaving a very wide portion of the top of the ridges open for the deposition of the metallic surface-plasmon carrying layer. For the latter we tested different metals with gold (~ 300 nm thick) to give the best results.

The dielectric profile of a $\lambda \sim 17$ μm laser structure with the corresponding surface plasmon mode calculated using the transfer matrix method are reported in Fig. 1A together with the values of the confinement factor Γ (defined as the normalized integral of the optical mode over the active material), waveguide attenuation α_w, and modal effective refractive index n_{eff}. For comparison we also show in the same plot the profile of the mode obtained with the conventional semiconductor-cladding waveguide of Ref. 6. With the present solution the thickness of the epitaxial growth is reduced from almost 9 μm to less than 4 μm, at the same time concentrating the field intensity in the active material, thereby raising Γ from 0.47 to 0.81. In Fig. 1B the mode profile obtained exchanging gold with titanium, which has a much less negative dielectric constant (Re $\varepsilon_{Ti} \sim -1\times10^3$, Re $\varepsilon_{Au} \sim -1.1\times10^4$ at $\lambda \sim 17$ μm), is reported; the penetration depth in the metallic layer (and therefore the waveguide losses) are strongly increased. The ridges were then cleaved into lasers bars (0.75-1.5 mm long) leaving the facets uncoated, soldered to copper plates, wire bonded and mounted in a He-flow cryostat for the measurements.

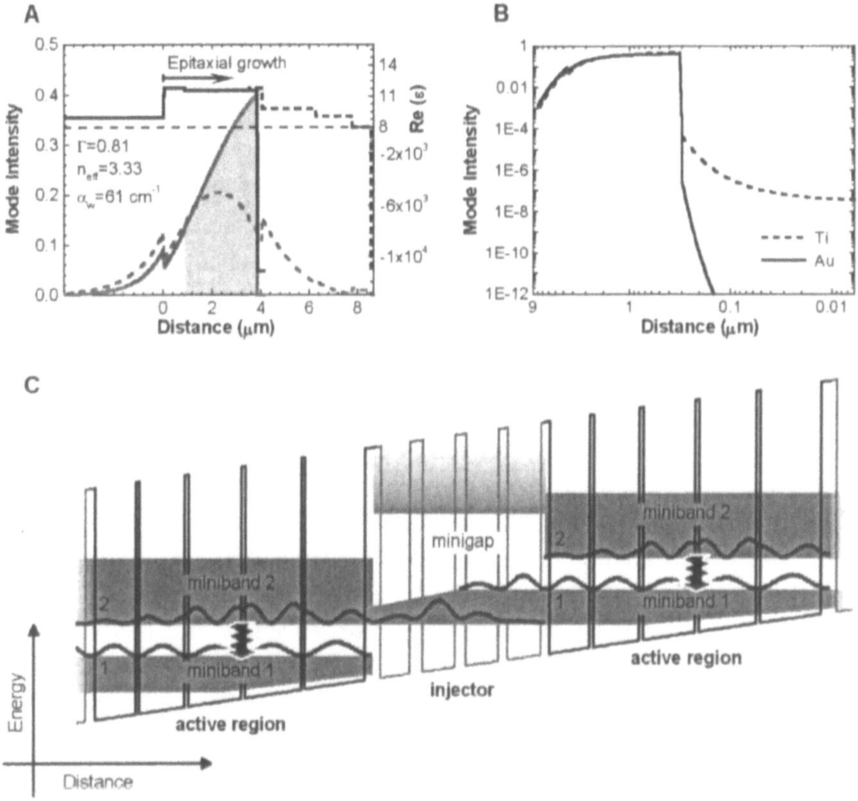

Figure 1. (A) Real part of the dielectric constants (black) of the various device layers and the resulting normalized intensity profile of the optical mode (gray). Solid lines refer to the present surface plasmon waveguide, dotted lines to the corresponding dielectric waveguide with semiconductor claddings of Ref. 6. The shaded area indicates the stack of active regions and injectors. (B) Optical mode intensity of the same surface plasmon waveguide of (A) as computed for two different metals and plotted on a logarithmic scale. The origin of the abscissa is at the air-metal surface. (C) Electronic band structure of a portion of the $\lambda \sim 17$ μm laser under an applied bias of 20.5 kV/cm. The actual layer thicknesses and compositions are reported in Ref. 8. The moduli squared of the relevant wavefunctions in the active and injector regions and the laser transition (arrow) are also shown.

3. EXPERIMENTAL RESULTS

The peak power as a function of current emitted by a $\lambda \sim 17$ μm gold surface plasmon laser is reported in Fig. 2A for various temperatures. It was measured with f/1 KBr optics and a cooled HgCdTe detector calibrated against the average power reading at high duty-cycles (100 kHz) of a sensitive thermopile power meter. An intensity attenuator was used in conjunction with the cooled detector to prevent response non-linearities. The dependence of the threshold current density J_{th} on the temperature is shown in Fig. 2B. At low temperature $J_{th} \sim 6$ kA/cm^2, approximately a factor of 2 lower than in previous QC lasers of comparable wavelength [6]. Using the calculated values of the waveguide losses and of the active material parameters, we estimate from the well-known expression for the gain [6] $J_{th} \sim 1.2$ kA/cm^2. The discrepancy originates from several factors: first the simple free-carrier absorption model for doped semiconductors neglects various effects (like direct two-phonon absorption) and is strongly dependent on the choice of a phenomenological scattering time[1]. Second, the injection efficiency in our structure is clearly non-ideal: as one can see from the I-V curve of Fig. 2, injection of electrons into the second miniband of the structure (marked by a strong reduction in differential resistance) takes place when the device is already conducting ~ 3 kA/cm^2. Third, the actual nature of the metal-semiconductor interface may be non-ideal. Taking into account the first two effects one already arrives at an estimated J_{th} = 5-6 kA/cm^2, in basic accordance with the experiment. The low threshold translates in record high powers (38 mW at 5 K) and operating temperatures T_{max}= 240 K for semiconductor lasers of this wavelength. The somewhat high voltage and resistance of the device can be attributed to the small size of the contacts with respect to the total area of the ridge. The resulting non-uniformity of the current profile can also be at the origin of the marked kink in the I-V at ~ 18 kA/cm^2, which is accompanied by a broadening towards longer wavelengths of the laser multi-mode emission and by a sudden increase in power. The laser line was centered at ~ 16.7 μm at low currents and at ~ 17.1 μm at the highest ones (see Fig. 2B).

[1] There are also large uncertainties in the dielectric constants of metals in this wavelength range, for which a large spread of values exist in the literature.

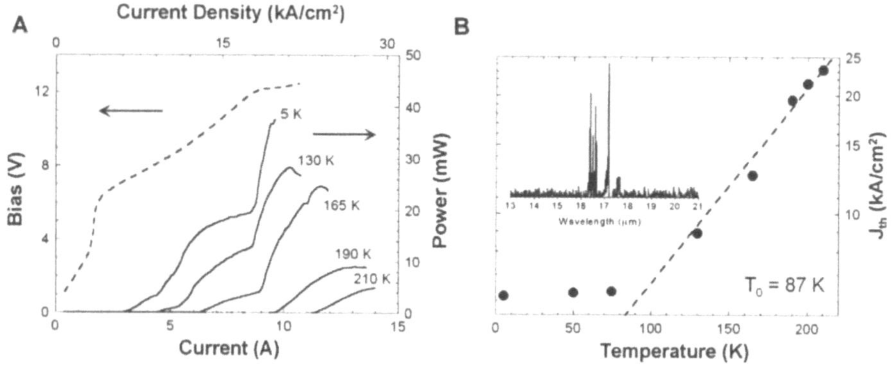

Figure 2. (A) Light-output (L)-current (I) (solid lines) and current-voltage (V) (dashed line) characteristics of a 1.4 mm long and 34 μm wide, deep etched, ridge waveguide λ ~ 17 μm laser operated in pulsed mode (50 ns pulse width, 5 kHz duty-cycle). The optical power is measured from a single facet with ~ 50% collection efficiency. The I-V characteristic has been measured at 5 K. L-I curves are shown at temperatures as indicated in the graph. (B) Threshold current density as a function of temperature. The dashed line is a fit with $J_{th}=J_0\exp(T/T_0)$ over the last five points. A typical laser emission spectrum at high currents is reported in the inset.

Making use of the same design philosophy for the waveguide and the active region we have realized the longest wavelength III-V semiconductor laser to date. A laser emission spectrum at a temperature of 80 K is reported in Fig. 3A. It shows multiple lines corresponding to the longitudinal modes of the Fabry-Perot cavity, with the intensity peaked at λ ~ 19 μm. The spacing of the fringes is 1.80 cm^{-1} which compares well with the 1.98 cm^{-1} obtained from the computed n_{eff} of the waveguide. The slightly lower value suggests a larger penetration in the metallic layer, which would increase the effective refractive index. To obtain an estimate for the linewidth of the transition, we measured the electroluminescence emission spectrum from a shorter device (375 μm long) far below threshold. The result is shown in the inset of Fig. 3A: a full-width-at-half-maximum (FWHM) 2γ ~ 5 meV is observed.

The light output (L) as function of drive current (I) is reported in Fig. 3B for a laser 1.54 mm long and 31 μm wide. Up to approximately 14 mW of peak output power was obtained at 10 K, with ~ 12 mW at liquid nitrogen and a maximum operating temperature of 145 K. The threshold current density at 10 K is 4.5 kA/cm^2, even lower than for the previous λ ~ 17 μm surface plasmon lasers.

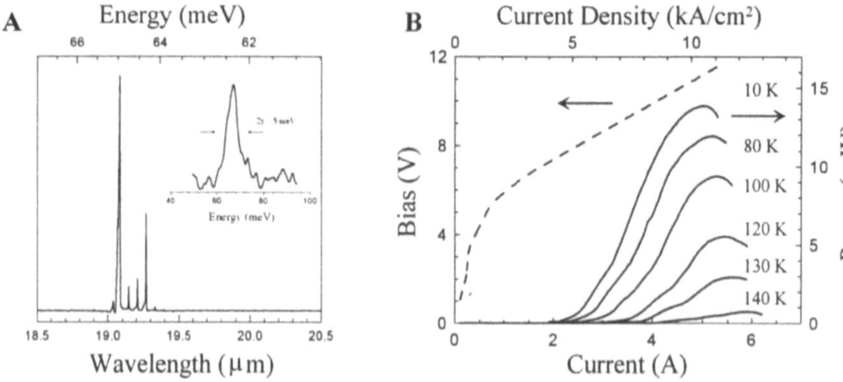

Figure 3. (A) Pulsed (50 ns width, 84.2 kHz repetition rate) laser emission spectrum of a device 35 μm wide and 760 μm long at 80 K under 2.8 A drive current. The spectrum was measured in rapid scan using a Nicolet Fourier transform infrared spectrometer (FTIR) and a cooled HgCdTe detector. The resolution of the set-up is 0.125 cm^{-1}. Inset: electroluminescence spectrum as collected in step-scan from the facet of a device 31 μm wide and 375 μm long at 5 K for a drive current of 450 mA. (B) L - I characteristics of a 1.54 mm long 31 μm wide ridge waveguide laser operated in pulsed mode (50 ns pulse width, 5 kHz duty-cycle) at various heat sink temperatures. The optical power emitted from a single facet was recorded. The 10 K current (I) - voltage (V) curve is also shown as a dashed line.

Mid-to-far-infrared semiconductor lasers are typically employed in gas sensing applications, where spectroscopic techniques with high resolution and sensitivity must be implemented [18]. For this purpose purely single-mode devices are highly desirable. This is usually accomplished for QC [19] and diode lasers in general [20] by implementing a DFB resonator by incorporating a grating of appropriate period and strength in the semiconductor material. This introduces a modulation of the refractive index n_{eff} and of the attenuation coefficient α_w (or of the effective net gain) favoring the optical mode which best matches the grating period over all other possible longitudinal modes. In our case the surface plasmon nature of the electromagnetic mode provides an elegant way of realizing a DFB laser without the need for any etching procedure. As already noticed in Fig. 1B, the metal chosen for carrying the plasmon strongly affects the penetration depth of the mode and with it its refractive index and loss factor. Deposition of 10 nm of titanium before the final gold layer already results in a variation of the refractive index $\Delta n_{eff}/n_{eff} \sim 1.8\text{x}10^{-3}$ and of the losses $\Delta\alpha_w/\alpha_w \sim 1.5\text{x}10^{-2}$. To spatially modulate this variation along the ridge, we have fabricated by optical contact lithography a first-order Bragg grating (nominal duty-cycle 50 %) of Ti stripes 10 nm thick. The subsequent evaporation of 300 nm of gold effectively resulted in an alternate sequence of Ti/Au and pure Au stripes along the direction of the surface plasmon propagation. A scanning electron microscope (SEM) image of an actual device is reported in Fig. 4. For a grating period $\Lambda = 2.05$ μm, we compute a complex coupling coefficient for the DFB mode [19]:

$$\kappa = \left| \frac{\pi \Delta n}{4n\Lambda} + i\frac{\Delta \alpha_w}{4} \right| = 6.9 \text{ cm}^{-1}, \qquad (4)$$

which for our case is apparently dominated by the real part. For the same reasons discussed above concerning the waveguide losses, we might somewhat have underestimated this value. Anyway, for typical cavity lengths $L_{cav} \sim 1.5$ mm, this leads to coupling products $\kappa L_{cav} \sim 1$, which is generally considered an optimum value to achieve good side mode suppression [19,20].

Figure 4. SEM picture of a surface plasmon DFB ridge laser. The stack of alternated active regions and injectors appears lighter on the front facet, compared to the darker substrate. The insulation oxide is almost transparent but is visible on the edges of the top of the ridge. Several periods of the Ti/Au grating can be seen in the upper portion, together with the contact layers on the sides. The last section of the ridge has been left uncovered by metal to show the flat, unmodified semiconductor surface.

As can be deduced from the spacing of the longitudinal modes in the emission spectrum of a regular Fabry-Perot surface plasmon laser shown in Fig. 5A, an experimental estimate of the effective refractive index of the waveguide is $n_{eff} \sim 3.95$. Based on the latter and on the grating period we expect a Bragg wavelength $\lambda_B = 2n_{eff}\Lambda \sim 16.4$ μm, although a precise value should also depend on the real width of the Ti stripes, i.e. on the lateral extension of the mode compared to the ridge width, etc. In Fig. 5B the single-mode emission spectrum of a surface-plasmon DFB laser is reported. The side-mode suppression ratio is at least 30 dB, which clearly reflects the high modal purity of the device. The wavelength of 16.15 μm is fairly close to what is expected. The width of the laser line is basically determined by the "chirping" of the emission wavelength during the current pulse, because of thermal effects.

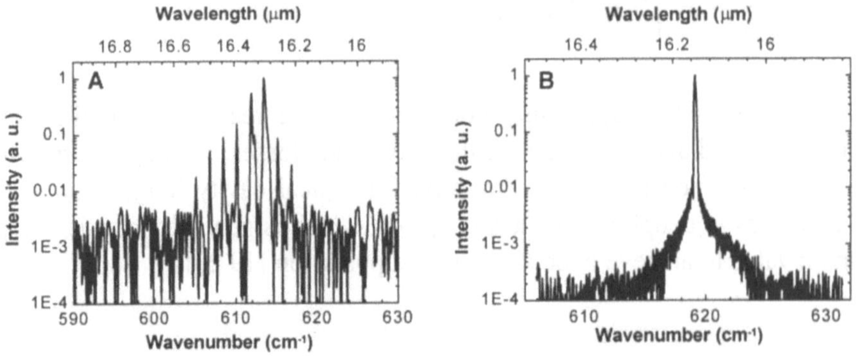

Figure 5. (A) Pulsed (50 ns width, 84.2 kHz repetition rate) laser emission spectrum in logarithmic scale at 5 K of a gold surface-plasmon laser (26 μm wide, 750 μm long) under 2 A drive current. The Fabry-Perot comb of longitudinal modes is evident with a spacing $\Delta\nu \sim 1.68$ cm^{-1}. (B) Pulsed laser emission spectrum in logarithmic scale of a DFB device (25 μm wide and 800 μm long) at 5K under 2 A drive current. The spectra were measured using a Nicolet Fourier transform infrared spectrometer (FTIR) and a cooled HgCdTe detector. The resolution of the set-up is 0.125 cm^{-1}.

The change of the refractive index with temperature can be used as a simple effective way to tune the laser frequency of DFB QC lasers [19]. In Fig. 6 we plot the tuning curve of a DFB surface-plasmon laser. Although the dependence is evidently non-linear, a linear tuning coefficient can be defined at the highest temperatures [19]. A value of 1 nm/K is obtained for the present device.

At lower frequencies of only a few THz (corresponding to wavelengths in the range 50 - 100 μm) a waveguide based on a single surface plasmon would face a considerable reduction of the optical confinement factor Γ, unless one could dramatically increase the thickness of the active material. The obvious solution is to resort to full planar metallic waveguides, which feature a metallic layer on both sides of the semiconductor. They offer the advantage of a near unity Γ and the inexistence of a cut-off frequency for the fundamental TM mode (which is basically given by two coupled surface plasmons). These characteristic are very important because they lift any major requirement on the thickness of the epitaxial growth that has to be performed. It must be noted however that the closer the two metallic layers the more the optical mode will be pushed inside them, resulting in a net increase of the attenuation factor.

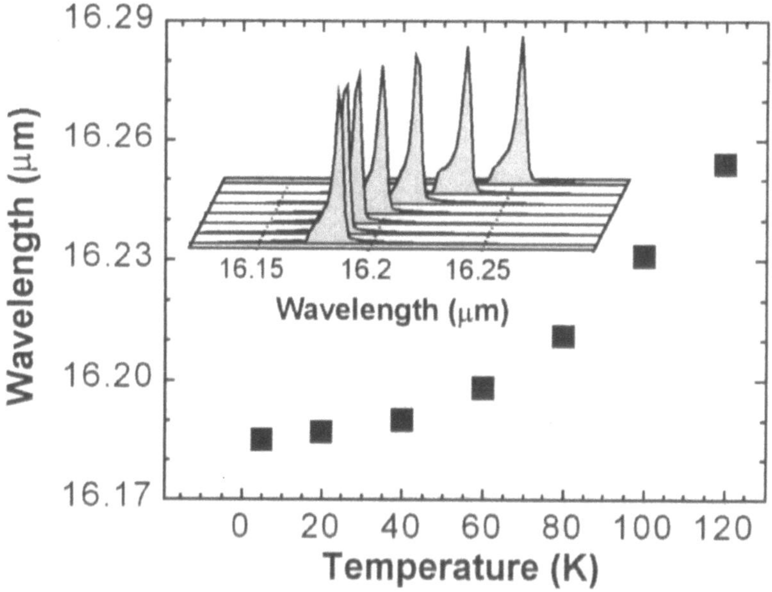

Figure 6. Emission wavelength as a function of temperature of a single-mode DFB laser (23 μm wide, 1.5 mm long). The period of the Ti grating is $\Lambda = 2$ μm. In the inset the spectra recorded at the various wavelengths are reported. The current level was varied from ~ 4.8 A to ~ 6.4 A increasing the temperature to keep the output power roughly at the same levels. This reflects in an broadened linewidth at the higher temperatures.

Deposition of metal on both sides of a QC structure implies a somewhat complicated processing, since the substrate has to be completely removed from the epitaxial material. However a much simpler possibility is to substitute the bottom metallic layer with a highly n-doped semiconductor (in our case InGaAs doped to ~ 5×10^{19} cm^{-3}); operating then far below its plasma frequency the semiconductor will present a strongly negative dielectric constant and behave as a metal for our waveguiding purposes. The price to pay in terms of optical losses depends of course both on the wavelength and on the thickness of the active material. We have employed this waveguide scheme in the design of a QC structure for emission at λ ~ 80 μm (see the inset of Fig. 7).

The active region superlattices are sandwiched between special injectors which allow electron relaxation through optical phonon emission without showing any resonant absorption at the wavevelength of the optical transition. Forty stages were grown in the InGaAs/AlInAs system, for a total thickness of the active material of about 6 μm. A spontaneous emission spectrum as detected from the facet of one such ridge device is reported in Fig. 7. It shows a thermal background of black-body emission resulting from the heating due to the current pulses with an evident peak on top (approximately 2 meV wide) at the energy of the superlattice minigap (~ 15 meV). The output power is estimated to be of the order of a few pW.

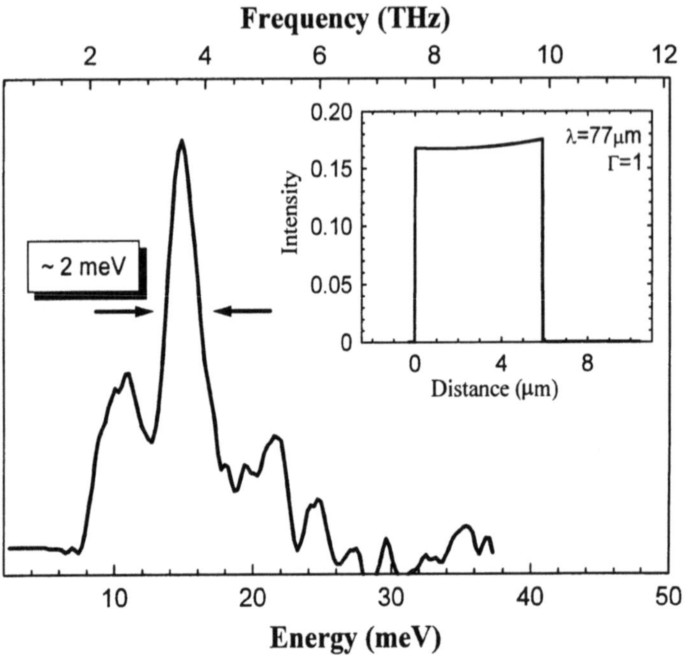

Figure 7. Pulsed emission spectrum of a λ ~ 80 μm QC superlattice device at a drive current of ~ 1 kA/cm^2. It was collected with a FTIR spectrometer using f/1 refractive optics and a cooled doped-Ge photoconductive detector. In the inset we plot the mode profile of the device waveguide: the zero of the abscissa is at the metal-semiconductor interface.

4. CONCLUSIONS

We have presented QC lasers operating at λ ~ 17 μm and λ ~ 19 μm, the longest wavelengths to date for III-V semiconductor lasers. They rely on interminiband transitions in semiconductor superlattices and on a new waveguide design based on surface plasmon interface modes. The emission can be made single-mode realizing a dual-metal grating that provides distributed feedback without etching the semiconductor material. This waveguide solution presents a great potential for the development of injection lasers at even longer wavelengths, and offers a means to implement DFB resonators in material systems where controlled etching is critical and laborious. Preliminary spontaneous emission data from a superlattice QC structure featuring a double surface-plasmon waveguide designed for operation at ~ 3.5 THz have been reported as well.

ACKNOWLEDGMENTS

This work was supported in part by DARPA/ARO through contract number DAAG55-98-C-0050.

REFERENCES

[1] F. Capasso, C. Gmachl, D. L. Sivco, and A. Y. Cho, Physics World **12**, 27 (1999), and references therein; F. Capasso, C. Gmachl, A. Tredicucci, A. L. Hutchinson, D. L. Sivco, and A. Y. Cho, Optics and Photonics News. **10** n.10, 31 (1999), and references therein.

[2] C. Sirtori, P. Kruck, S. Barbieri, P. Collot, J. Nagle, M. Beck, J. Faist, and U. Oesterle, Appl. Phys Lett. **73**, 3486 (1998); I. Vurgaftman, J. R. Meyer, F. H. Julien, and L. R. Ram-Mohan, Appl. Phys, Lett. **73**, 711 (1998); L. Friedman, R. A. Soref, and G. Sun, J. Appl. Phys. **83**, 3480 (1998).

[3] J. Faist, F. Capasso, D. L. Sivco, A. L. Hutchinson, S. G. Chu, and A. Y. Cho, Appl. Phys. Lett. **72**, 680 (1997).

[4] M. Rochat, J. Faist, M. Beck, U. Oesterle, and M. Ilegems, Appl. Phys. Lett. **73**, 3724 (1998); B. S. Williams, B. Xu, Q. Hu, and M. R. Melloch, Appl. Phys. Lett. **75**, 2927 (1999); J. Ulrich, R. Zobl, K. Unterrainer, G. Strasser, and E. Gornik, Appl. Phys. Lett. **76**, 19 (2000).

[5] See, for instance, P. Y. Yu and M. Cardona, *Fundamentals of Semiconductors* (Springer-Verlag, Berlin, 1996).

[6] A. Tredicucci, C. Gmachl, F. Capasso, D. L. Sivco, A. L. Hutchinson, and A. Y. Cho, Appl. Phys. Lett. **74**, 638 (1999).

[7] A. Tredicucci, F. Capasso, C. Gmachl, D. L. Sivco, A. L. Hutchinson, and A. Y. Cho, Appl. Phys. Lett. **73**, 2101 (1998); S. Slivken, A. Matlis, A. Rybaltowski, Z. Wu, and M. Razeghi, Appl. Phys. Lett. **74**, 2758 (1999).

[8] A. Tredicucci, C. Gmachl, F. Capasso, A. L. Hutchinson, D. L. Sivco, and A. Y. Cho, Appl. Phys. Lett. **76**, 2164 (2000).

[9] F. Capasso, A. Tredicucci, C. Gmachl, D. L. Sivco, A. L. Hutchinson and A. Y. Cho, IEEE J. Selected Topics in Quantum. Electron. **5**, 792 (1999).

[10] G. Strasser, S. Gianordoli, L. Hvozdara, W. Schrenk, K. Unterrainer, and E. Gornik, Appl. Phys. Lett. **75**, 1345 (1999).

[11] C. Sirtori, C. Gmachl, F. Capasso, J. Faist, D. L. Sivco, A. L. Hutchinson, and A. Y. Cho, Opt. Lett. **23**, 1366 (1998).

[12] A. Tredicucci, C. Gmachl, F. Capasso, D. L. Sivco, S. N. G. Chu, A. L. Hutchinson, and A. Y. Cho, Electron. Lett. **36**, 328 (2000).

[13] See, for instance, P. Yeh, *Optical Waves in Layered Media* (John Wiley & Sons, New York, 1988).

[14] N. W. Ashcroft and N. D. Mermin, *Solid State Physics*, Saunders College, Philadelphia, 1988. For device applications of surface plasmons see, for instance, G. I. Stegeman and J. J. Burke, Appl. Phys. Lett. **43** (1983) 221; A. Köck et al., Appl. Phys. Lett. **63**, 1164 (1993).

[15] D. M. Pozar, *Microwave Engineering* (John Wiley & Sons, New York, 1988).

[16] M. Helm, Semicond. Sci. Technol. **10**, 557 (1995), and references therein.

[17] H. Kroemer, RCA Rev. **18**, 332 (1957); F. Capasso, Annu. Rev. Mater. Sci. **16**, 263 (1986).

[18] M. W. Sigrist, *Air Monitoring by Spectroscopic Techniques* (John Wiley & Sons, New York, 1994).

[19] C. Gmachl, F. Capasso, J. Faist, A. L. Hutchinson, A. Tredicucci, D. L. Sivco, J. N. Baillargeon, S. G. Chu, and A. Y. Cho, Appl. Phys. Lett. **72**, 1430 (1998), and references therein.

[20] G. Morthier and P. Vankwinkelberge, *Handbook of Distributed Feedback Laser Diodes*, Artech House, Boston, 1997.

Terahertz emission from semiconductor nanostructures

J. Ulrich, R. Zobl, G. Strasser, K. Unterrainer

Institut für Festkörperelektronik, Technische Universität Wien,
A-1040 Wien, Austria

Abstract

We present a study on spontaneous THz emission from semiconductor quantum structures. THz emission from parabolically graded AlGaAs/GaAs quantum wells driven by an in-plane electric field has been measured. We have analyzed the grating coupled radiation in the temperature range from 20 K to 240 K. The peak emission corresponds to the intersubband plasmon in the parabolic potential. From a quantum cascade structure we observe spontaneous THz emission, based on carrier injection into the excited state of a 26 nm quantum well. We have measured luminescence at a photon energy of 17.3 meV with a full linewidth at half maximum of 1.3 meV.
In the presence of a magnetic field (B = 7.2 T) applied normal to the epitaxial layers the emission efficiency doubles. This effect is attributed to the suppression of non-radiative Auger-intersubband transitions caused by Landau-quantization of the in-plane electron motion.

Keywords

Terahertz, far-infrared, electroluminescence, parabolic quantum well, quantum cascade structure

1. Introduction

For the development of electrically driven semiconductor sources emitting in the far-infrared wavelength region several concepts have been pursued. A far-infrared laser based on radiative transitions of holes in p-doped Germanium between Landau levels was demonstrated /1/. Sources operating without a magnetic field can be constructed employing intersubband transitions in heterostructures. THz intersubband electroluminescence has been observed in parabolic quantum wells /2/, in superlattices /3/, and more recently in quantum cascade structures /4,5/.

R.E. Miles et al. (eds.), Terahertz Sources and Systems, 115–124.

We studied the electroluminescence from two different structures. The first structure is a parabolically graded quantum well and the second one is quantum cascade strucutre

2. THz emission from parabolic quantum wells

Parabolic quantum wells are promising candidates for THz sources operating above liquid nitrogen temperature, because their emission frequency is independent of the electron distribution and concentration in the well. In absorption and emission spectroscopy /6/, coupling between the radiation and the electron system has been observed at only one frequency. This is the harmonic oscillator frequency, solely determined by the width and the energetic depth of the quantum well. These observations confirmed the validity of the generalized Kohn theorem /7/ in parabolic quantum wells, stating that long wavelength radiation (as compared to the width of the well) couples only to the center of mass co-ordinate of an electron system confined in a parabolic potential.

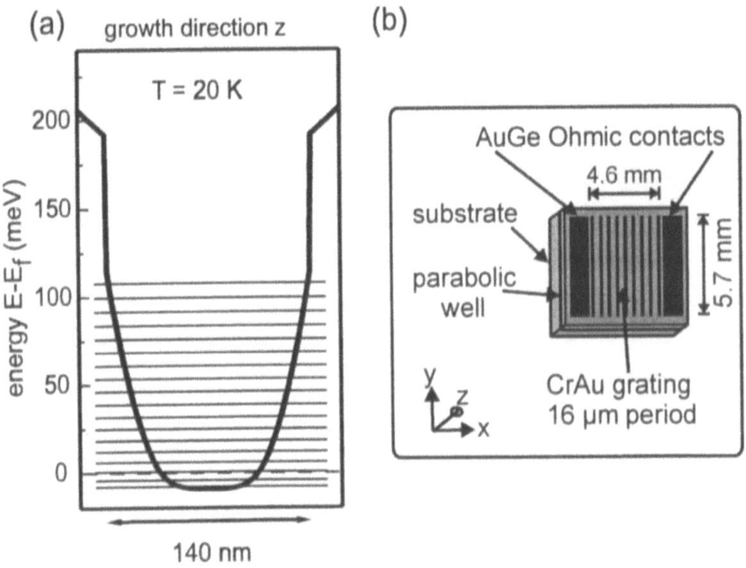

Figure 1: (a) Self-consistent calculation of the band diagram of the 140 nm well. The horizontal lines denominate the energy levels in the graded part of the well. (b) Device geometry of the 140 nm well sample.

Two samples were examined: one with 140 nm, the other one with 200 nm well width, both with 167 meV energetic well depth. The samples were grown by molecular beam epitaxy using a digital alloy technique. The electron concentration in both wells were determined by Hall measurements.

In the 140 nm well, the electron concentration varies from $3.8x10^{11}$ cm^{-2} at 20 K to $5.1x10^{11}$ cm^{-2} at 240 K. In the 200 nm well, the corresponding values are $2.7x10^{11}$ cm^{-2} at 20 K, $3.4x10^{11}$ cm^{-2} at 240 K. The energy band diagram of the 140 nm well is displayed in Fig. 1 (a).

Fig. 1 (b) shows the device geometry schematically. The wells are contacted with two AuGe Ohmic contact stripes. In order to couple out the intersubband radiation, that is polarized with its electrical field perpendicular to the layers, a metallic CrAu grating of 16 μm (50 μm) period, with its stripes parallel to the contacts, was evaporated between the contacts.

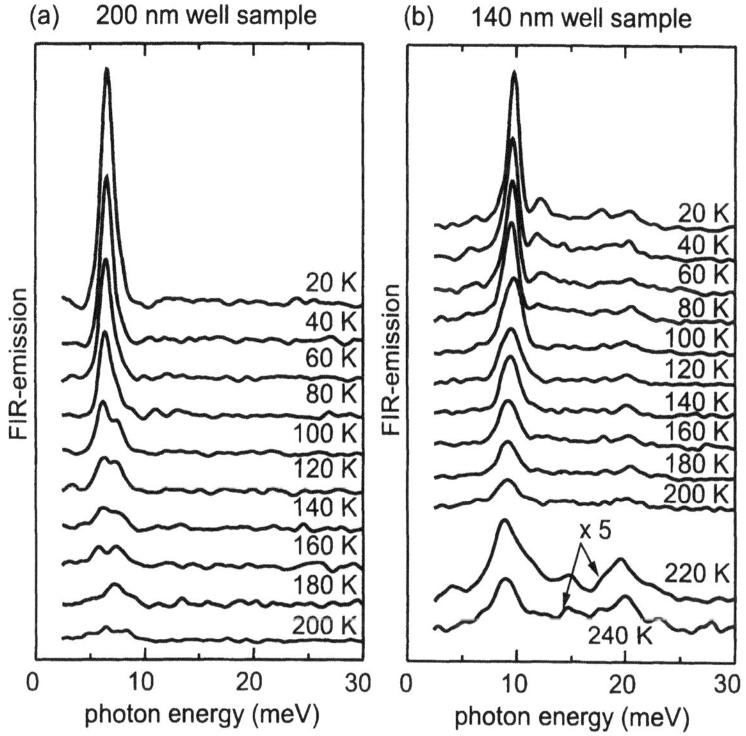

Figure 2: Emission spectra (a) of the 200 nm well sample and (b) of the 140 nm well sample for various temperatures as indicated. The spectra of the two highest temperatures in (b) are shown enlarged by a factor of five.

We measured the electroluminescence of the parabolic quantum wells using a Fourier-transform infrared spectrometer in the step scan mode with a spectral resolution of 0.5 meV. The sample was mounted on the cold finger of a helium-flow cryostat. The emitted radiation was collected by an off-axis parabolic mirror with an f/2 aperture, transmitted through the spectrometer, and then focussed on a helium cooled Si-bolometer. The whole beam path was purged with dry nitrogen gas to minimize the far-infrared absorption of water vapor. The electron gas in the well was excited by application of a pulsed electric field between the Ohmic contacts at a frequency of 411 Hz and 50 % duty cycle. A lock-in amplifier was employed to detect the bolometer signal.

In Fig. 2 (a), spectra of the 200 nm well at various temperatures are displayed. The peak emission is observed at a photon energy of 6.6 meV for all temperatures. This value corresponds to the harmonic oscillator energy, calculated from the well dimensions as 6.0 meV. In agreement with the generalized Kohn theorem, the emission energy is unaltered by the temperature-induced variation of the electron distribution in the well. The spectra of the 140 nm well in Fig. 2 b show a 20 K emission peak at 9.8 meV (calculated as 8.4 meV) that is slightly shifted to lower energies (9.1 meV), as the temperature is raised up to 240 K. The full width at half maximum of the emission line of both samples ranges from 1 meV at low temperatures ($T < 100$ K) to 2 meV in the high temperature regime ($T > 100$ K).

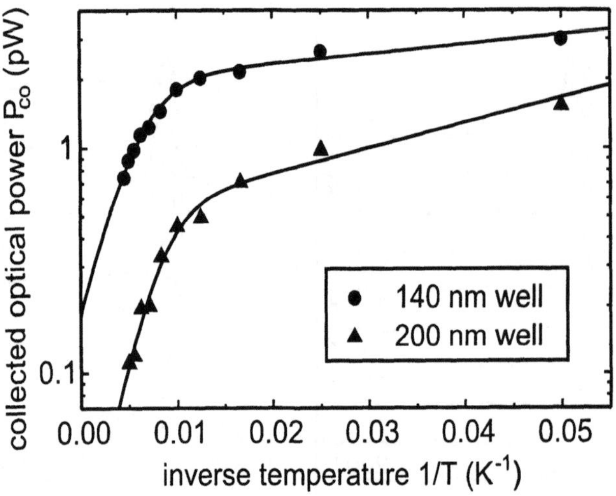

Figure 3: Collected optical power P_{co} versus inverse temperature 1/T of the 200 nm well sample driven at an electrical power P_{el} of 70 mW (solid triangles) and of the 140 nm well sample driven at 65 mW (solid circles). The solid lines are inverse two stage exponentials fitted to the data.

The optical power collected by the bolometer P_{co} of both samples is depicted in Fig. 3 as a function of inverse temperature. It was determined as the integrated area of the emission spectra divided by an amplification factor and the bolometer responsivity. The collected optical power P_{co} is assumed to be proportional to the electrical input power P_{el} and to the ratio of non-radiative lifetime τ_{nr} and radiative lifetime τ_r (in case of $\tau_r << \tau_{nr}$):

The difference in the efficiency P_{co}/P_{el} between the two samples can be qualitatively understood, looking at the radiative lifetimes τ_r. They were calculated in the model of a classical electron oscillator as 59 μs for the

140 nm well and 130 μs for the 200 nm well. With the simplifying assumption of similar non-radiative lifetimes, grating-coupler efficiencies and excitation efficiencies, one would expect the efficiency P_{co}/P_{el} of the 140 nm well sample to be approximately twice the one of the 200 nm well sample.

The decrease of (collected) optical power is thermally activated. The solid lines in Fig. 3 are fits of the inverse sum of two exponential functions in 1/T yielding in activation energies of Δ_1 = 34 meV, Δ_2 = 0.8 meV for the 140 nm well sample and Δ_1 = 41 meV, Δ_2 = 2.2 meV for the 200 nm well sample. The higher energy Δ_1 is probably the activation barrier for the emission of optical phonons, where the energy spacing ΔE between the two lowest subbands is 4 meV in the 140 nm well [see Fig. 1 (a)] and 2.5 meV in the 200 nm well. We have no interpretation of the low energy Δ_2 activation process. The intersubband transition rate $1/\tau_{nr}$ shows a similar behavior of thermal activation, as described by Heyman et al. /8/. We may conclude that the temperature dependence of the optical power is governed by the non-radiative lifetime τ_{nr}. Here, we disregard any changes in the excitation process. At temperatures above ~100 K, the emission of optical phonons limits the optical power. Using a non-radiative transition rate τ_0 for of 2.2 ps^{-1} as estimated by Rochat et al. /5/, the phonon relaxation time at the temperature of 100 K is determined as 23 ps for the 140 nm well sample and 53 ps for the 200 nm well sample.

The above times represent lower limits for the relaxation times below 100 K which show only small temperature dependence. The recombination below 100 K is most probably caused by electron-electron or interface assisted scattering processes . /9/

3. Terahertz-electroluminescence from quantum cascade structures

Investigations of intersubband electroluminescence /3,4,5/ have been performed aiming at an electrically driven coherent terahertz emitter based on intersubband transitions. In the mid-infrared wavelength region a quantum cascade laser has been demonstrated by Faist et al. /10/. We are trying to downscale the concept for emission of radiation with a photon energy below the LO-phonon energy in GaAs (36 meV). In that case, the upper state lifetime is not limited by LO-phonon scattering.

The described structure is made up of 50 periods of a chirped $Al_{0.15}Ga_{0.85}As$/GaAs superlattice, as seen in Fig. 4. The radiative transition occurs in the widest (26 nm) quantum well. At an appropriate electric field (22 mV/period), electrons are resonantly injected from the first miniband of the injector (4 wells) into the second subband |2> of the transition well. The electrons escape from the ground subband |1> via fast resonant tunneling whereas tunneling from |2> downstream is inhibited by the minigap of the adjacent injector. The second miniband at the operation bias is Wannier-Stark localized and does not contribute to carrier transport. The two center wells of the injector are lightly doped ($n = 1x10^{16}$ cm^{-3}) to insure a constant electric field across the structure.

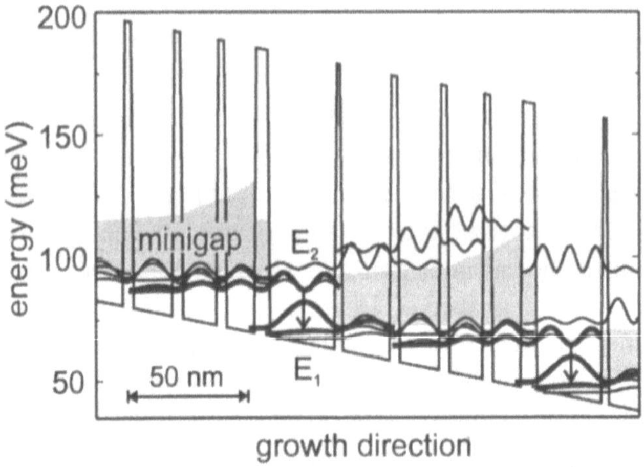

Figure 4: Part of the band structure of a quantum cascade structure designed for emission of terahertz radiation. It has been calculated by selfconsistently solving Schrödinger's and Poisson's equation.

The layer system is grown by molecular beam epitaxy on an n+ doped substrate that is used as the electrical bottom contact. One sample (1300 μm x 250 μm) was prepared with a 16 μm period Au-grating on top that serves as the electrical top contact and as the optical outcoupler at the same time. In a second sample with a larger period grating (50 μm) the layer system was removed between the Au-stripes by reactive ion etching, leaving 7 ridges of 1300 μm length and 25 μm width emitting radiation from the edges.

In order to detect the terahertz luminescence of the samples we have employed two different spectrometers. The first one is the Fourier spectrometer described in the previous section. Our second electroluminescence set-up contains an InSb-cyclotron resonance detector which is tuned in a magnetic field. The sample is located in a second magnet in order to screen the field. The radiation is guided from the sample to the detector in a closed light pipe. The whole optics is immersed in liquid He, so that room temperature background radiation cannot reach the detector nor the sample.

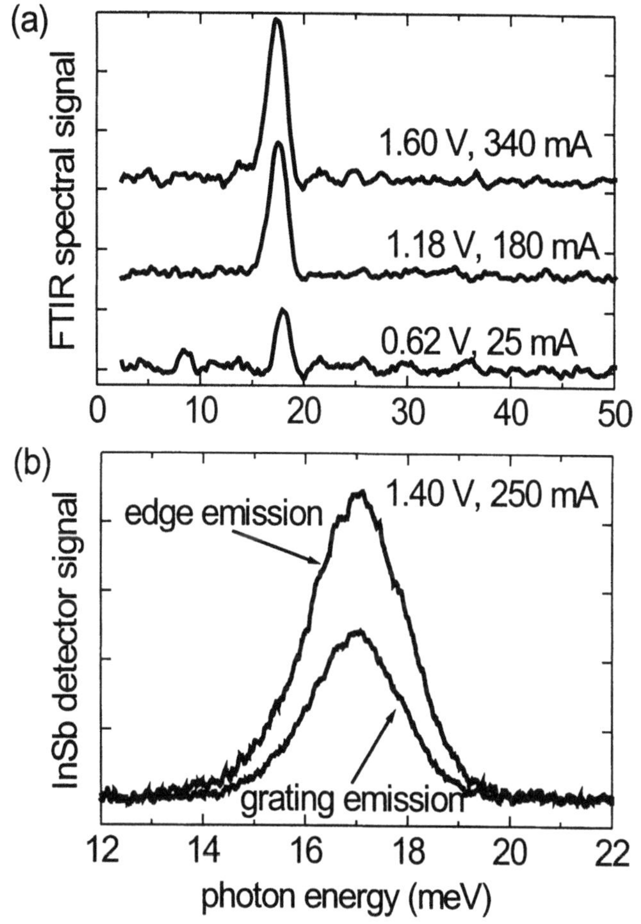

Figure 5: (a) Emission spectra recorded with the Fourier-transform spectrometer at three different biases. (b) Emission spectra recorded with the InSb-detector spectrometer, comparing grating coupled and edge emission at the same current.

We observe terahertz-emission at 17.3 meV as displayed in Fig. 5. This is consistent with the calculated energy difference E_2 - E_1 of 17.1 meV. The linewidth is 1.3 meV, see Fig. 5 (a). The linewidth of 2.2 meV in the InSb-detector spectra in Fig. 5 (b) is not the true linewidth of our sample because the InSb-detector has an intrinsic linewidth of 1.5 meV. We do not observe emission from transitions other than E_2 - E_1. There is a monotonous increase of signal with the injection current up to the breakup of the first miniband. The increase of light output with current is sublinear. This is an indication /5/ that the non-radiative scattering rate increases with the injection current and with the population of the excited state |2>, as it is the case for electron-electron- or Auger-scattering processes.

In spite of the advantages of the FTIR-spectrometer over the InSb-setup, i. e. the high spectral resolution (0.25 meV) the large spectral range (0 - 50 meV), we have used the InSb-setup for its higher signal to noise ratio, for the absence of distortion of the signal by room temperature background radiation, and for the possibility to apply a magnetic field to the sample. The comparison of the two differently processed samples shown in Fig. 5 (b) yields: The edge outcoupling is about twice as efficient as the grating coupled surface emission. We attribute this difference to strong absorption in the highly doped top contact layer /11/.

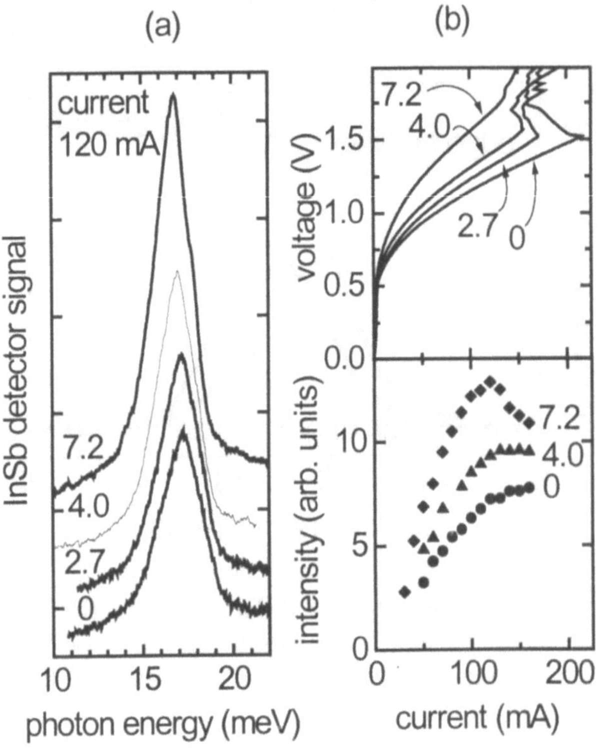

Figure 6: (a) Intersubband electroluminescence spectra at a constant current for various magnetic fields given in Tesla. The spectra are plotted with arbitrary offsets. (b) Voltage versus current characteristics (top panel) and light intensity versus current characteristics (bottom panel). The intensity has been extracted from the spectra as the integrated area below the emission peak.

Application of a magnetic field enhances the emission intensity by a factor of two at B = 7.2 T [see Fig. 6 (a)] /12/. This effect is understood as a consequence of the discretization of the energy spectrum. In the parabolic dispersion relation in the absence of a magnetic field the energy conservation requirement for intersubband Auger-scattering processes is fulfilled for a continuum of energy changes. However, for a non-vanishing magnetic field the only allowed energy changes are multiples of the cyclotron resonance energy. Since the Auger-scattering rate scales inversely with the associated change in energy and momentum /13/, it is more and more suppressed with increasing Landau-level splitting. The emission intensity is proportional to the internal quantum efficiency. The latter is governed by the non-radiative Auger-scattering rate. Hence, the observed enhancement of emission intensity is a direct measure of the suppression of the intersubband scattering rate from subband $|2\rangle$ to $|1\rangle$.

Examining the intensity versus current characteristics given in Fig. 6 (b) we find an enhancement in a magnetic field by about the same factor for currents below 120 mA. In this region the intensity increases sublinearly, indicating that for all magnetic fields Auger-scattering is the governing non-radiative transition process /5/. At the highest magnetic field of 7.2 T the intensity drops when the current is increased above 120 mA. This is a consequence of the misalignment of the structure. If the $|2\rangle$-$|1\rangle$ intersubband scattering rate is quenched by means of a magnetic field, a higher voltage must be applied to maintain the same current. This can be seen in the voltage versus current characteristics [Fig. 6 (b)]. Eventually, the bias reaches the region of negative differential conductance where the injection into subband $|2\rangle$ and therefore also the emission break down.

The emission peak is significantly narrowed in the presence of a magnetic field. The line narrowing in a magnetic field cannot solely be explained by the change in intersubband lifetime. The in-plane dephasing time, which presumably determines the linewidth, is also reduced by the Landau-quantization.

4. Conclusion

We have demonstrated electrically driven THz emission of parabolic quantum wells up to a temperature of 240 K. At high temperatures, the thermal energy k_BT exceeds the photon energy by a factor greater than two. However, we observe single frequency emission, and the impact of high temperatures on the frequency and the line shape is small. The emission efficiency is limited by the temperature dependent decrease of intersubband life time.

From the quantum cascade structures we observe a narrow (1.3 meV) spontaneous emission line at 17.3 meV. The application of a magnetic field enhances the efficiency and reduces the linewidth of terahertz intersubband electroluminescence. These two effects together could facilitate stimulated terahertz-emission in a quantum cascade structure.

Acknowledgments

This work has been supported by the Austrian Science Foundation (START Y47) and by the EU-TMR Program (INTERACT).

References

/1/ K. Unterrainer, C. Kremser, E. Gornik, C. R. Pidgeon, Y. L. Ivanov, E. E. Haller; Phys. Rev. Lett. **53**, 1714 (1988).

/2/ K. D. Maranowski, A. C. Gossard, K. Unterrainer, E. Gornik; Appl. Phys. Lett. **69**, 3522 (1996).

/3/ M. Helm, P. England, E. Colas, F. DeRosa, S. J. Allen, Jr.; Phys. Rev. Lett. **63**, 74 (1988).

/4/ B. Xu, Q. Hu, M. R. Melloch, Appl. Phys. Lett. **71**, 440 (1997).

/5/ M. Rochat, J. Faist, M. Beck, U. Oesterle, M. Ilegems, Appl. Phys. Lett. **73**, 3724 (1998).

/6/ A. Wixforth, M. Kaloudis, C. Rocke, K. Ensslin, M. Sundaram, J. H. English, A. C. Gossard; Semicond. Sci. Technol. **9**, 215 (1994).

/7/ L. Brey, N. F. Johnson, B. I. Halperin Phys. Rev. B **40**, 647 (1989).

/8/ J. N. Heyman, K. Unterrainer, C. Craig, B. Galdrikian, M. S. Sherwin, K. Campman, P. F. Hopkins, A. C. Gossard; Phys. Rev. Lett. **74**, 2682 (1995).

/9/ J. Ulrich, R. Zobl, K. Unterrainer, G. Strasser, E. Gornik, K.D. Maranowski, A.C. Gossard; Appl. Phys. Lett. **74**, 3158 (1999).

/10/ J. Faist, F. Capasso, D. L. Sivco, C. Sitori, A. L. Hutchinson, A. Y. Cho, Science **264**, 553 (1994).

/11/ J. Ulrich, R. Zobl, N. Finger, K. Unterrainer, G. Strasser, E. Gornik, Physica B **272**, 216 (1999).

/12/ J. Ulrich, R. Zobl, K. Unterrainer, G. Strasser, E. Gornik, Appl. Phys. Lett. **76**, 19 (2000).

/13/ P. Hyldgaard, J. W. Wilkins, Phys. Rev. B **53**, 6889 (1996).

Theme 4

CW Sources, Optical and Electronic

PERSPECTIVES OF CONTINUOUS–WAVE OPTOELECTRONIC THZ IMAGING

K.J. SIEBERT, H. QUAST AND H.G. ROSKOS
Physikalisches Institut der Johann Wolfgang Goethe–Universität, Robert–Mayer–Str. 2-4, D-60054 Frankfurt (M), Germany, WWW address: physik@uni-frankfurt.de

1. Introduction

During the last fifteen years, tremendous advances in the ability to generate and detect THz (far-infrared) radiation optoelectronically have led to an increased interest in the application of this part of the electromagnetic spectrum in areas as diverse as solid–state physics, chemistry, microelectronics, and medical imaging. The progress of THz optoelectronics is based mainly on the efficiency of the photomixing techniques developed for optical femtosecond (fs) pulses which allow the generation and detection of broadband, coherent THz pulses with a frequency content from a few GHz to many THz. A great advantage of these techniques is the fact that both the generation and detection process are synchronized to the same laser source via optical gating thus enabling the sensitive measurement of both the amplitude and phase of the electric THz field in a quasi–stroboscopic way. The photoconductive or electrooptic photomixers are operated at room temperature thus ridding far–infrared spectroscopy of the burden of detection at cryogenic conditions. Tracing of the THz wave–form in the time domain permits the extraction of both the real and imaginary part of the dielectric function of the object under investigation.

The very success of the time–domain approach in THz optoelectronics has led to the question whether it is actually necessary to employ fs laser pulses at all. A replacement of pulsed lasers by continuous–wave (cw) lasers providing two beams for photomixing promised advantages for a number of applications. A very obvious and general one is that of cost reductions by replacing expensive fs lasers by comparatively cheap cw semiconductor diode lasers. This replacement could also possibly lead to a reduction in complexity and robustness of a measurement setup. There are also a num-

R.E. Miles et al. (eds.), Terahertz Sources and Systems, 127–143.

ber of technical issues which were of significance for the growing interest in cw THz optoelectronics. Several important aspects concern the bandwidth of the radiation. (i) Although THz pulses provide radiation over a wide frequency band simultaneously, it is tedious to perform measurements with them when a very high spectral resolution is required. In order to raise the resolution in a time–domain measurement, one has to scan over a longer time delay between pump and probe pulse. Although this is principally possible, the alignment efforts and the duration of measurements rapidly increase with the required scan length. The spectral resolution of cw THz systems, on the other hand, is only determined by the bandwidth and spectral stability of the difference frequency of the two cw laser beams. If the THz source provides a narrow bandwidth, all measurements are automatically high–resolution measurements. (ii) Another bandwidth issue is the spectral intensity of the radiation. In THz–pulse systems, the energy is distributed over the entire bandwidth of the pulse. As a consequence, the signal–to–noise ratio (SNR) over a narrow band of frequencies is generally not impressive, although the ratio of the peak signal amplitude to the noise level of the time–domain data can be very large (values as high as 10^5 are not uncommon). Cw THz systems were expected to have an inherent advantage with respect to the SNR and the achievable spectral dynamic range. (iii) A narrow bandwidth makes cw systems also suitable for the application as local oscillators in heterodyne THz detection thus promising the opening of a whole new area of applications for THz optoelectronics, e.g. the detection of extraneous THz signals with the help of solid–state lasers operating in the visible or near–infrared.

Because of the heterodyning potential of cw THz optoelectronics, a strong push for the development of a cw technology came and continues to come from radioastronomy and airborne far–infrared atmospheric spectroscopy. More recently, potential applications in ultra–high–frequency telecommunications (such as local–area networking) and in medical imaging have contributed to a rising interest in cw THz optoelectronics. As all of these fields are of interest and relevance for society, one can expect a growth of funding for THz optoelectronics in forthcoming years.

In the following, we will discuss the status of research in general before reviewing the development of dual–color lasers and advanced photoconductive mixers as key components of cw THz measurement systems. We will then focus on the potential of cw systems for two specific applications, as local oscillators in radio–astronomy and for THz imaging.

2. Status of cw optoelectronic THz spectroscopy systems

As recently as 1998, the only detectors used for optically generated cw THz radiation were bolometers which inherently have to be operated at cryogenic temperatures and yet are too noisy and unreliable in their calibration to be of interest for future applications [1].

Many in the THz community believed that neither photoconductive sampling nor electrooptic sampling — both of which have been instrumental for the success of time–domain THz spectroscopy with high-repetition–rate laser systems— would be applicable for room–temperature detection of cw THz radiation thus confining cw THz optoelectronics to a side–show role. There were good reasons for this pessimism. An important feature of photoconductive sampling is the temporal–windowing effect by optical gating which contributes much to limiting the noise picked up by the photoconductive antenna. The windowing effect is to be understood with respect to the cavity–round–trip time of the laser pulses typically lasting up to ten nanoseconds. During this time, the antenna is only sensitive upon excitation by the laser pulse for less than a picosecond thus rejecting noise quite effectively. In contrast, detection of cw THz radiation requires the antenna to be active equally long as it is inactive. Hence, noise is picked up much more easily. With respect to electrooptic sampling, the peak intensity of optoelectronically generated cw THz radiation.was believed to be too low for electrooptic detection with sufficiently high SNR.

Thus, it came quite as a surprise when Verghese et al. at MIT Lincoln Laboratories demonstrated in 1998 that photoconductive detection is feasible and effective after all [2]. In fact, they established benchmarks for cw THz systems which right away challenged pulsed THz systems. In the words of the authors: "Such a system can outperform systems based on time–domain photoconductive sampling in frequency resolution, spectral brightness, system size, and cost." Their cw THz system was based on two uncoupled Ti:sapphire lasers pumping both an emitter and a receiver planar log–spiral antenna on LT-GaAs with 35 mW of optical power for each antenna. The maximum THz output power ever achieved was 1 μW at 1 THz [3]. The smallest linewidth was below 1 MHz [4]. The accessible frequency range of the photomixers used in the photoconductive detection scheme extended beyond 2 THz. The SNR was better than 100:1 up to 600 GHz with a strong roll–off to 3:1 at 2 THz. The authors explicitly compare the spectral brightness to that of pulsed THz measurement systems and come to the conclusion that the cw power available in any 1–MHz band is 10^4 times higher than in a high–repetition–rate pulsed system. Key parameters of this system which represent the present state of the art are summarized in Tab. 1.

TABLE 1. State of the art THz systems

THz power	1 μW at 1 THz
Frequency	1 GHz up to 3.8 THz
Linewidth	< 1 MHz
Conversion efficiency	10^{-5}
Signal to Noise ratio	≈ 100:1 (at 600 GHz)

The rather unexpected success of cw THz detection may be explained by the homodyning character of the receiver action which is characteristic for both cw and pulsed THz optoelectronics. It involves narrow–band detection of electric fields instead of intensities which leads to cancellation of much of the noise by summation over noise contributions with varying sign. Apparently, time windowing is not as important as previously assumed. A natural advantage of cw THz systems, however, is the higher radiation and detection efficiency of antennas in cw mode than under δ–pulse excitation. The fact that the generation of THz radiation by photomixing with cw lasers achieves higher conversion efficiencies (10^{-5}) than THz generation with high–repetition–rate laser pulses (conversion efficiency: 10^{-6}), is indicative for the fact that resonances are only taken advantage of in full under cw operation.

Electrooptic detection has also made good — albeit not so spectacular — progress. In 1999, Nahata et al. demonstrated electrooptic detection of photoconductively generated cw THz radiation for the first time [5]. The experimental set–up is shown in Fig. 1. THz radiation was generated with a LT–GaAs dipole antenna optimized for maximum output power by the design of the switch metallization (triangular–shaped tips) and by modulation of the applied bias voltage (+25 V to -25 V). The emitter was illuminated with 35 mW of laser radiation mixed from two single–mode DBR lasers operating at wavelengths around 850 nm. The THz radiation was detected with the help of a 6.7–mm–long <110> ZnTe crystal after coupling the optical radiation needed for the nonlinear mixing process into the THz beam path by a pellicle beamsplitter. Care was taken to avoid the formation of standing waves both in the optical and the THz beams by applying various anti–reflection coatings. With these measures, THz signals up to 1.3 THz could be detected with a maximum SNR of 100 at 0.5 THz.

In spite of this successful demonstration, electrooptic detection still has a long way to go before it can be employed as a standard high–SNR detection technique. In the measurements of Nahata et al., the maximum SNR of 100 could only be achieved with an unusual null–bias setting of the

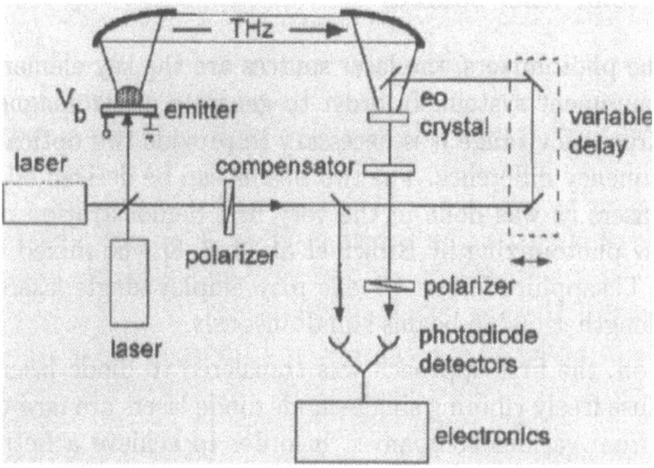

Figure 1. Set–up for electrooptic detection of cw THz radiation. From Nahata et al. [5].

polarizers normally avoided in electrooptic sampling because of a limited linearity range. In the standard quarter–wave polarizer setting, the SNR was reduced by a factor of five. The difficulty of reaching high SNR values clearly results from the low THz peak power (in the experiments of Nahata et al., the THz field amplitude at 0.5 THz was estimated to be only 30 mV/cm in the crystal). More than for pulsed THz systems it is necessary to raise the coherence length (interaction length) of the optical and the THz beams. There are ways to achieve this, however, which are not amenable to pulsed systems. One potential route is quasi–phase–matching which is based on the periodic resetting of the relative phases of the optical and THz waves in the electrooptic detector. Such techniques are not available yet in the THz frequency range because they have not been developed for pulsed THz systems where the large bandwidth of the optical and THz pulses denies application of quasi–phase–matching. The small linewidth of cw radiation not only promises interesting developments in this respect but also opens an additional avenue to noise suppression by spectral filtering. The electrooptic effect is a frequency–mixing effect generating signals at the sum and the difference frequency of the incoming optical and THz radiation. Laser noise can be reduced much by filtering out laser radiation at the original wavelengths behind the electrooptic crystal and in front of the detector photodiodes [6].

3. Two–color laser sources

Besides the photomixers, the laser sources are the key elements of any cw THz measurement system. In order to generate electromagnetic waves in the THz frequency range it is necessary to provide two optical beams with a THz frequency difference. The two beams can be derived either from two disjunct lasers as was done in the very first demonstration of frequency–tunable cw photomixing by Brown et al. [3, 7, 8] who mixed the radiation from two Ti:sapphire lasers. Or one may employ single lasers which emit two wavelength–tunable beams simultaneously.

Early on, the first approach was transfered to diode lasers [9, 10] not least because freely running single–mode diode lasers are now commercially available from various companies. In order to achieve a better frequency stability and thus a better frequency resolution of the spectroscopic system, Chen et al. frequency–locked two distributed–Bragg–reflector diode lasers [11]. They achieved a linewidth of the beat node of < 2 MHz on a time scale of 10 s. On short time scales, the linewidth of the beat node is even better (on the order of 50 kHz for 50–ms integration time).

Working with a single laser emitting two beams appears to be both more economical and quite a bit more elegant than running a two–laser THz system. It has been shown that multimode laser diodes can generate THz radiation with the emission frequencies given by the longitudinal–mode spacings of the laser diode [12]. Such a system is hampered by the limited tunability and the need for an external wavelength selector to obtain single–frequency THz radiation. An improved approach employs laser diodes which emit exactly two longitudinal modes [13, 14], but the tunability is still rather restricted. More advanced is the external–cavity approach by Wang et al. [15] who employ an intra–cavity grating and a V–shaped mirror for the selection of two lasing frequencies. By moving the mirror up and down, the difference frequency can be altered continuously from 250 GHz up to 2.1 THz. The output power is 8 mW, sufficient for the generation of about 2 nW of THz radiation at 250 GHz [16].

In 1995, Goyette et al. [17] suggested to utilize femtosecond pulses from a mode–locked Ti:sapphire laser for the generation of narrow–band cw THz radiation by filtering out two longitudinal modes from the pulse spectrum. This approach promises wide tunability, low noise and high frequency stability of the difference signal. The penalty for these advantages, however, is quite severe because the low power of each mode makes it necessary to amplify them in diode amplifiers thus reducing the stability and increasing the noise again and in addition raising the costs of the final system. The need for amplification of the laser radiation appears to be a general feature of all diode–laser–based approaches known to us at present. The amount of

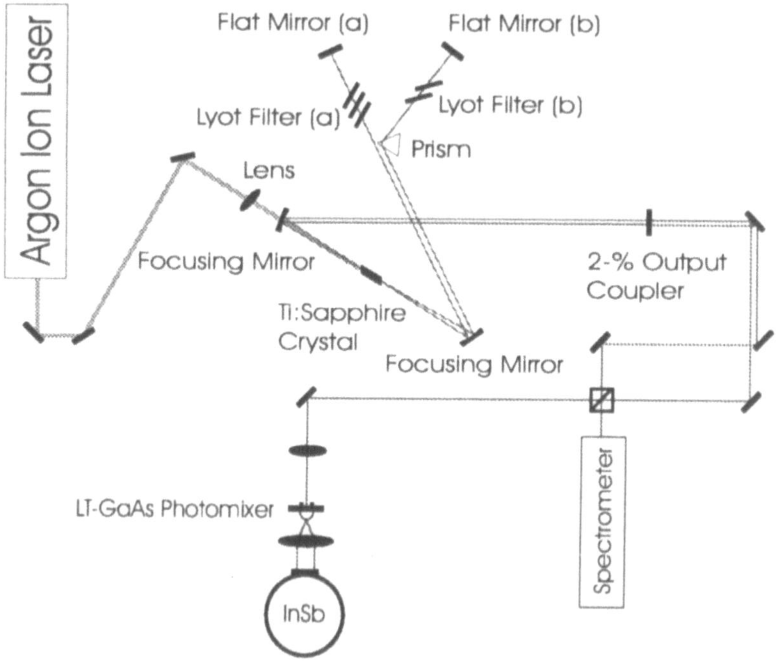

Figure 2. Two–color Ti:sapphire laser with photomixing setup. The laser emits two non–collinear but parallel beams which are combined in a single beam for photomixing.

total optical power (sum of both beams) wanted for a state–of–the–art THz system is on the order of 150–200 mW, mainly determined by the damage threshold of the LT–GaAs photomixers which is on the order of 60 mW [18]. Often, diode lasers can provide this power without amplification but then the mode quality and the noise properties appear to be so problematic that operation of the lasers at lower power and subsequent amplification e.g. in a tapered amplifier is prefered.

As the performance of diode lasers is continually improving it is likely that the need for amplifiers will diminish over the next few years. As this is not yet the case, however, we have decided to develop for our own laboratory purposes a laser source which is easy to handle, provides sufficient output power, is easily and widely tunable and exhibits good stability at the difference–frequency signal. All of these requirements are met by the dual–color Ti:sapphire laser whose resonator lay–out is displayed in Fig. 2 [6, 19]. The α–shaped linear cavity consists of five mirrors, three of them — as well as the Ti:sapphire gain medium — being shared by both beams while two mirrors in the wavelength–selection segment are used by either beam only. The Ti:sapphire crystal is pumped by a single beam from an argon–ion laser operated in all–lines mode at an output power of 6 W. The

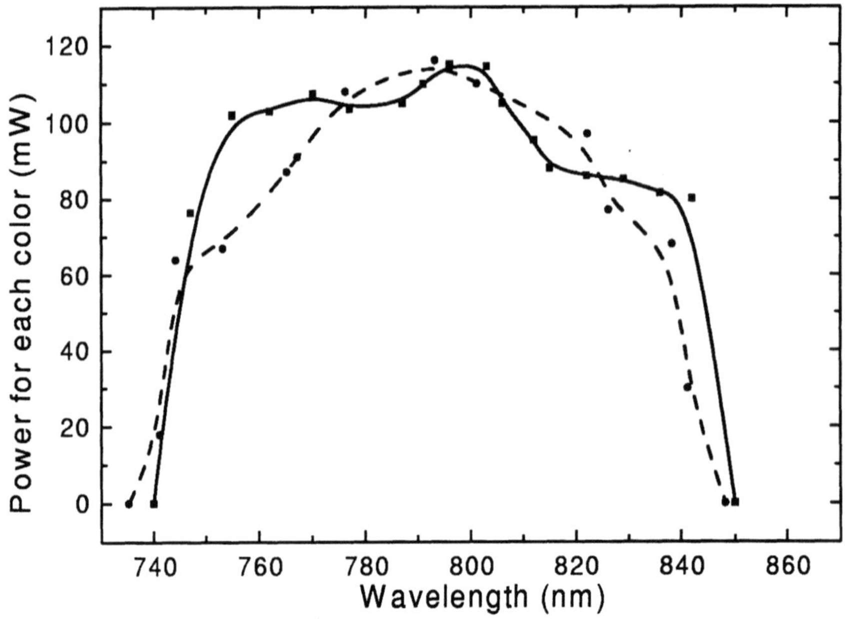

Figure 3. Wavelength tuning curve of the two–color Ti:sapphire laser. The solid line denotes the tuning characteristics for the case that one beam is tuned while the other is kept fixed at 800 nm. The dashed line displays the case when both colors are tuned simultanously with a fixed wavelength difference of 7 nm between them.

wavelength tuning of the Ti:sapphire laser beams is achieved with the help of two Lyot filters consisting of two respectively three birefringent quartz plates. Figure 3 displays the tuning characteristics of the laser. The two beams can be tuned independently from each other over a range of more than 100 nm allowing to set the frequency difference anywhere between less than 1 GHz and 50 THz.

After the data of Fig. 3 were taken, the 2–% output–coupling mirror was replaced by a 4–% output coupler leading to a slight increase in the power which can be extracted from the laser. We now achieve a total output power (sum over both beams) of up to 250 mW for a pump power of 6 W. The laser action is as stable as that observed under single–beam operation with no indications for instabilities induced by crosstalk of the beams.

Before concluding this chapter it should be mentioned that recently a new trend emerged towards laser sources operating in the telecommunication wavelength windows of 1.3 μm and 1.55 μm. There are three reasons for this development. First, the stringent requirements of telecommunication technology have led to the development of high–power single–

mode diode–laser systems with excellent frequency stability which are now available commercially. Such lasers principally have sufficient tunability for THz photomixing because tunability is a prerequisite for the application of such lasers for wavelength–division multiplexing in telecommunications. Secondly, the 1.55–μm wavelength region offers a variety of other types of highly developed lasers such as fiber lasers as interesting alternatives to diode lasers. The third reason for the shift to telecommunication wavelengths stems from some prospective applications where it will be necessary to transmit the laser radiation to the photomixers over larger distances via optical fibers. Such applications are envisaged both in radio–astronomy and in ultra–high–frequency local–area networking (see below). In this case it is unavoidable to move away from the 800–nm wavelength range towards longer wavelengths in order to overcome the problem of high losses in optical fibers around 800 nm. The interest in the telecommunication wavelength region has already led to the development of novel dual–color laser sources such as that realized by Alouini et al. [20] who developed an Er:Yb–glass–based two–color laser at 1.5 μm for THz frequency generation. The difference frequency is tunable from DC to 2.7 THz with a remarkably small spectral linewidth of less than 10 kHz.

The problem faced by the wavelength shift is the lack of suitable photoconductive materials for THz photomixing. There is an urgent need for further developments in this area. We address this issue in the next chapter.

4. Advanced photoconductive photomixers

The idea of mixing two optical frequencies in a semiconductor device dates back at least to 1974 when a radio frequency signal was produced for the first time by mixing two visible laser frequencies with a tungsten-nickel point–contact diode [21]. However it took about twenty years until E. Brown et al. came up with the first photomixers based on the ultrafast photoconducting material low–temperature–grown–GaAs which now is the standard semiconductor material in photoconductive antenna mixers [7].

To obtain a THz modulation of the current within the antenna, the dynamics of the charge carriers in the optical switch must be extremely fast. This puts severe constraints on the properties of the semiconductor material. The mobility lifetime, determined by the trapping time of the charge carriers which have been excited optically in the semiconductor must be less than 1 ps in order to cut off the electric currents, which are to be modulated at THz frequencies, sufficiently rapidly. The photoconductive materials LT–GaAs and ion–implated SOS (silicon–on–sapphire)which represent the standard for pulsed THz systems both fulfill this first requirement.

There is, however, a second requirement which is unique for cw systems and does not affect typical pulsed THz systems. This second prerequisite concerns screening of the electric bias field, respectively of the electric field of the radiation which is to be detected, by trapped charge carriers. The repetition rate of femtosecond laser pulses is typically on the order of 80 MHz which leaves much time — more than 10 ns — for the carriers to relax from the trapped states into the ground states until the next pulse hits the switch. As the relaxation out of the traps occurs via recombination of electrons and holes, this has the effect of removing charge from the photoconductive switch and of terminating any screening effects. For cw systems, the situation is much more severe as the next maximum of the beat signal follows much more rapidly after a first one. For a 1–THz signal, the maxima are separated by only 1 ps. Ideally, one would like all charge carriers or at least a large fraction of them to disappear out of the traps on the same time scale or faster in order to avoid field screening with certainty. However, state–of–the–art ion–implanted SOS has a carrier recombination lifetime of more thn 10 ns [22] while LT–GaAs is assumed to have a recombination lifetime on the order of 10 ps [23] [1]. It is hence unavoidable that trapped charge carriers accumulate over several or many periods of the THz signal. It depends on the average density of carriers whether field screening is significant or not. Both photoconductive materials are not optimal for ultrafast cw photomixing but LT–GaAs clearly is much better than ion–implanted SOS. It is not by accident that all the successful cw photomixing experiments have been performed so far with photomixers based on LT–GaAs where charge accumulation seems to be sufficiently small not to cause screening problems. Whether ion–implanted SOS can be employed for cw THz photomixing at all is unknown to us. We are not aware of any systematic study of this issue nor of any general study of screening in cw photoconductive switches.

In the last chapter we have mentioned the trend towards optoelectronic THz system based on 1.3–μm and 1.5–μm optical technology. This trend meets a big obstacle, which is the lack of suitable photoconductive materials. Criteria for good media are a high carrier mobility, a subpicosecond trapping time, a picosecond recombination lifetime, a high breakthrough voltage, good thermal conductivity and thermal breakdown characteristics, and mechanical robustness. In the past, there have been sporadic attempts to develop such materials but with insufficient success [25]. More systematic studies have started at several laboratories in the world and there seems to be no fundamental reason why they should not be able to develop good pho-

[1]It must be pointed out that the figures for the carrier recombination liefetime in LT-GaAs are controversially dicussed in the literature [23, 24]. Further work has to be carried out to clarify the influence of screening on the generation and detection of cw THz radiation.

toconductors for this wavelength range. It should be pointed out that the search for appropiate materials is not limited to low–temperature growth, but also to other techniques of material synthesis and modification such as ion implantation [26, 27, 28, 29]. However, to our knowledge, no reports of successful photomixing have been published yet.

Another key feature of photoconductive photomixers which we want to address briefly is their geometry. They normally consist of antenna structures with the photoconductive switch being embedded as a metal–semiconductor–metal contact with a typical size of 10 μm × 10 μm. Higher efficiencies are obtained if the metal contacts are laid out as interdigitated finger structures. A number of antennas have been tested over the years. Concentrating on the emission efficiency, simple dipole antennas give reasonable output powers at their resonance frequency, but show a rather narrow bandwidth of operation [12]. Broadband devices such as bow–tie antennas [12] and spiral antennas [3] have a broader bandwidth, but correspondingly don't allow to reach as high peak powers as the resonant structures. These rather simple types of emitter antennas are restricted to THz powers of several hundred nW. More complex structures, the so–called resonant double–dipole or double–slot antennas, allow for output powers of about 1 μW [30]. This is the highest cw THz power ever achieved with photomixing. In addition, they provide the radiation in a very symmetrical emission cone which is favorable for the further handling of the radiation, notably its focussing.

In principle, the THz output power grows quadratically with the optical power [18] and superquadratically with the applied bias voltage [8]. Due to electric breakthrough one can't increase the bias indefinitely. Due to thermal failure, the optical pump power is restricted to $\approx$ 75 mW [18].

Solving this problem leads away from lumped–element photoconductive switches to a new generation of photomixers currently under investigation. These so called travelling–wave photomixers distribute the optical pump power over a larger area of the switch. In the devices studied by the group at MIT Lincoln Laboratories, the optical beat signal is guided by a dielectric ridge waveguide from where it couples evanescently into the photoconductor which is placed on top of the waveguide [31]. The THz signal generated in the switch is guided with coplanar strips and copropagats with the optical signal before being coupled into an antenna for emission. Phase matching between the optical and the THz wave is necessary and is achieved by slowing down the THz wave by loading the coplanar strips with thin interdigitated metal electrodes. Another travelling–wave approach which so far seems to be more successful is pursued by a group at CALTECH/JPL [32]. The experimental set–up is shown in Fig. 4. The photomixer is a coplanar stripline ending in a dipole antenna, all fabricated on LT-GaAs. Optical

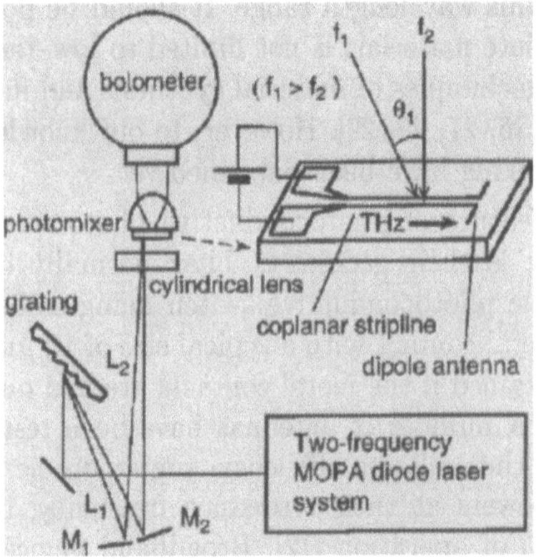

Figure 4. Travelling–wave device developed by Matsuura et al. [32].

excitation occurs with two beams each focussed with a cylindrical lens into the gap between the metal strips of the waveguide. The two beams (with different optical frequencies) are tilted with respect to each other producing a moving interference pattern. Phase matching between the travelling interference pattern and the THz wave is achieved by choosing the right tilt angle between the two optical beams. So far, this device generates output powers of about 1 μW while pumped with an optical power of 300 mW. The emission efficiency (THz output power to incident optical power) is still inferior to that of the older–generation single–point photomixers and needs to be improved. Output powers of about 10 μW appear to be feasible for 300 mW pump power.

5. Perspectives for cw THz imaging

THz optoelectronics has now entered a phase where potential applications are explored seriously. One area of special interest is that of THz imaging, a field which gained much momentum by two developments introduced nearly simultaneously for pulsed THz measurement systems. Until 1995 it was believed that taking images with THz pulses would be too time–consuming to be practical. Images were obtained by placing objects into the focus of a THz beam, recording the time–domain data, Fourier–transforming them in order to extract spectral information, and raster scan the object through the focus of the beam repeating the procedure described before at each spatial point. Taking a single image took hours. Then, in 1955, Hu et al. introduced a high–repetition–rate pulsed THz measurement system which reduced this time to minutes [33]. A further improvement was achieved by high–speed data acquisiton and by introducing pattern–recogition techniques which replaced the time–consuming Fourier–transformation algorithms [34]. A second development gave THz imaging an even better outlook. Zhang et al. showed that THz images could be taken very rapidly with the help of low–repetition–rate amplifier laser systems and electrooptic detection employing conventional Si–chip cameras for data recording [35]. The key to this success was the high power of the THz pulses which proved to be sufficient for taking images without the need for raster scanning. Even single–shot imaging may be feasible for some applications [36].

Since then, many papers have been published on pulsed imaging systems [37, 38, 39]. Promising areas of application are in medical and biological imaging. Arnone et al. recently showed first applications which use the unique features of electromagnetic radiation in the THz frequency range in dentistry [40]. They demonstrated that they can detect caries in a tooth in the first stage of the decay when the outer surface of the tooth still remains intact. Different absorption of healthy enamel and demineralized tissue in the tooth gives a contrast which can't be observed with other methods like X-rays.

All THz imaging systems presented in the literature to date have been based on THz time–domain spectroscopy systems. With cw THz measurement systems having improved in the way described in the preceeding chapters, and taking into account that they, too, allow to detect both amplitude and phase changes, one can argue that the time has come to introduce cw THz imaging systems as serous alternative to pulsed systems because of the following reasons. (i) Cw THz systems can be operated with laser diodes. For imaging and inspection applications this will induce a significant reduction in cost, size and complexity. (ii) Compared to high–repetition–rate pulsed THZ systems, cw THz systems have a fundamental advantage in terms of the time which is needed for data acquisition. In pulsed systems, time scans with subsequent Fourier transformation or pattern recognition are needed at each spatial point of the sample in order to obtain the frequency and phase information from the time transient of the detected electric field. If one considers that for the identification of many substances it is only necessary to obtain information at certain characteristic absorption lines, pulsed systems deliver more information than necessary while consuming much time for data acquisition. (iii) An additional advantage of cw systems is their higher spectral brightness and better frequency resolution. For pulsed systems the length of the delay stage determines the achievable frequency resolution whereas for cw systems the bandwidth of the THz radiation is restricted by the bandwidth of the laser source.

Research in cw THz imaging has only started recently. In fact, as this article is being written, there has not been any report in the literature yet about the demonstration of cw THz imaging. One has to see what the future will bring.

Before concluding this chapter, we want to point out an interesting THz system which in a sense combines the advantages of pulsed and cw systems. In 1996, Kawase et al. introduced an optoelectronic THz source based on a Q–switched Nd:YAG laser. It delivers 25 ns long light pulses. The THz radiation generated by parametric down–conversion in a $LiNbO_3$ crystal is tunable and has a rather small bandwidth but at the same time is characterized by a high intensity. With this system, it may be feasible to perform imaging with unprecedented speed because the small bandwidth makes temporal scanning unnecessary while the high intensity principally allows to employ electrooptic detection with a camera for recording images over larger areas thus reducing the need for spatial raster scanning. The only disadvantage of the system is the wavelength of the laser (1.064 μm) which prevents the straightforward application of Si–based cameras.

6. THz optoelectronic local oscillators

Cw THz optoelectronics presently receives a strong technological push from an unexpected side, radio–astronomy. The background is the following. For radio–astronomers, far–infrared radiation from the sky provides essential information about the distribution of matter and about chemical processes in the universe. The radiation is detected by down–conversion with the help of radiation derived from cw local oscillators (LOs).

At present, radio–astronomy aims for higher spatial resolution by building phased arrays of telescopes. This requires one LO for each telescope with all LOs being phase–locked with each other. The most ambitious project of this kind is the Atacama Large Millimeter Array (ALMA) endeavor, a collaboration of the European, American and Japanese space organizations and to become perative in the Atacama desert in Chile. (ALMA) [41]. It is planned to combine 64 or more antennas distributed over a maximum distance of 10 km and is designed to achieve an angular resolution approaching 10 milliarcseconds. The plan aims for a coverage of frequencies up to 1 THz to be reached at later stages of the project. With rising frequency it becomes harder and harder to phase–lock the LOs because the ultrahigh–frequency signals have to be distributed over many kilometers of distance to achieve coupling of the LOs. It may become unavoidable to employ optoelectronic techniques where THz signals are distributed as beat signals of optical waves transmitted through optical fibers followed by

Figure 5. Animation of the Atacama Large Millimeter Array site. The furthest distance between two antennas is about 10 km [41].

down–conversion at each telescope site employing photonic LOs.

The requirements specified for each LO are that it should be able to generate 100 μW of THz power at 1 THz. This condition cannot be met at present. A second requirement concerns the optical wavelength the system has to be operated at. In order to cover the rather long distances between telescopes, the wavelength has to be in the telecommunication windows at 1.3 μm respectively 1.55 μm. Only there are absorption losses in optical fibers low enough. As pointed out above, however, a viable photomixer technology does not yet exist for optical wavelengths in the 1.3–1.55 μm range.

The need to find solutions for this problem has led to a strong push to develop a high–efficiency photomixing technology for the 1.3–1.55–μm–wavelength regime. This situation is clearly of interest beyond the realm of radioastronomy. Any technological advance in this area can be adopted in a straightforward manner by other THz applications thus improving the technological outlook of the field as a whole.

References

1. Richards, P.L. (1994) Bolometers for Infrared and Millieter Waves, *J. Appl. Phys.*, **76**, 1-24.
2. Verghese, S., McIntosh, K.A., Calawa, S., Dinatale, W.F., Duerr, E.K. and Molvar, K.A. (1998) Generation and Detection of Coherent Terahertz Waves Using two Photomixers, *Appl. Phys. Lett.*, **73(11)**, 3824-3826.
3. Brown, E.R., McIntosh, K.A., Nichols, K.B. and Dennis, C.L. (1995) Photomixing up to 3.8 THz in Low-Temperature-Grown GaAs, *Appl. Phs. Lett.*, **66**, 285-287.

4. Matsuura, S., Chen, P., Blake, G.A., Pearson, J.C., Picket H.M. (2000) A tunable Cavity-Locked Diode Laser Source for Terahertz Photomixing, *IEEE Transactions on Microwave Theory and Techniques*, **48(3)**, 380-387.
5. Nahata, A., Yardley, J.T. and Heinz, T.F. (1999) Free–Space Electro–Optic Detection of Continuous–Wave Terahertz Radiation, *Appl. Phys. Lett.*, 75(17), 2524-2526.
6. Siebert, K., Siebe, F., Thomson, M., Baghbidi, J.Z., Leonhardt, R., Roskos, H.G. (1999) Advances in Continous–Wave THz Generation, *Terahertz Spectroskopy, Applications–II,J.M. Chamberlain (Editor), SPIE*, **3828**, 234-243.
7. Brown, E.R., McIntosh, K.A., Smith, F.W., Manfra, M.J. and Dennis, C.L. (1993) Measurements of Optical-Heterodyne Conversion in Low-Temperature-Grown GaAs, *Appl. Phys. Lett.*, **62(11)**, 1206-1208.
8. Brown, E.R., McIntosh, K.A., Smith, F.W., Nichols, K.B., Manfra, M.J., Dennis, C.L. and Mattia, J.P. (1994) Milliwatt Output Levels and Superquadratic Bias Dependence in a Low-Temperature-Grown GaAs Photomixer, *Appl. Phys. Lett.*, **64(24)**, 3311-3313.
9. McIntosh, K.A., Brown, E.R., Nichols, K.B., McMahon, O.B., DiNatale, W.F. and Lyszczarz, T.M. (1995) Terahertz Photomixing with Diode Lasers in Low–Temperature–Grown GaAs, *Appl. Phys. Lett.*, **67**, 3844-3846.
10. Matsuura, S., Tani, M. and Sakai, K. (1997) Generation of Coherent Terahertz Radiation by Photomixing in Dipole Photoconductive Antennas, *Appl. Phys. Lett.*, **70**, 559-561.
11. Chen, P., Blake, G.A., Gaidis, M.C., Brown, E.R., McIntoch, K.A., Chou, S.Y., Nathan, M.I. and Williamson, F. (1997) Spectroscopic Applications and Frequency Locking of THz Photomixing with Distributed–Bragg–Reflector Diode Lasers in Low-Temperature–Grown GaAs, *Appl. Phys. Lett.*, **71(12)**, 1601-1603.
12. Tani, M., Matsuura, S., Sakai, K. and Hangyo, M. (1997) Multiple–Frequency Generation of Sub–Terahertz Radiation by Multimode LD Excitation of Photoconductive Antenna, *IEEE Microwave and Guided Wave Lett.*, **7**, 1-3.
13. Gu, P., Tani, M., Hyodo, M., Sakai, K. and Hidaka, T. (1998) Generation of cw-Terahertz Radiation Using a Two–Longitudinal–Mode Laser Diode, *Jpn. J. Appl. Phys.*, **37**, L976-L978.
14. Pelusi, M.D., Liu, H.F., Novak, D. and Ogawa, Y. (1997) THz Optical Beat Frequency Generation from a Single Mode Locked Semiconductor Laser, *Appl. Phys. Lett.*, **71**, 449-451.
15. Wang, C.-L. and Pan, C.-L. (1995) Tunable Multiterahertz Beat Signal Generation from a Two–Wavelength Laser–Diode Array, *Optics Lett.*, **20(11)**, 1292-1294.
16. Gu, P., Chang, F., Tani, M., Sakai, K. and Pan, C.-L. (1999) Generation of Coherent Cw-Terahertz Radiation using a Tunable Dual-Wavelength External Cavity Laser Diode, *Jpn. J. Appl. Phys.*, **38**, L1246-L1248.
17. Goyette, T.M., Guo, W., De Lucia, F.C., Swartz, J.C., Everitt, H.O., Guenther, B.D. and Brown, E.R. (1995) Femtosecond Demodulation Source for High–Resolution Submillimeter Spectroscopy, *Appl. Phys. Lett.*, **67(25)**, 3810-3812.
18. Verghese, S., McIntosh, K.A. and Brown, E.R. (1997) Optical and Terahertz Power Limits in the Low–Temperature–Grown GaAs Photmixers, *Appl. Phys. Lett.*, **71(19)**, 2743-2745.
19. Siebe, F., Siebert, K., Leonhardt, R. and Roskos, H.G. (1999) A Fully Tunable Dual–Color CW Ti:Al_2O_3 Laser, *IEEE Journ. Quantum Electronics*, **35(11)**, 1731-1736.
20. Alouini, M., Brunel, M., Bretenaker, F., Vallet, M. and Le Floch, A. (1998) Dual Tunable Wavelenggth Er:Yb:Glass Laser for Terahertz Beat Frequency Generation, *IEEE Photonics Technology Lett.*, **10(11)**, 1554-1556.
21. Sakuma, E. and Evenson, K.M. (1974) Characteristics of Tungsten–Nickel Point Contact Diodes Used as Laser Harmonic–Generator Mixers, *IEEE Journ. Quantum Electronics*, **Qe-10(8)**, 599-603.
22. Brorson, S.D., Zhang, J., Keiding, S.R. (1994) Ultrafast Carrier Trapping and slow

Recombination in Ion–Bombarded Silicon on Spphire Measured via THz Spectroscopy, *Appl. Phys. Lett.*, **64(18)**, 2385-2387.
23. Dekorsy, T., Segschneider, G., Nagel, M., Heiliger, H.M., Kurz, H., Hey, R., Ploog, K., Luysberg, M. (1998) Characterisation and Application of Low Temperature Grown-GaAs Based on Femtosecond Laser, Symposium On *Non-Stoichiometric III-V Compounds*, Edt. T. Marek, University of Erlangen **6**, 85-91.
24. Allam, J., Baynes, N. de B., Cleaver, J.R.A., Ogawa, K., Mishima T., Ohbu, I. (1996) Monolithically–Integrated Optoelectronic Circuit for Ultrafast Sampling of a Dual–Gate FET, *Optical and Quantum Electronics*, **28**, 875-896.
25. Gupta, S., Whitaker, J.F., Mourou, G.A. (1992) Ultrafast Carrier Dynamics in III-V Semiconductors Grown by Molecular–Beam Epitaxy at Very Low Substrate Temperatures, *IEEE Journ. Quantum Electronics*, **28(10)**, 2464-2472.
26. Krotkus, A., Marcinkevicius, S., Jasinski, J., Kaminska, M., Tan, H.H., Jagadish, C. (1995) Picosecond Carrier Lifetime in GaAs Implanted with High Doses of As Ions: An Alternative material to low–temperature GaAs for Optoelectronic Applications, *Appl. Phys. Lett.*, **67(23)**, 3304.
27. Ganikhanov, F., Lin, G.-R., Chen, W.-C., Chang, C.-S., Pan, C.-L. (1995) Subpicosecond Carrier Lifetimes in Arsenic–Ion–Implanted GaAs, *Appl. Phys. Lett.*, **67(23)**, 3465.
28. Jagadish, C., Tan, H.H., Jasinski, J., Kaminska, M., Palczewska, M., Krotkus, A., Marcinkevicius, S. (1995) High Resistivity and Picosecond Carrier Lifetime of GaAs Implanted with MeV Ga Ions at high fluences, *Appl. Phys. Lett.*, **67(12)**, 1724.
29. Kostoulas, Y., Waxer, L.J., Walmsley, I.A., Wicks, G.W., Fauchet, P.M. (1995) Femtosecond Carrier Dynamics in Low–Temperature–Grown Indium Phosphide, *Appl. Phys. Lett.*, **Vol. 66, no. 14**, pp. 1821.
30. Duffy, S.M., Verghese, S., McIntosh, K.A., Jackson, A., Gossard, A.C. and Matsuura, S. (2000) Accurate Modeling of Dual Dipole and Slot Elements used with Photomixers for Coherent Terahertz Output Power, *To be published.*.
31. Duerr, E.K., McIntosh, K.A. and Verghese, S. (2000) Distributed Photomixers, *CLEO 2000 Proceedings*, **CWU6**, 382-383.
32. Matsuura, S., Blake, G.A., Wyss, R.A., Pearson, J.C., Kadow, C., Jackson, A.W. and Gossard, A.C. (1999) A Travelling–Wave THz Photomixer based on Angle-Tuned Phase Matching, *Appl. Phys. Lett.*, **74(19)**, 2872-2874.
33. Hu, B.B. and Nuss, M.C. (1995) Imaging with Terahertz Waves, *Optics Lett.*, **20**, 1716-1718.
34. Mittleman, D.M., Jacobsen, R.H., Nuss, M.C. (1996) T-Ray Imaging, *IEEE Journ. of Selected Topics in Quantum Electronics*, **2(3)**, pp. 679-692.
35. Wu, Q., Hewitt, T.D., Zhang, X.-C. (1996) Two-Dimensional Electro-Optic Imaging of THz Beams, *Appl. Phys. Lett.*, **69(8)**, 1026-1028.
36. Jiang, Z., Zhang, X.-C. (1998) Electro-Optic Measurement of THz Field Pulses with a Chirped Optical Beam, *Appl. Phys. Lett.*, **72(16)**, 1945-1947.
37. Jacobsen, R.H., Mittleman, D.M. and Nuss, M.C. (1996) Chemical Recognition of Gases and Gass Mixtures with Terahertz Waves, *Optics Lett.*, **21**, 2011-2013.
38. Mittleman, D.M., Gupta, M., Neelamani, R., Baraniuk, R.G., Rudd, J.V. and Koch, M. (1999) Recent Advances in Terahertz Imaging, *Appl. Phys. B*, **B 86**, 1085-1094.
39. Lu, Z.G., Zhang, X.-C. (1998) Real Time THz Imaging System Based on Electro-Optic Crystals, *SPIE-Applications of Photonic technology III*, **3491**, 334.
40. Arnone, D., Ciesla, C. and Pepper, M. (2000) Terahertz Imaging comes into View, *Physics World*, **April**, 35-40.
41. http://www.mma.nrao.edu.

CONTINUOUS-WAVE TERAHERTZ GENERATION USING PHOTOMIXERS

S. Verghese, K. A. McIntosh, S. M. Duffy, and E. K. Duerr
Lincoln Laboratory, Massachusetts Institute of Technology
Lexington, MA 02420-9108
simonv@ll.mit.edu

Keywords: photomixer, GaAs, terahertz, THz, transceiver, local oscillator

Abstract Optical heterodyne conversion, or photomixing, is a frequency agile technique that generates continuous-wave radiation at THz frequencies using thin films of low-temperature-grown GaAs. Optimizing photomixers for maximum output power requires careful design of the epitaxial growth sequence, and detailed analyses of the radio-frequency (RF) circuitry and of the optical feed. Control of the LTG-GaAs epitaxy leads to material with short photocarrier lifetime and robustness to thermal failure. Key tradeoffs are discussed for optimizing the RF and optical feeds for THz output power. Some promising applications for photomixers include local oscillators for THz heterodyne detectors based on superconductors, and high-resolution spectrometers useful for rotational spectroscopy of airborne molecules. Two experiments are described that demonstrate the viability of photomixers for these applications.

Introduction

Photomixers are compact, all-solid-state sources that use a pair of single-frequency tunable diode lasers to generate a THz difference frequency by photoconductive mixing in low-temperature-grown (LTG) GaAs [1, 2, 3]. The output frequency is tuned over several THz by temperature or current detuning of the two diode lasers by a few nanometers in wavelength. As sources, photomixers have been used for local oscillators with cryogenic THz heterodyne detectors [4] and for high-resolution gas spectroscopy when used in conjunction with liquid-helium-cooled bolometers [5, 6]. On the most basic level, optical heterodyne conversion with a photoconductive switch (photomixing) is analogous to the operation of a transistor amplifier. In a transistor, a small RF signal applied to the gate modulates the conductance of a switch under a relatively large

R.E. Miles et al. (eds.), Terahertz Sources and Systems, 145–165.

DC bias. The modulated output power is drawn from the source providing the DC bias. In photomixing, the photoconductance is modulated by the optical beating of the two laser diodes and the output THz power is generated from current drawn from the battery that provides a DC bias between the photoconductor electrodes. Photomixing is fundamentally different than difference-frequency generation using a $\chi^{(2)}$ process in a material such as $LiNbO_3$. In this type of difference-frequency generation, the output THz power is generated from the optical photons, and only one THz photon can be created per each pair of optical photons. The fact that two hot photons are required to create one cold photon results in an efficiency penalty that makes photomixing more favorable between 1 and 3 THz. At frequencies above several THz, however, the $\chi^{(2)}$-mixing process is more efficient than photomixing, since photomixers then suffer from parasitic impedances and the $\chi^{(2)}$-mixing process gets more efficient. This trade-off has been discussed in more detail by Brown *et al.* [7].

1. DEMONSTRATED CW TECHNOLOGY FOR THZ GENERATION

Many sources are available for generating coherent continuous-wave THz radiation, including backward-wave oscillators, molecular lasers, Schottky diode upconverters, and parametric downconverters. If we restrict the discussion to solid-state devices, the most widely used sources are varactor multipliers [8] or fundamental sources such as negative-resistance diodes and, more recently, high-electron mobility transistors (HEMTs). In general, electronic technologies that use electron transport for THz generation are limited by parasitic resistance and capacitance as frequencies exceed 1 THz. A developing technology is the THz quantum cascade laser—although only electroluminescence has been observed to date, this technology represents a fundamentally different approach to THz generation than the electronic technologies [9, 10]. Quantum-effect devices such as Bloch oscillators and quantum cascade lasers are generally limited to operation at cryogenic temperatures since thermal energy kT at 300 K otherwise smears out the quantum levels.

Photomixers can be thought of as combining some of the attributes of electronic technologies and optical technologies. Although their output power is modest (1 μW typical), this power is suitable or nearly suitable for applications like local oscillators for cryogenic heterodyne receivers and for spectrometers. Two of the photomixer's most attractive features are: its wide tuning range (25-3000 GHz), and that it operates at room temperature.

2. BASIC DESIGN TRADE-OFFS

The basic design tradeoffs for lumped-element photomixers are constrained by the following: optical spot size, thermal limits from optical and ohmic heating, capacitance, and the photoconductor's RF resistance.

A photomixer comprises a thin photoconducting film of low-temperature-grown GaAs, submicron electrodes, and a radiating antenna also used to provide a dc bias. In the photoconducting film, the incident light modulates the conductivity through the generation of photocarriers. The photomixing process occurs during illumination of an electrode region with two single-mode lasers with average powers P_1 and P_2 and frequencies ν_1 and ν_2, respectively. The instantaneous optical power incident on the photomixer contains mixing terms (beats) between the two optical frequencies:

$$P_i = P_1 + P_2 + 2\sqrt{P_1 P_2}\left(\cos 2\pi(\nu_1 - \nu_2)t + \cos 2\pi(\nu_1 + \nu_2)t\right) , \quad (2.1)$$

where $f = \nu_1 - \nu_2$ is the difference frequency which can be easily adjusted by changing the bias current or operating temperature of the diode lasers. The $\nu_1 + \nu_2$ term occurring at the sum frequency doesn not couple efficiently through the THz metal circuitry and can be neglected.

The DC photocurrent generated by the optical power is

$$I_o = \eta_e \frac{e}{h\nu} P_o , \quad (2.2)$$

where $e/h\nu$ is a constant of 0.69 for an optical wavelength near 860 nm (the bandgap of GaAs) and η_e is the external quantum efficiency. Using the above two equations, the amount of THz power transmitted out of the radiating antenna is then

$$P_\omega = \frac{0.5 I_o^2}{1 + \omega^2 \tau_e^2} \eta_{ant} \mathrm{Re}(Z_{\mathrm{ant}}) , \quad (2.3)$$

where $\omega = 2\pi f$, τ_e is the photocarrier lifetime, η_{ant} is an antenna efficiency that accounts for rf absorption, and $\mathrm{Re}(Z_{\mathrm{ant}})$ is the real part of the antenna impedance and includes all the parasitic reactances associated with the electrodes. Note that the effective THz current is only a fraction of the DC photocurrent

$$I_\omega = \frac{I_o}{\sqrt{1 + \omega^2 \tau_e^2}} . \quad (2.4)$$

A detailed analysis of the effect of photocarrier lifetime on photomixer performance is available in [11, 12, 13]. Here, we present a simplified argument that explains the importance of a short carrier lifetime. From (1.2), it is clear that the dc photocurrent is proportional the external quantum efficiency, which in a simplified one-dimensional case can be written as

$$\eta_e = \eta_{\text{opt}} \frac{\mu E \tau_e}{L(1 + (\mu E)/(v_{\text{sat}}))}, \tag{2.5}$$

where η_{opt} is the fraction of the absorbed optical power, μ is the effective carrier mobility, E is the electric field, L is the distance between electrodes, and v_{sat} is the saturated velocity of the carrier. The second factor in (1.5) is the photoconductive gain. In a more accurate model, the strength of E decays with depth into the sample and only a fraction of the carriers travel at v_{sat}. This has led some [14, 12] to design photomixers with film thickness ($\sim 0.2\,\mu$m) considerably thinner than the absorption depth ($\sim 1\,\mu$m) so that all of the carriers move at v_{sat}. In this case, η_e and therefore I_o are proportional to τ_e and inspection of (1.3) suggests that the photomixer output power is independent of τ_e. Figure 1(a) depicts this relationship by plotting THz output power versus frequency for various values of τ_e.

The above argument does not take into account the dissipated power limit imposed on photomixers by their failure mechanism [15, 12, 14]. Measurements at different labs suggest that photomixers fail thermally because of the combination of optical heating and ohmic dissipation. Ohmic dissipation is particularly severe since the power deposited per unit volume is proportional to $j \cdot E$ (where j is the current density) and both j and E are highly concentrated near the edge of a given electrode. The main point is that while the THz power can be independent of τ_e , the DC ohmic power deposited near the electrodes is proportional to τ_e. Therefore, a given electrode geometry with a given voltage bias has a maximum sustainable value of I_o imposed by thermal limits. By reducing τ_e, the optical power and dc bias can be increased thereby increasing the THz output power while staying below the maximum sustainable value of I_o. Figure 1(b) shows the maximum THz power available given the constraint on total I_o. Note that if one is willing to increase the diode laser power, a gain of 40 dB is achievable by reducing τ_e by a factor 0.01 (e.g. 25 ps instead of 2.5 ns). The optimum lifetime τ_e for a given operating frequency f is approximately the value of τ_e that satisfies the equation $2\pi f \tau_e \approx 1$. This represents a near optimal tradeoff between reduced DC power dissipation and efficient generation of THz photocurrent.

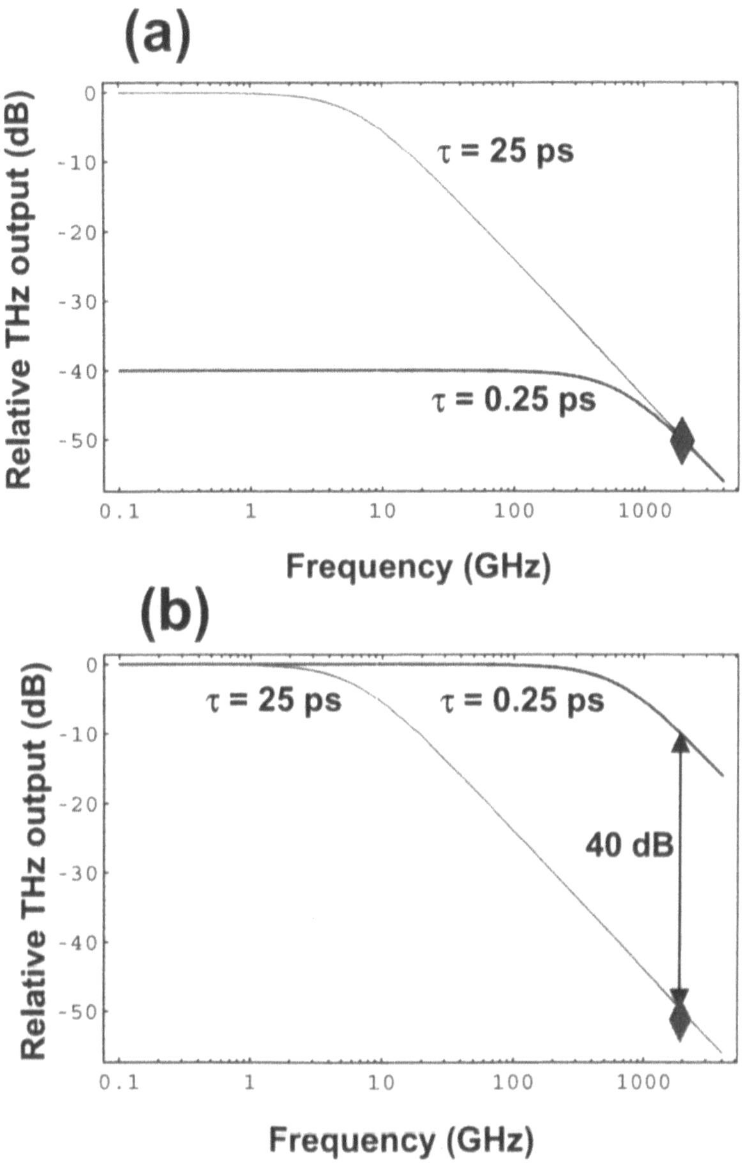

Figure 1 Calculated THz output power comparisons. (a) Two photomixers using photoconductors with different values of carrier lifetime τ_e and illuminated by the same laser power. (b) Two photomixers with different τ_e operating at the same DC photocurrent but with different laser illumination powers. These photomixers have similar amounts of ohmic heating due to DC photocurrent but have very different THz output powers.

3. ROLE OF PHOTOCARRIER LIFETIME

The value of τ_e for a particular wafer is determined by the growth conditions in the molecular-beam epitaxy apparatus. Figure 2 shows the photocarrier lifetime—measured by optical pump-probe reflectometry—as a function of substrate temperature during the molecular-beam-epitaxy process [16]. The lifetime measured with the optical pump-probe technique was consistent with the value inferred at low voltage bias from autocorrelating two $\sim$ 100 fs laser pulses on a LTG-GaAs photoconductor [17].

According to the argument in the above section, a photomixer designed to operate at a particular frequency has a corresponding optimum carrier lifetime. Since τ_e is a material property, this suggests that a given LTG-GaAs wafer is best suited for one operating frequency. Experimental results, however, suggest that LTG-GaAs with short τ_e that produces high output power at high frequencies can still produce high output power at lower frequencies. A possible explanation of this effect is based on the measurements of Zamdmer [11, 18] that showed how the photocarrier lifetime is lengthened at high values of DC electric field. These measurements were performed in the time-domain by measuring the decay time of the photoconductive response of a LTG-GaAs switch to a mode-locked laser pulse. Figure 3 shows measured lifetime as a function of DC electric field. This effect was simulated assuming a barrier-lowering model in a material with many electron traps. Figure 4 shows how this effect manifests itself in the frequency domain in photomixer output power. In a simple model, the THz power should scale with the DC photocurrent as $P_\omega \sim I_o^2$. The measured THz power at frequencies above 326 GHz, however, no longer increases with applied DC voltage and DC photocurrent. This suggests that $2\pi f \tau_e > 1$ for $f > 326$ GHz in this particular LTG-GaAs sample when biased at 30 V.

The above data suggest that one possible strategy is to grow LTG-GaAs wafers with a short lifetime, and then optimize the DC voltage bias for photomixers with antennas optimized for different frequencies—for example, use a larger voltage bias to operate a photomixer with an antenna designed for lower frequencies.

4. RESONANT ANTENNA DESIGN

The section above described how to optimize the properties of the photoconductor. This section describes the basic antenna design problem of extracting THz power from the photoconductor using metallic electrodes. While illuminated, the effective RF resistance of the photoconductor is estimated to be $> 10\,\mathrm{k}\Omega$ for typical operating conditions.

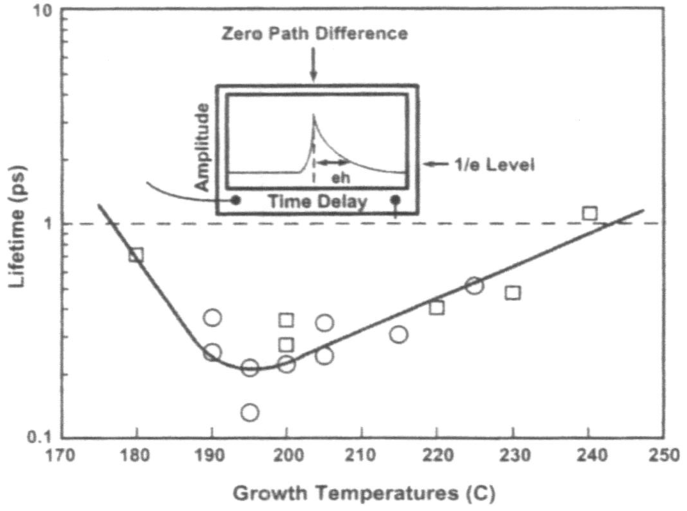

	MBE GaAs	LTG GaAs
Growth Temperature	600 C	200 C
Stoichiometry	Stoichiometric GaAs	1 at. % Excess As $Ga_{0.495}As_{0.505}$
Trap Density	< 10^{15} cm^{-3}	~ 10^{19} cm^{-3}
Breakdown Field	~ 3 10^{4} V/cm	~ 5 10^{5} V/cm
Photo-Excited Carrier Lifetime	~ 1 ns	≳ 0.150 ps

Figure 2 Optical pump-probe measurements of the photocarrier lifetime for different LTG-GaAs wafers. Also shown are typical values of LTG-GaAs electronic properties compared to conventional epitaxial GaAs.

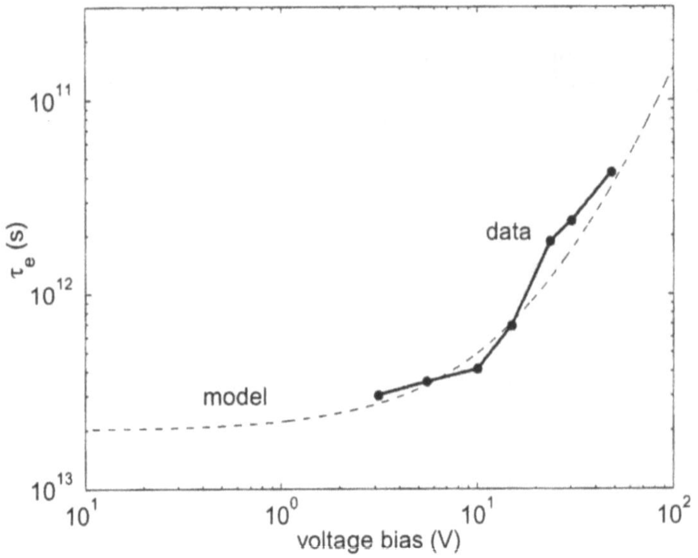

Figure 3 Measured values of photocarrier lifetime inferred from the decay time resulting from correlating two fs laser pulses on a LTG-GaAs photoconductor. The increase in leftime with electric field is attributed to a barrier-lowering model in which the deep traps become less efficient at trapping carriers under high electrical bias.

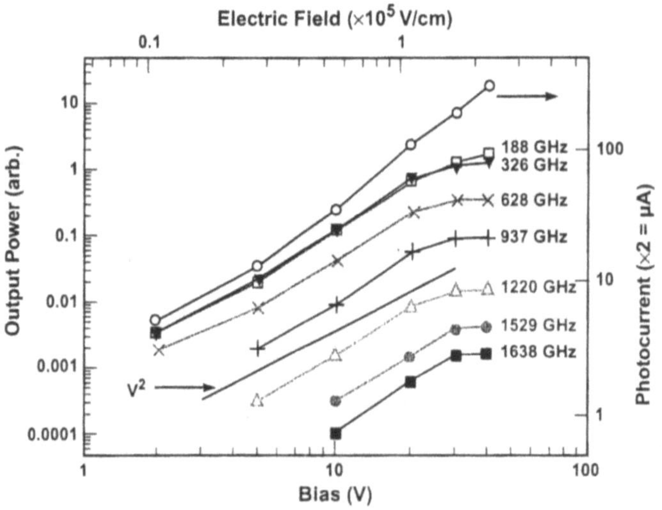

Figure 4 Measured THz output power at various frequencies as a function of DC voltage bias. Note how the output saturates above 30 V for frequencies above 300 GHz.

To good approximation, (1.3) describes the output power of the photomixer where the effect of the electrodes is modeled as part of the complex impedance Z_{ant}. Ideally, the antenna would present a $\sim 10\,\mathrm{k\Omega}$ load impedance to the photoconductor to optimally transfer power out of the antenna. In practice, a broadband antenna—such as a log-spiral or log-periodic—on GaAs typically has a real impedance $\sim 70\,\Omega$ as well as capacitive reactance arising from the fringing fields between the interdigitated electrodes that collect the photocurrent. Resonant antennas are useful if the application of interest allows one to trade octaves of tuning bandwidth for increased power over a narrow band (typically $\sim 20\%$ fractional bandwidth). In principle, a resonant antenna could present a $> 10\,\mathrm{k\Omega}$ impedance to the photoconductor. In practice, metallic loss at THz frequencies preclude the use of tuning circuits with the necessary Q-factor to achieve such a high antenna impedance.

The work by Duffy discusses a detailed design procedure for photomixer antennas as well as the computational techniques used to verify them [13]. Here, we describe in words some of the critical trade-offs used to design planar antennas for photomixers. Single-element antennas are relatively easy to design and understand. Dual element antennas have a performance advantage, although their equivalent circuits are more complex.

4.1. SINGLE ELEMENT ANTENNAS

The design goal is a metallic structure comprising interdigitated electrodes for collecting photocurrent, planar radiating elements, and a transmission line for biasing the photoconductor. The interdigitated electrodes should cover the illuminated area (typically between $4 \times 4\,\mu\mathrm{m}$ and $8 \times 8\,\mu\mathrm{m}$) without obscuring it or adding more parasitic capacitance than can be tuned out. The radiating elements (dipoles or slots) should present a large antenna resistance to the equivalent current source used to model the illuminated photoconductor. Then, the radiated fields should propagate into the GaAs substrate with an approximately Gaussian pattern that can be coupled out by mounting the substrate on a high-resistivity Si lens. The transmission line must provide a DC bias path while preventing parasitic leakage of the THz power down the line. Full-wave dipoles and half-wave slot antennas on a semi-infinite substrate both offer a high antenna resistance [19]. The parasitic reactances from the interdigitated electrodes are more easily tuned out in the dipole structures, so here we only discuss dipoles. Slot antennas are discussed in [20, 13]. Figure 5 shows a scanning electron micrograph (SEM) and the performance of a full-wave dipole antenna designed to operate at

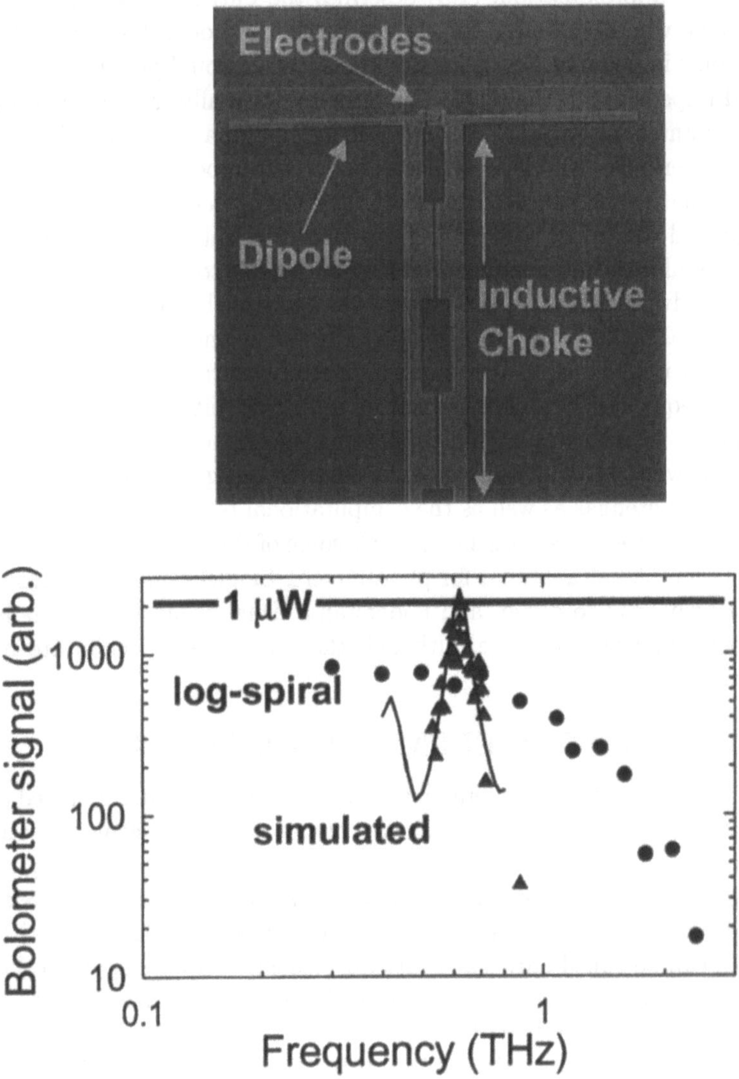

Figure 5 Full wave dipole antenna. Scanning electron micrograph and measured ▲ versus calculated performance - . Measured data from a photomixer • with a log-spiral antenna is included for comparison.

630 GHz. The total length of the dipole is approximately one wavelength long. The electrodes shunt the drive point of the dipole and the bias lines perform two functions in addition to DC biasing. First, they prevent leakage of the 630 GHz signal by using a quarter-wave choke. Second, the first high-impedance section of the choke is slightly shorter than quarter-wave so that an effective inductance appears in parallel connection with the electrode capacitance and goes into resonance at 630 GHz. The resulting impedance presented to the photomixer equivalent current source is real. The output power was measured with a liquid-helium-cooled bolometer while the laser diodes were detuned to sweep their difference frequency. The agreement between experiment and theory was close when using a planar electromagnetic solver based on the method of moments.

While the single-element full-wave dipole meets the design goal stated above, there are several disadvantages to this structure that are exacerbated when designing photomixers for operation above 1 THz. First, because the electrodes are situated at the drive point of the antenna, they interact with the near-field distribution and degrade the dipole's radiation resistance. Second, the choke must present a shunt inductance while also preventing THz leakage down the bias lines. At higher frequencies, more effective inductance is needed to increase the LC resonance frequency. Simulations show that increasing the amount of inductance also increases the parasitic leakage through the choke. The net effect is an additional reduction in the antenna resistance. Last, because there is only a single antenna, the H-plane pattern is broad and some of the available THz power is not coupled into the Si lens.

4.2. TWIN-ELEMENT ANTENNAS

Twin-element photomixers correct some of these disadvantages and offer superior performance between 1-3 THz. Twin dipoles and slots have been analyzed for photomixers by Duffy *et al.* in [13]. Here we only discuss the twin dipoles. The twin elements allow more modular control of the various design aspects than single elements. For example, the electrodes are situated between the two antennas and away from their near-field distributions. Also, the capacitance is tuned out with two lengths of line that function as single-section impedance transformers. This requires the chokes to only block THz leakage down the bias lines. Last, the twin dipole elements are driven in phase and are separated by roughly a half-wavelength. This improves the antenna gain of the H-plane pattern since they now act as a two-element phased array. Figure 6 shows the equivalent circuit and a SEM. The photoconductor and its interdigitated electrodes are modeled as a current source with a capacitance C_{elect}. Between the photoconductor and each antenna element is approximately a quarter-wavelength of high-impedance transmission line Z_{CPS}. The drive-point impedance of the dipoles and the input impedance of the chokes are appear in parallel connection as Z_{dipole} and Z_{CPS}. As discussed by Duffy, twin slot antennas can be analyzed in a similar fashion, but with the equivalent circuit elements in series connection.

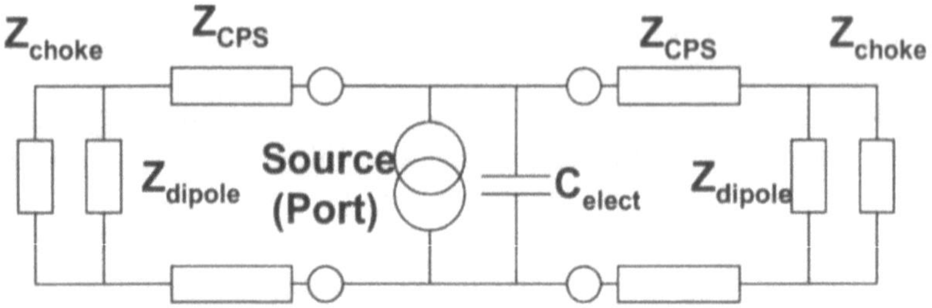

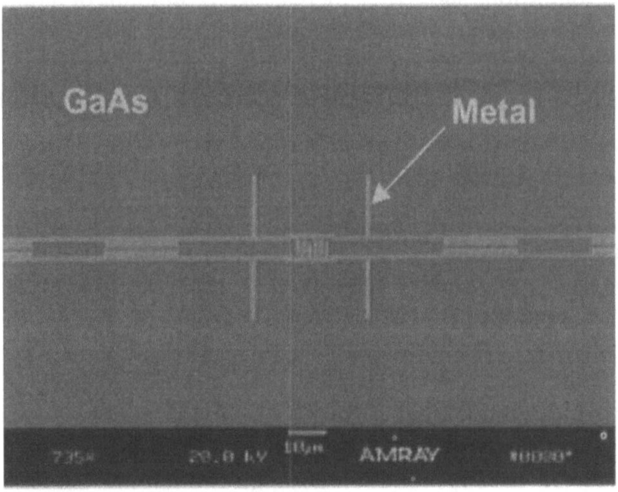

Figure 6 Twin half-wave dipole antennas and their equivalent circuit. This design outperforms single-element designs at frequencies above 1 THz.

Figure 7 shows the measured and calculated performance of several twin-dipole photomixers designed for various operating frequencies [13, 12]. A broadband spiral antenna is shown for comparison with its -12 dB/octave rolloff caused by the combination of parasitic capacitance and photocarrier lifetime. The key indication that the twin-dipole photomixers were properly designed is that the line indicating a -6 dB/octave rolloff passes along the maximum outputs of the various antennas. This indicates that the twin-dipole photomixers were designed with the parasitic capacitance optimally tuned out at each frequency. We speculate that the 2.7 THz design was less optimal because the THz loss in the Ti:Au metallization limited the quality-factor Q available for resonating out the electrode capacitance. These data represent state-of-the-art performance for these types of photomixers.

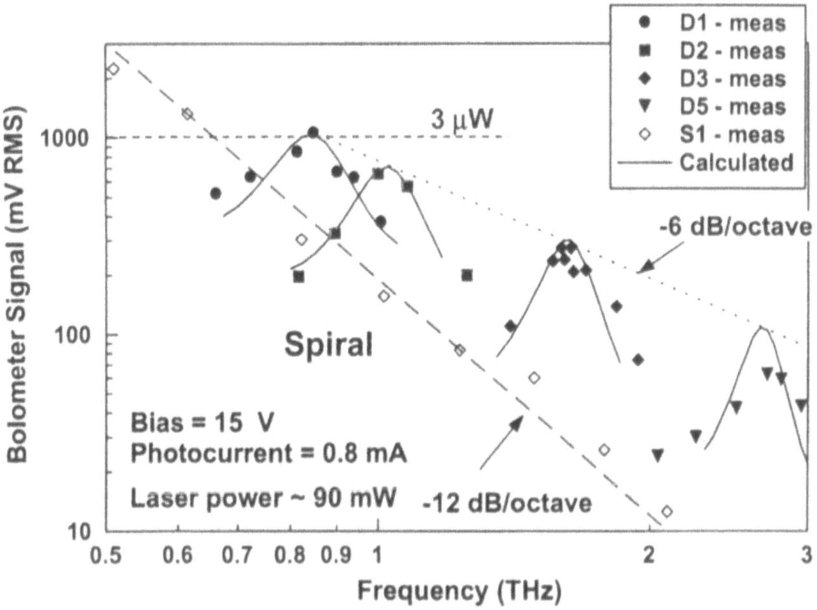

Figure 7 Simulations and measurements for four different twin-dipole photomixers. The 6 dB/octave roll-off shows that the pole resulting from capacitive loading by the electrodes has been efficiently cancelled at the resonance frequency of each antenna. A log-spiral antenna that rolls off at 12 dB/octave is shown for comparison.

4.3. DISTRIBUTED PHOTOMIXERS

An alternative approach to photomixer design is the distributed photomixer. These devices can potentially improve the bandwidth and output power of photomixers operating above 2 THz. The relevant analogy is the success of traveling-wave transistor amplifiers and nonlinear transmission lines in enhancing the high-frequency performance of transistors and Schottky diodes, respectively. Two approaches look encouraging. A surface-illuminated approach has been developed by Matsuura *et al.* that has demonstrated record output power above 1 THz [21]. This approach has the advantage of relatively simple fabrication. Velocity matching of the optical group velocity to the THz phase velocity is achieved by controlling the angle of incidence of the optical waves. A second approach has been described by Duerr *et al.* that is similar to the design of high-power photodiodes used for analog optical communication links [22].

Figure 8 shows two SEM's of a distributed photomixer that is terminated by a full-wave dipole antenna [22]. Here, an AlGaAs ridge waveguide couples the optical beat signal to a thin layer of LTG-GaAs photoconductor that is patterned on its surface. Interdigitated electrodes collect the photocurrent onto a coplanar stripline (CPS) and add just enough capacitance per unit length to velocity match the THz and optical waves. Figure 9 shows initial results from a velocity-matched distributed photomixer. Such a structure coupled to a broadband bow-tie antenna is consistent with the model and is limited only by the photocarrier lifetime and RF loss in the CPS. The structure coupled to a dipole antenna has its output spectrum band-limited by the antenna impedance. In theory, this device should outperform lumped-element photomixers at frequencies above 2 THz. In practice, engineering details such as efficient thermal management, optical facet coatings, and RF antenna design need to be optimized before this device realizes its potential.

5. APPLICATIONS

5.1. LOCAL OSCILLATORS

Recent advances in superconducting heterodyne detectors promise significant scientific payoff in submillimeter-wave astronomy and the study of planetary atmospheres. Hot-electron bolometers (HEB) and superconductor-insulator-superconductor mixers (SIS) can achieve almost quantum-limited sensitivity at frequencies above 600 GHz. Such systems presently use local oscillators (LO's) consisting either of frequency-multiplied diode oscillators, vacuum tubes, or molecular lasers. The pho-

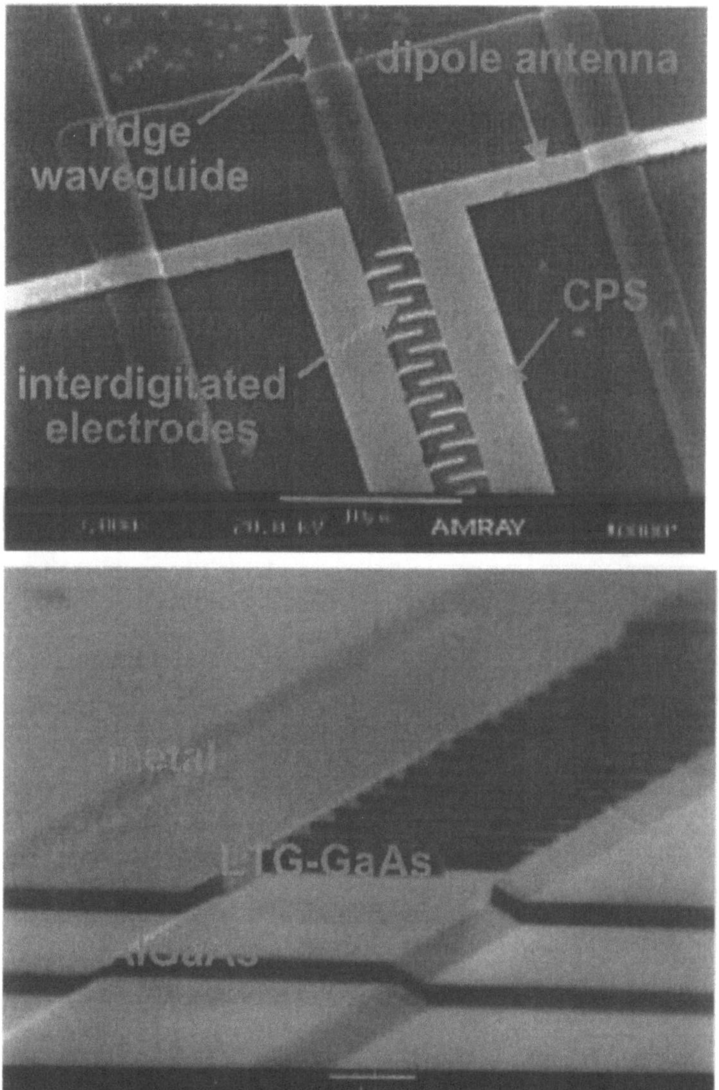

Figure 8 Scanning electron micrographs of a distributed photomixer. Top view shows the AlGaAs ridge waveguide, the electrodes and coplanar strips (CPS), and the full-wave dipole antenna. Side view shows the LTG-GaAs film on top of the AlGaAs waveguide and the metal electrodes.

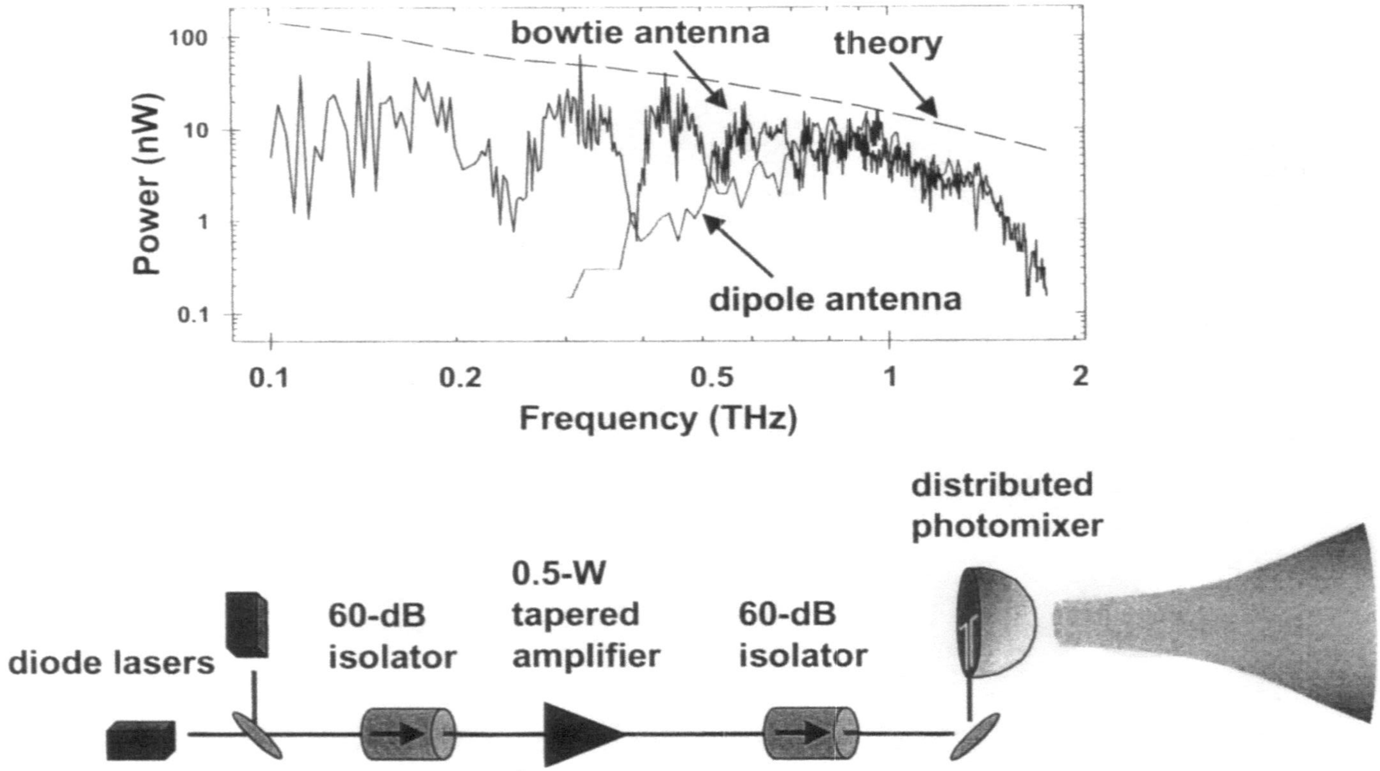

Figure 9 Measured data for two distributed photomixers with different antennas. Simulations of the expected performance are also shown. A bolometer was used to detect the output power.

tomixer is an interesting alternative LO technology. It can be assembled with no moving parts, is relatively low power, has a wide tuning range, and is based on commercial lasers and a custom photoconductor. For array applications, multiple photomixer LO's can be remotely located from the lasers, thereby easing signal distribution to remote antennas.

The main challenges in making a viable LO out of photomixers are: generating a coherent tone with minimal frequency jitter and amplitude noise, and generating sufficient power to overcome diplexer and other losses incurred when coupling a LO to a cryogenic heterodyne detector. An experiment in collaboration with Harvard-Smithsonian Center for Astrophysics was performed which demonstrated the use of a photomixer as a local oscillator at 630 GHz [4]. The heterodyne detector was a Nb SIS mixer mounted in a waveguide mixing block. The photomixer was a single-element full-wave dipole. Using the photomixer as an LO, approximately $0.2\,\mu$W of RF power was coupled to the SIS mixer and the resulting double sideband noise temperature was 331 K—in good agreement with the 323 K noise temperature obtained when a multiplied Gunn oscillator coupling in $0.25\,\mu$W was substituted for the photomixer. The twin-element photomixers described above have higher output power and operate at higher frequencies. These device should be adequate for laboratory experiments that integrate HEB mixers with photomixer LO's. The photomixer technology should be a reliable LO technology as it becomes more robust against thermal failure and as the LO-power consumption of the HEB mixers drops.

5.2. TRANSCEIVERS

Photomixers are also attractive sources for high-resolutions spectroscopy of gases. Pine *et al.* used a photomixer in conjunction with a liquid-helium-cooled bolometer to measure the rotational broadening parameters of sulfur dioxide (SO_2) [5] Chen *et al.* measured the rotational spectrum of acetonitryl (CH_3CN) [6]. Gas spectrometers based on photomixers and bolometers exhibit high sensitivity, as well as frequency resolution that is unmatched by systems such as Fourier-transform spectrometers that do not use coherent sources. A practical disadvantage for fielded systems, however, is the requirement of a helium-cooled bolometer. The photomixer transceiver addresses this limitation [23].

The transceiver uses of a pair of photomixers that are pumped by the same pair of diode lasers. One photomixer is a coherent transmitter of THz radiation. Its input consists of a DC bias voltage and the optical beat signal produced by the combined output of the two lasers. Its output is a coherent THz beam. The THz beam passes through a gas cell and is detected by the second photomixer. The second photomixer is a coherent homodyne receiver of THz radiation. Its input consists of a THz waveform and the optical beat signal. Its output is a DC current that is proportional to the amplitude of the incident THz electric-field strength. Two photomixers with twin slot antennas with a center frequency of 1.4 THz were used to perform a demonstration of high-resolution spectroscopy on water vapor. Figure 10 shows successive measurements of the transmission at 1.411 THz through the gas cell with the water vapor pressure in a 50-cm long gas cell varying between 0.1 and 1 Torr (from top to bottom). Note the narrowing of the line that occurs as the pressure is reduced and the ability of the transceiver to resolve it. This particular set of photomixers had sufficient THz power to measure the water-vapor spectrum from 1.15 to 1.5 THz (26% fractional bandwidth).

Recently, an on-chip photomixer transceiver was demonstrated, where the two photomixers were coupled by a coplanar waveguide rather than through free space [18]. The on-chip transceiver is especially useful for studying the frequency dependence of the photoconductor and interdigitated electrodes since propagation along a well-designed transmis-

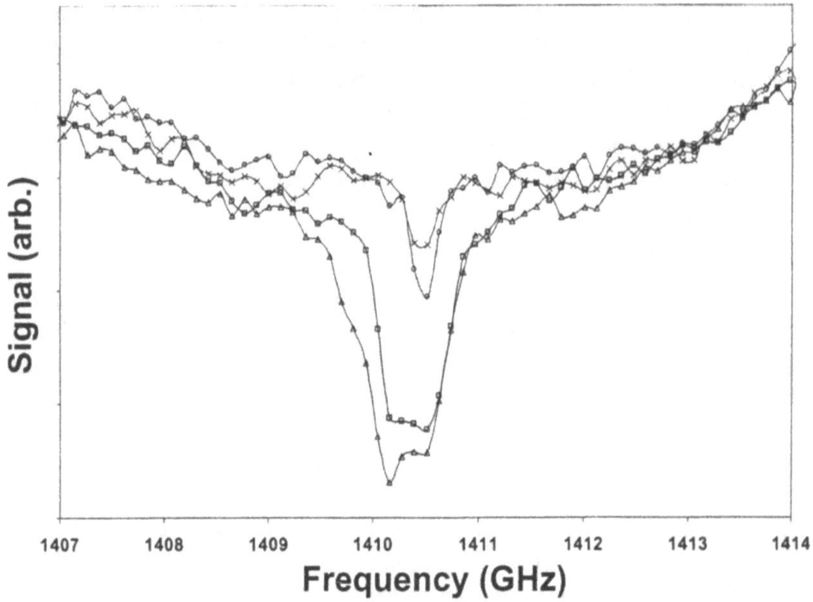

Figure 10 Measurements of the 1.4 THz absorption line in water vapor. From the bottom trace to the top, the pressure in the gas cell was reduced from 1 Torr to 0.1 Torr. Further resolution in frequency can be obtained by stabilizing the diode lasers.

sion line is mostly linear phase and can be deembedded more accurately than with quasioptical systems. Analysis of the data from the on-chip transceiver confirmed the predictions of the model described by Zamdmer *et al.* for the electric-field dependence of the photocarrier lifetime [11].

6. SUMMARY

Photomixers are a frequency agile technology for generating continuous-wave THz radiation. An understanding of photocarrier lifetime and of antenna design is necessary to fabricate high-performance photomixers. Newer structures that use a distributed design may ultimatedly outperform conventional structures for frequencies above 2 THz. Potential applications of photomixers include local oscillators and gas spectroscopy.

Acknowledgments

This work was sponsored by the Department of the Air Force under United States Air Force Contract No. F19628-95-C-0002. Opinions, interpretations, conclusions, and recommendations are those of the author and are not necessarily endorsed by the United States Air Force.

References

[1] E. R. Brown, F. W. Smith, and K. A. McIntosh, "Coherent millimeter-wave generation by heterodyne conversion in low-temperature-grown GaAs photoconductors," *J. Appl. Phys.*, vol. 73, pp. 1480-1484, 1993.

[2] K. A. McIntosh, E. R. Brown, K. B. Nichols, O. B. McMahon, W. F. Dinatale, and T. M. Lyszczarz, "Terahertz photomixing with diode lasers in low-temperature-grown GaAs," *Appl. Phys. Lett.* vol. 67, pp. 3844-3846, 1995.

[3] S. Verghese, K. A. McIntosh, and E. R. Brown, "Highly tunable fiber-coupled photomixers with coherent terahertz output power," *IEEE Trans. Microw. Theory Tech.*, vol. 45, pp. 1301-1309, 1997.

[4] S. Verghese, K. A. McIntosh, S. D. Calawa, C.-Y. E. Tong, R. Kimberk, and R. Blundell, "A photomixer local oscillator for a 630-GHz heterodyne receiver," *IEEE Microwave and Guided Wave Lett.*, vol. 9, pp. 245-247, 1999.

[5] A. S. Pine, R. D. Suenram, E. R. Brown, and K. A. McIntosh, "A terahertz photomixing spectrometer: Application to SO2 self broadening," *J. Mol. Spectrosc.* vol. 175, pp. 37-47, 1996.

[6] P. Chen, G. A. Blake, M. C. Gaidis, E. R. Brown, K. A. McIntosh, S. Y. Chou, M. I. Nathan, and F. Williamson, "Spectroscopic applications and frequency control of submillimeter-wave photomixing with distributed-Bragg-reflector diode lasers in low-temperature-grown GaAs," *App. Phys. Lett.* vol. 71, pp. 1601-1602, 1997.

[7] E. R. Brown, S. Verghese, and K. A. McIntosh, "Terahertz photomixing in low-temperature-grown GaAs," *Proc. SPIE Conference on Advanced Technology MMW, Radio, and Terahertz Telescopes*, vol. 3357, pp. 132-142, 1998.

[8] N. Erickson, "Diode frequency multipliers for THz local oscillator applications," *Proc. SPIE Conference on Advanced Technology MMW, Radio, and Terahertz Telescopes*, vol. 3357, pp. 75-84, 1998.

[9] B. Xu, Q. Hu, and M. R. Melloch, "Electrically pumped tunable terahertz emitter based on intersubband transition," *Appl. Phys. Lett.*, vol. 71, pp. 440-442, 1997.

[10] M. Rochat, J. Faist, M. Beck, U. Oesterle, and M. Ilegems, "Far-infrared ($\lambda = 88\,\mu$m) electroluminescence in a quantum cascade structure," *Appl. Phys. Lett.*, vol. 73, pp. 3724-3726, 1997.

[11] N. Zamdmer, Q. Hu, K. A. McIntosh, and S. Verghese, "Increase in response time of low-temperature-grown GaAs photoconductive switches at high voltage bias," *Appl. Phys. Lett.*, vol. 75, pp. 2313-2315, 1999.

[12] A. W. Jackson, *Low-temperature-grown GaAs photomixers designed for increased terahertz output power*, Ph.D. Dissertation, Dept. of Materials Science, Univ. of California, Santa Barbara, 1999.

[13] S. M. Duffy, S. Verghese, K. A. McIntosh, A. W. Jackson, A. C. Gossard, and S. Matsuura, "Accurate modeling of dual dipole and slot elements used with photomixers for coherent terahertz output power," *IEEE Trans. Microw. Theory Tech.*, to be published.

[14] E. R. Brown, "A photoconductive model for superior GaAs THz photomixers," *Appl. Phys. Lett.*, vol. 75, pp. 769-771, 1999.

[15] S. Verghese, K. A. McIntosh, and E. R. Brown, "Optical and terahertz power limits in low-temperature-grown GaAs photomixers," *Appl. Phys. Lett.*, vol. 71, pp. 2743-2745, 1997.

[16] K. A. McIntosh, K. B. Nichols, S. Verghese, and E. R. Brown, "Investigation of ultrashort photocarrier relaxation times in low-temperature-grown GaAs," *Appl. Phys. Lett.*, vol. 70, pp. 354-356, 1997.

[17] S. Verghese, N. Zamdmer, E. R. Brown, A. Förster, Q. Hu, "An optical correlator using a low-temperature-grown GaAs photoconductor," *Appl. Phys. Lett.*, vol. 69, pp. 842-844 (1996).

[18] N. Zamdmer, Q. Hu, K. A. McIntosh, S. Verghese, and A. Förster, "On-chip frequency-domain submillimeter-wave transceiver," *Appl. Phys. Lett.*, vol. 75, pp. 3877-3879, 1999.

[19] M. Kominami, D. M. Pozar, and D. H. Schaubert, "Dipole and slot elements and arrays on semi-infinite substrates," *IEEE Trans. Antennas Propag.*, vol. 33, pp. 600-607, 1985.

[20] K. A. McIntosh, E. R. Brown, K. B. Nichols, O. B. McMahon, W. F. Dinatale, and T. M. Lyszczarz, "Terahertz measurements of resonant planar antennas coupled to low-temperature-grown GaAs photomixers," *Appl. Phys. Lett.*, vol. 69, pp. 3632-3534, 1996.

[21] S. Matsuura, G. A. Blake, R. A. Wyss, J. C. Pearson, C. Kadow, A. W. Jackson, and A. C. Gossard, "A traveling-wave THz photomixer based on angle-tuned phase matching," *Appl. Phys. Lett.*, vol. 74, pp. 2872-2874, 1999.

[22] E. K. Duerr, K. A. McIntosh, and S. Verghese, "Distributed photomixers," in *Conference on Lasers and Electro-Optics*, OSA Technical Digest, (Optical Society of America, Washington DC), pp. 382-383, 2000.

[23] S. Verghese, K. A. McIntosh, S. D. Calawa, W. F. DiNatale, E. K. Duerr, and K. A. Molvar, "Generation and detection of coherent terahertz waves using two photomixers," *Appl. Phys. Lett.* vol. 73, pp. 3824-3826, 1998.

Theme 5

Emerging Technologies - Sources and Circuits

EMERGING SOLID STATE TERAHERTZ ELECTRONICS

M. S. SHUR
Rensselaer Polytechnic Institute
CII-9017, CIEEM, Troy NY 121180 USA
shurm@rpi.edu

V. RYZHII
Computer Solid State Physics Laboratory, University of Aizu
Aizu-Wakamatsu City
Fukushima, 965-8580 Japan

Abstract

We review emerging solid-state technologies that have promise for generation, detection, and modulation of terahertz radiation. Gunn, IMPATT, and TUNNETT diodes have achieved operating frequencies close to 300 GHz. Record maximum frequencies of oscillations of transferred-substrate HBTs have reached 800 GHz. Resonant Tunneling Diodes have also demonstrated operation at 700 GHz or so, albeit at very small power levels. We also discuss new ideas of solid-state devices for application in the terahertz band - from negative effective mass amplifiers and two-stream instability to cyclotron resonance MASERs and plasma wave electronics.

The terahertz range of frequencies is a new frontier of solid-state electronics. This frequency range is borderline between microwave electronics and photonics (see Figure 1) and corresponds to the frequency bands of molecular and lattice vibrations in gases, fluids, and solids.

Numerous potential and emerging applications of terahertz electronics include imaging and characterization, ranging from semiconductor to battlefield imaging, detection of poisonous and biological substances, medicine, environmental monitoring, radio astronomy, covert communications, antiterrorist devices, and industrial controls. As an example, Figure 2 shows the application of terahertz radiation for imaging ion-implantation layers in a semiconductor device. [1] Since terahertz radiation easily passes through a semiconductor wafer, it can detect and image the wafer internal structure. Using scanning techniques and computer data processing, it is possible to generate three-dimensional maps of semiconductors wafers, biological and live objects, etc. Figure 3 deals with another application of terahertz technology. It shows the distribution of chlorine monoxide in the upper atmosphere at 20 km height from sea level measured by Upper Atmosphere Research Satellite (detected using submillimeter Schottky diodes [2,3]). GaAs-based Schottky diodes have now reached the terahertz range, where new effects related to coupling with polar optical phonons become important. [4]

R.E. Miles et al. (eds.), Terahertz Sources and Systems, 169–185.

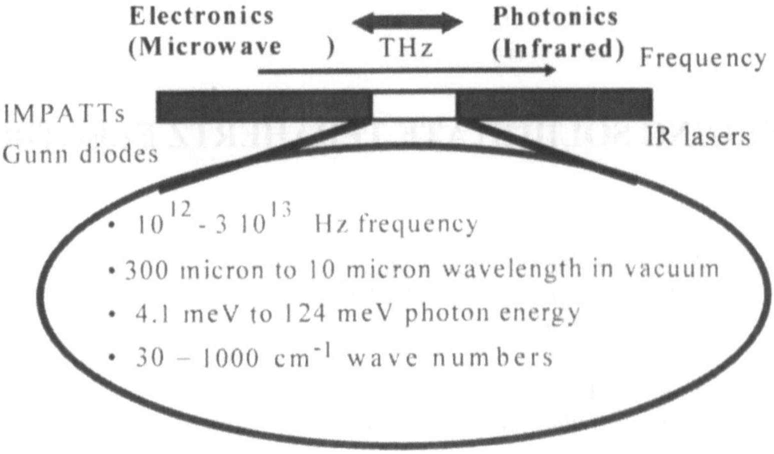

Figure 1. Terahertz range of frequencies.

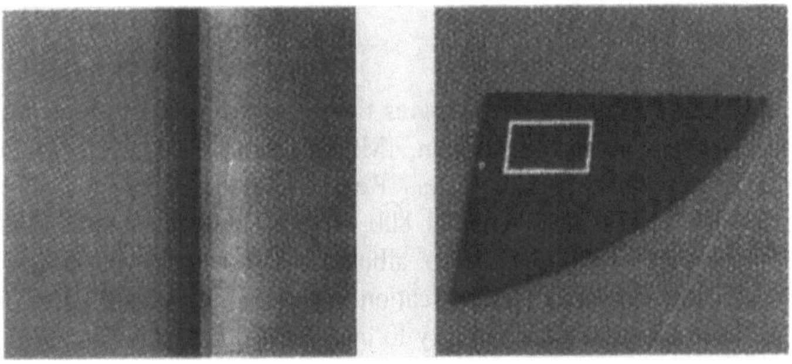

Figure 2. Terahertz (a) and visible image (b) of semiconductor wafer. Terahertz image shows p-n junction region. [1] *(Courtesy of Tochigi Nikon Corporation.)*

The most developed solid-state techniques for generation, detection, and modulation of the terahertz radiation are based on using femto second pulses for nearly instantaneous change in the conductivity of a semi-insulating semiconductor sample. Figure 4 illustrates this technique which allows one to obtain a broadband, pulsed coherent terahertz radiation in the frequency range from 0.3 to 30 THz at power levels on the order of microwatts. The generation of terahertz pulses using such a technique is associated with the ability of the photoexcited electrons to be accelerated by the electric field to fairly high velocities immediately after their emergence (the electron velocity overshoot effect). This effect can be used for the generation of terahertz radiation by ultra short optical pulses in planar interdigitated metal-semiconductor-metal photodiodes [5] and quantum well infrared photodetectors. [6] In the structures based on materials similar to low-temperature-grown GaAs, a fast response of the photoexcited electrons and holes can be associated with their very short lifetime [7,8,9]

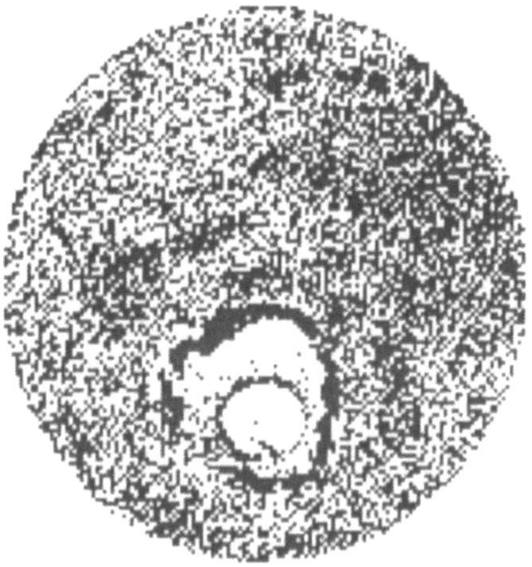

Figure 3. Distribution of chlorine monoxide in the upper atmosphere at 20 km height from sea level measured by Upper Atmosphere Research Satellite. Dark area indicates a high concentration of chlorine monoxide. (After G. Zarpette,[2] copyright © IEEE 1993.)

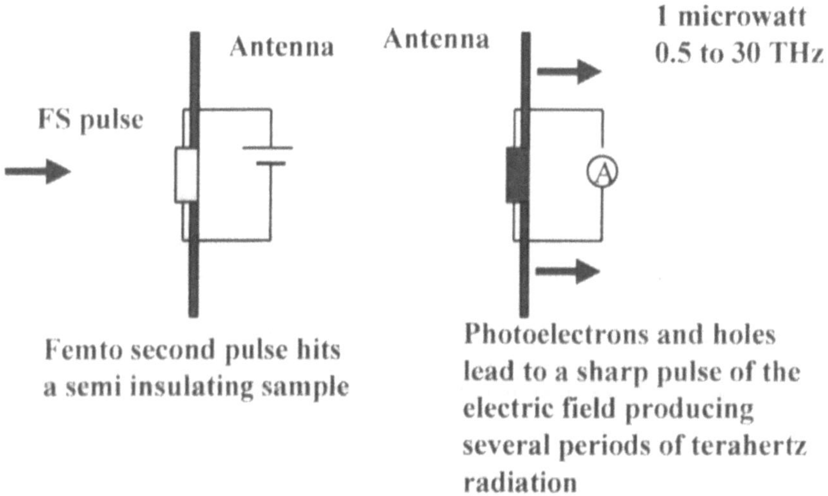

Figure 4. Terahertz system using femto second laser.

Kersting and co-workers developed another variation of this technique based on screening of the surface field by photogenerated carriers (see Figure 5).[10,11] This method yields emission in the range between 0.5 – 4.0 THz with emission power on the order of 100 nW. The dynamics of the pulse decays gives information about electron relaxation times.

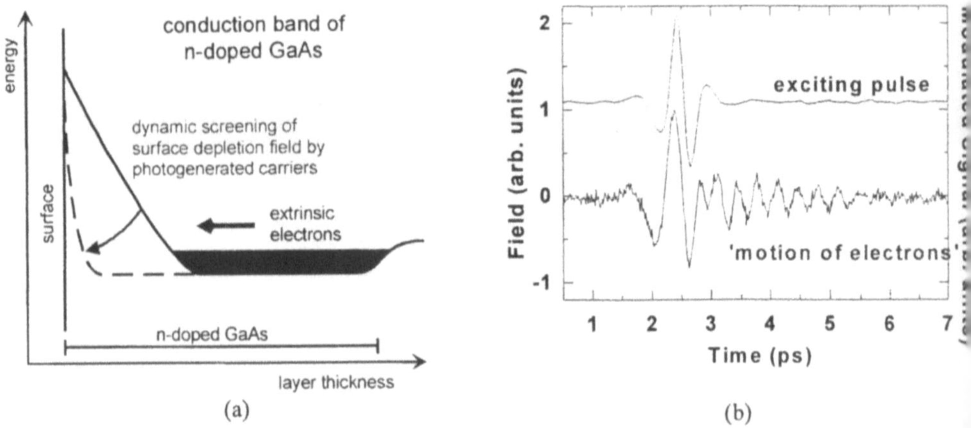

Figure 5. Dynamic screening of surface depletion field by photogenerated carriers (a) and electron response to exciting terahertz pulse. [10, 11]

Resonant Tunneling Diodes (RTDs) hold promise of generating CW terahertz power a frequencies up to several terahertz [12,13], even though parasitic contact resistances make difficult to obtain frequencies higher than 200 to 800 GHz. [14]

Figure 6 shows a schematic RTD structure and the band diagram along with a typic measured current voltage characteristic. [15] The electron transmission coefficient through such structure has a sharp peak when the component of the electron momentum perpendicular to th double-barrier structure is such that

$$\frac{p_{\perp}^{2}}{2m_n} = E_o - E_c, \qquad (\;$$

where E_o - E_c is the energy separation between the electron subband in the quantum well an the bottom of the conduction band. This corresponds to a resonance condition. In an RTD, th resonance condition is achieved at a certain bias voltage at which the diode current reaches peak value. Figure 7 shows power generated by RTDs as a function of frequency.

At the present time, one of the most practical ways to obtain CW terahertz radiation from solid-state source is to use Transferred Electron Oscillators (TEOs) in combination wi frequency multipliers based on Schottky diodes. Ridley and Watkins in 1961 and Hilsum i 1962 were first to suggest that such a negative differential mobility in high electric fields related to an electron transfer between different minima (valleys) of the conduction band i GaAs (intervalley transfer). When the electric field is low, electrons are primarily located in th central valley of the conduction band. As the electric field increases, many electrons ga enough energy for the intervalley transition into higher satellite valleys. The electron effecti mass in the satellite valleys is much greater than in the central valley. Also, the intervall transition is accompanied by an increased electron scattering. These factors result in a decrea of the electron velocity in high electric fields.

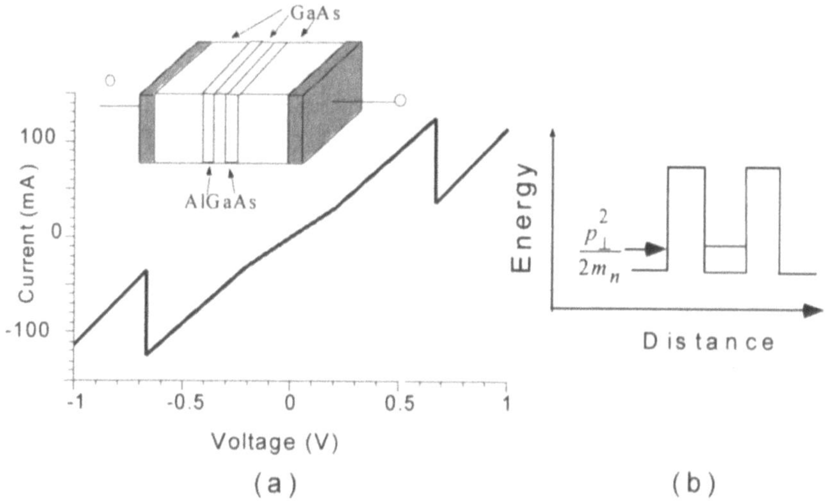

Figure 6. Schematic RTD structure, measured current voltage characteristic haracteristics of $Al_{0.8}In_{0.52}As/Ga_{0.57}In_{0.53}As_{0.47}$ RTD (a), and band diagram (b). (After [16]).

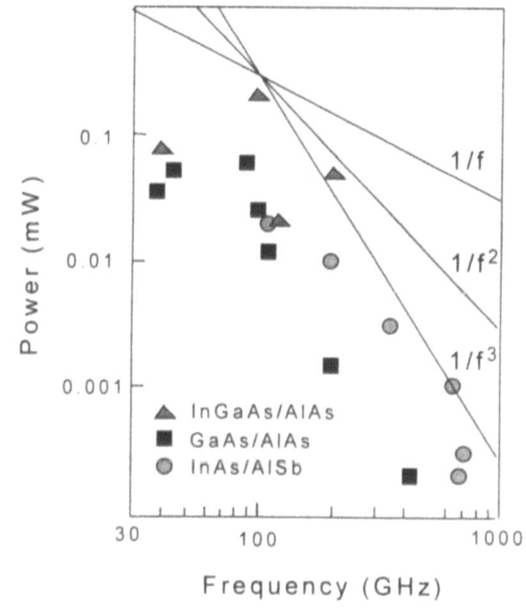

Figure 7. Power versus frequency for resonant tunneling diodes. (After [17])

Figure 8 shows the velocity-field curves for GaAs for two values of the low field mobility, = 0.85 m^2/Vs and μ = 0.5 m^2/Vs. In both cases, we observe a range of electric fields where e differential mobility $\mu_d = dv/dF$ is negative.

A simplified equivalent circuit of a uniformly doped semiconductor is a parallel combination of the differential resistance and capacitance (see Figure 9):

$$R_d = \frac{L}{q\mu_d n_o S}$$

$$C_d = \frac{\varepsilon S}{L}$$

Here S is the cross section of the sample, L is the sample length, and n_o is the electr concentration.

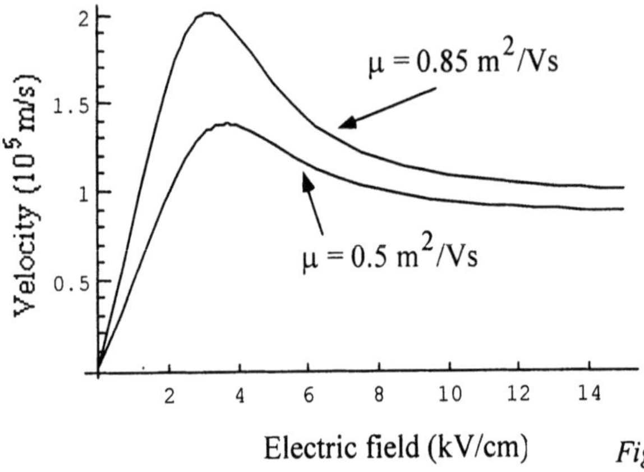

Figure 8. Velocity-field curves for GaAs.

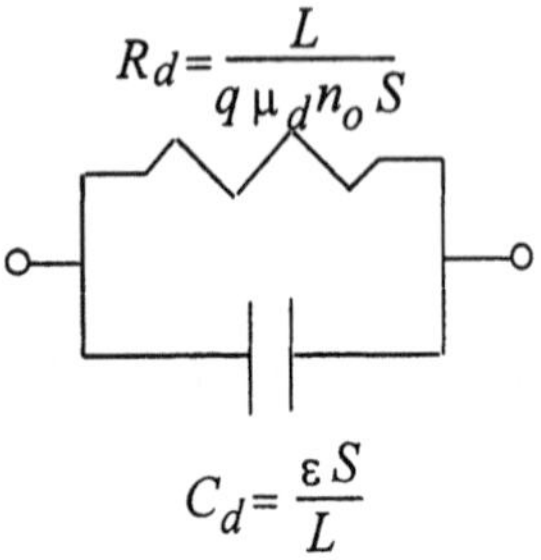

Figure 9. Equivalent circuit for a unifo piece of semiconductor.

The equivalent RC time constant determining the evolution of the space charge is equal to t Maxwell dielectric relaxation time:

$$\tau_{md} = R_d C_d = \frac{\varepsilon}{q\mu_d n_o}$$

In a material with a negative differential conductivity, the space charge fluctuation may grc with time. What actually happens depends on the relationship between τ_{md} and t_{tr}, where t_{tr} L/v the electron transit time, L is the sample length, and v is the electron drift velocity. If τ_{md}) << t_{tr} a space charge fluctuation grows tremendously during a small fraction of the tran time. In this case, it develops into a high field region (called a high field domain), whi propagates from the cathode toward the anode with the velocity that is approximately equal the electron saturation velocity, v_s.

John Gunn observed high field domains in GaAs in 1963. Each domain nucleates near t cathode, propagates through the sample (during the transit time L/v_s), and is destroyed near t anode. After the domain annihilation at the anode, the next domain nucleates near the catho and the process repeats itself. The current density in a sample with a domain is equal to j_s $qn_o v_s$. When the domain is extinct, the current density is equal to the peak current density, j_p

qn_0v_p where v_p is the peak velocity. During the domain annihilation and formation the current density rises from j_s to j_p.

Stable propagating domains exist in relatively long samples that operate as microwave oscillators at relatively low frequencies (10 to 40 GHz or less). In short samples, operating as Transferred Electron Oscillators (TEOs) at higher frequencies (50 GHz and above), the distribution of the electric field is quite different. In such samples, growing accumulation layers are nucleated. The propagation velocity of accumulation layers may be even larger than the peak velocity in GaAs. In both regimes, TEOs can operate as microwave or millimeter wave oscillators. (The frequency of oscillations increases approximately in inverse proportion to the diode length.)

Figure 10 shows the power obtained from TEOs versus frequency. As can be seen from the figure, power on the order of milliwatts can be obtained at frequencies around 300 GHz. [8,19,20] With frequency multiplication using Schottky diodes, one might reach the terahertz band.

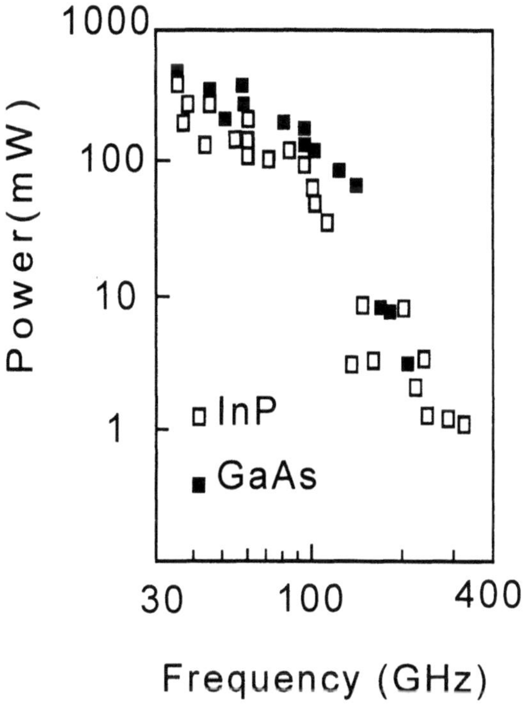

Figure 10. Power obtained from TEOs versus frequency. [21]

In a non-uniformly doped sample, the current waveform reproduces the $n_0(x)S(x)$ profile. Jones et al. [22] proposed to use this property for generating the current waveform that has a much more pronounced second harmonic in a TED that has a doping spike within the active region (see Figures 11 and 12). They called such diode a Modulated-Impurity-Concentration Transferred-Electron Device (MICTED).

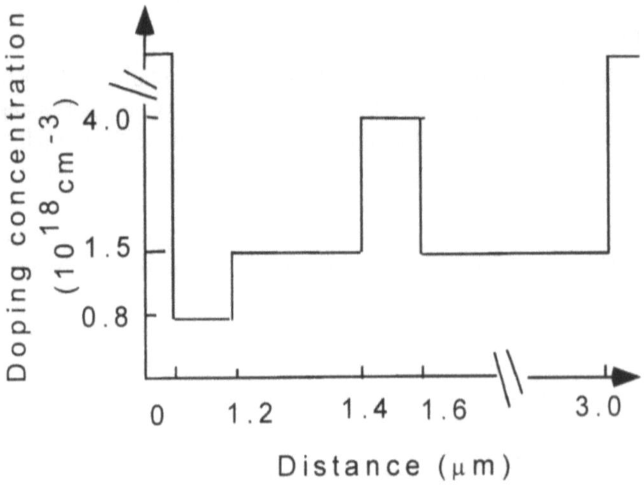

Figure 11. MICTED doping profile [22] *(Copyright © IEEE 1992)*

More recent computer simulations confirm that such a device should yield a higher power at very high frequencies. [23]

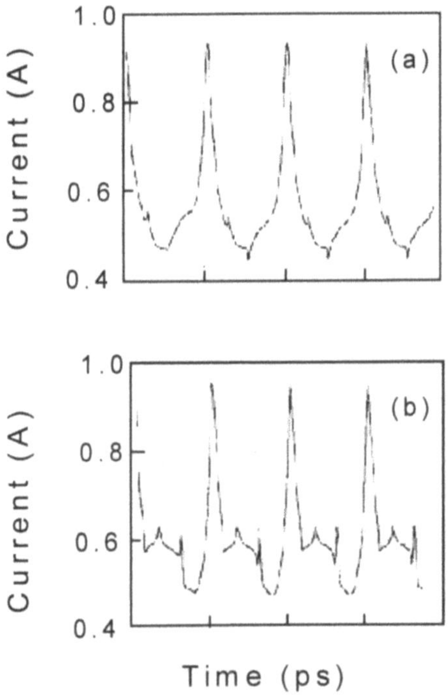

Fig.12. Comparison of waveforms of conventional TED and MICTED. [22] *(Copyright © IEEE 1992)*

Transit Time Diodes (which use a transit time delay by introducing a drift region for injected carriers in order to achieve a dynamic negative differential resistance) present another technology that is able to reach fairly high frequencies. An example is an **IMPact ionization Avalanche Transit Time (IMPATT) diode** (see Figure 13). In this device, avalanche breakdown (generating electron-hole pairs at the p^+-n interface where the electric field is the highest) and carrier drift across a special "drift" section of the device create a phase delay responsible for a dynamic negative resistance.

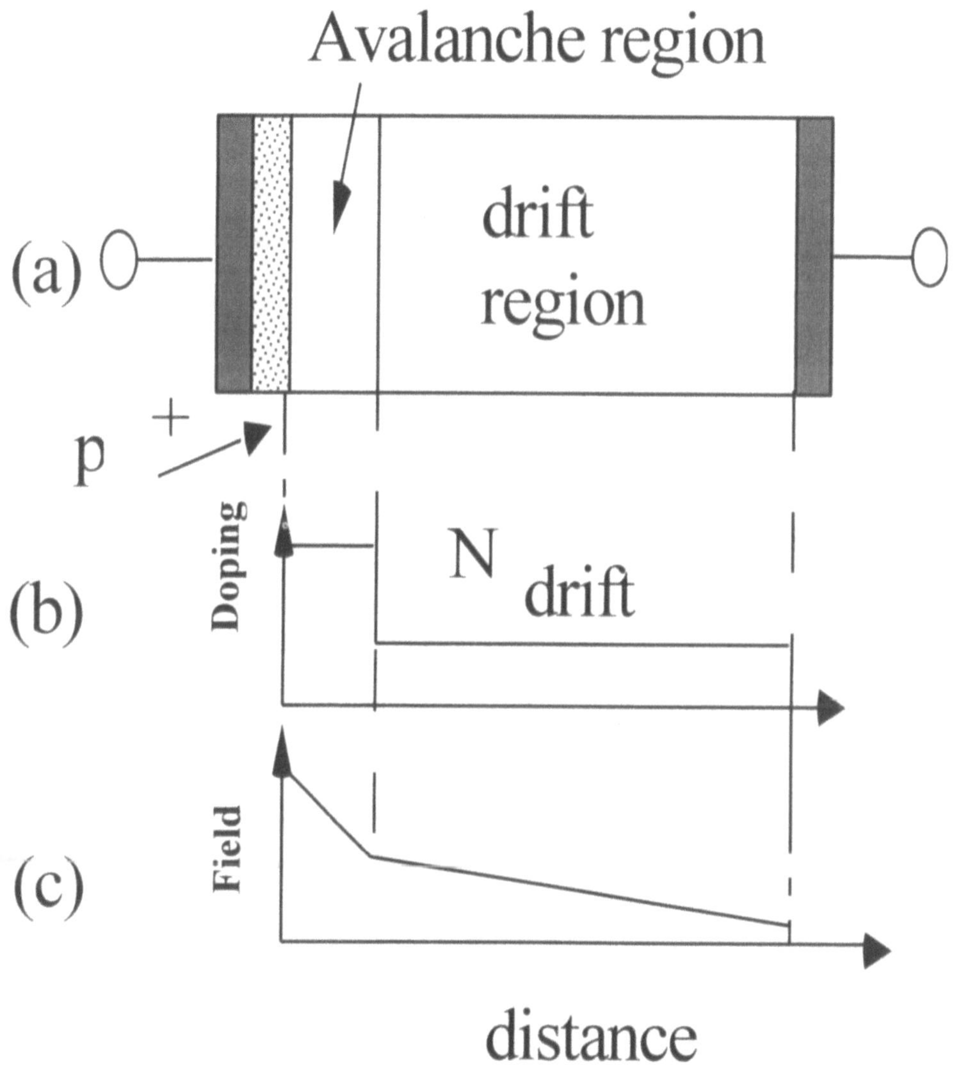

Figure 13. (a) Schematic diagram of an IMPATT diode along with (b) a typical doping profile and (c) electric field distribution. (After [16])

Tunnel Transit-Time (TUNNETT) diodes operate in a similar way, except that the carrier injection occurs via tunneling at the high field region interface. Figure 14 shows the power versus frequency achieved for different transit time diodes. Figure 15 presents such data for tunnel injection transit time diodes.

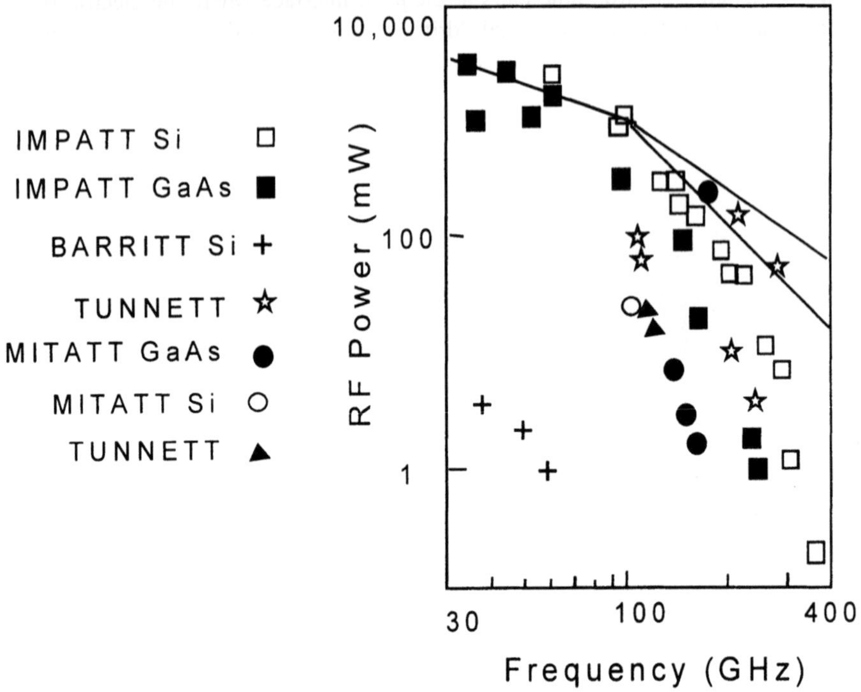

Figure 14. Power versus frequency achieved different transit time diodes. [17, 21]

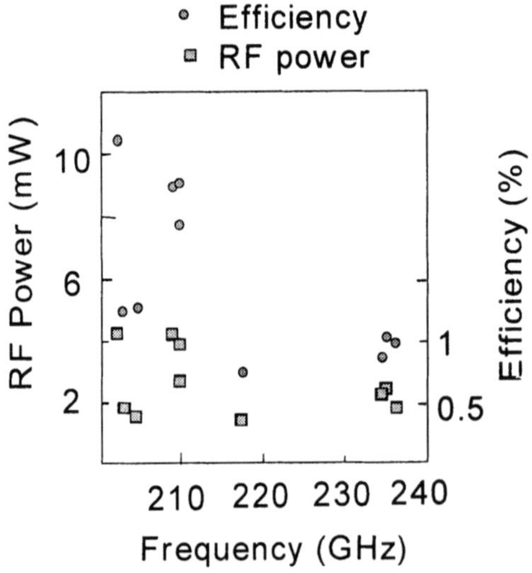

Figure 15. Power and efficiency versus for tunnel injection transit time diodes. [17, 21]

igures 10, 14, and 15 show that even though it is possible to use these diode sources for enerating power in the terahertz band by using frequency doublers or triplers, these devices all short as CW terahertz sources.

Many alternative ideas and device designs for terahertz devices explored in literature nclude improved resonant tunneling structures of various kind [24], Hot Electron Transistors [25], ransferred Substrate HBTs (see Fig. 16) [26], Quantum Cascade Lasers and other devices [27] ased on interband transitions in quantum wells, cyclotron resonance MASERs [28], [29], [30], diodes sing two-stream instabilities [31],[32] devices using negative effective mass [33], diodes based on ynamic negative differential mobility [34], superlattice devices using Bloch oscillations [35],[36] and lasma wave electronics devices.[37], [38],[39]

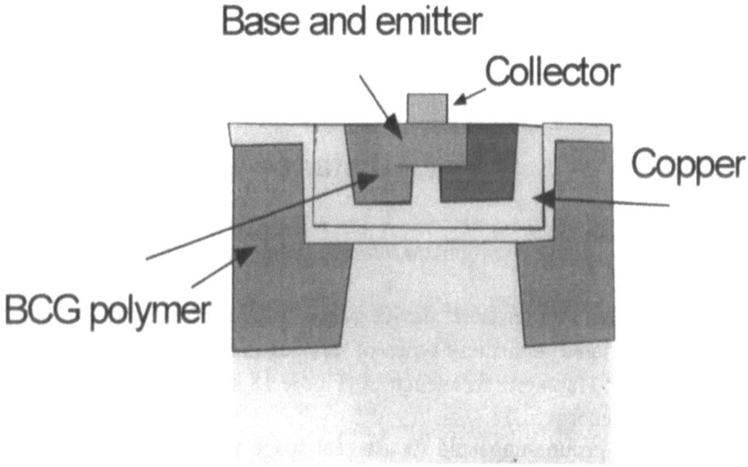

Figure 16. Schematic structure of Transferred Substrate HBT.

In 1994, Faist, Capasso, Sivco, Sirtori, Hutchnson, and Cho of AT&T Bell Labs reported n a new type of a far infrared laser called the Quantum Cascade Laser. [40] (The idea of this uperlattice device was proposed by R. Kazarinov and R. Suris of A. F. Ioffe Institute in 1971.) The band diagram of the active region of this device is shown in Figure 17. The laser onsisted of 25 active regions connected by superlattices with a very short period (much shorter an the electron de Broglie wavelength). Electrons are injected into the first $Ga_{0.47}In_{0.53}As$ uantum well (with a thickness of only 8 Å) through a 45 Å $Al_{0.48}In_{0.52}As$ barrier. (This aterial system was chosen to obtain lattice match to the InP substrate.) The laser transition is etween the third subband in the narrow 8 Å quantum well and the second subband (n = 2) in e adjacent, wider (35 Å) quantum well. The emission wavelength (λ = 4.26 μm) was in xcellent agreement with the calculated energy difference between the subbands. Since then e Bell Labs group has achieved a remarkable progress in these devices demonstrating recently quantum cascade laser operating at temperatures above 425 K at 8.4 micron with the output ower of 17 mW. [42] (This laser incorporated 75 stages of alternated active regions and jectors.)

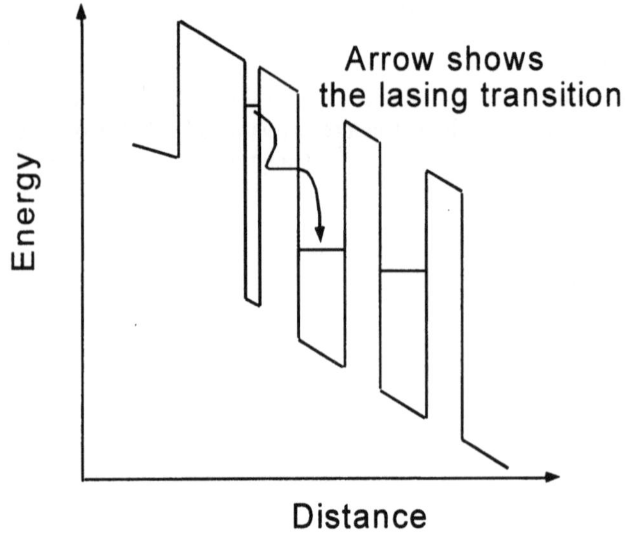

Figure 17. Active region of quantum cascade laser (after [40])

Buckle et al. proposed another terahertz device based on intersubband transitions. [43] Howeve general difficulty associated with these types of devices is a very long radiative lifetime, τ_r which sharply rises at the terahertz frequencies. (Figure 18 shows the dependence of τ_{rad} on t intersubband transition energy. [44])

Andronov and his group suggested to use real-space transfer intersubband transitions AlGaAs/GaAs quantum wells by hot electrons moving in the direction parallel to t heterointerfaces (see Figure 19). A big advantage of this approach is a possibility to use ma parallel layers for enhancing the intensity of the terahertz radiation. The difficulty is a possib high-field domain formation and other instabilities in the direction of the applied electric fie and hot electron drift. This serous problem has remained unsolved.

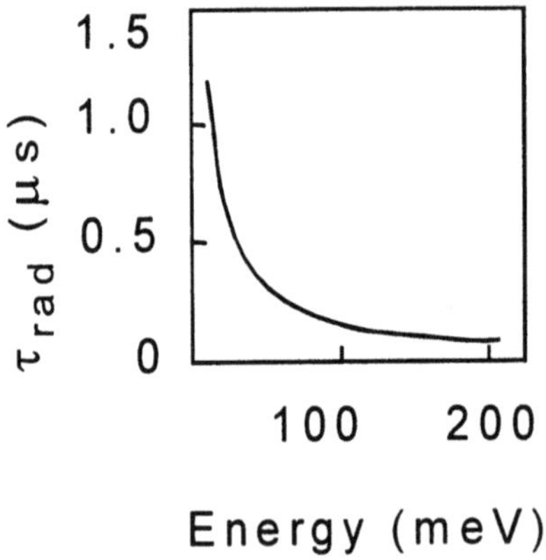

Figure 18. Radiative time versus transition energy. [44]

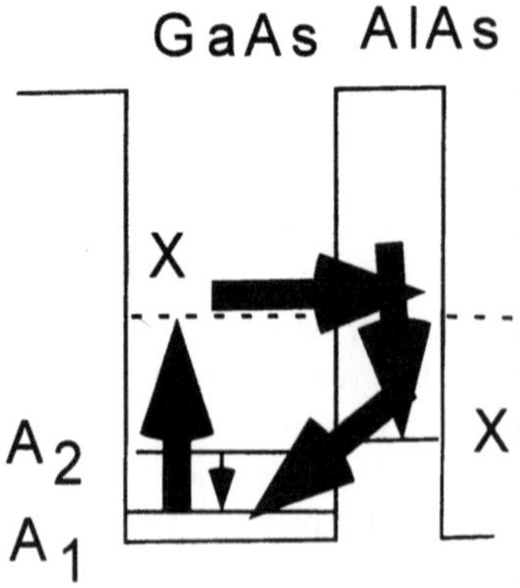

Figure 19. Schematic band diagram of interband transition terahertz device propose by Andronov et al. and designed for lasing at microns. [45]

Surface plasma wave instability in short-channel High Electron Mobility Transistors could be used for the generation of terahertz radiation.[37] The first measurement results by Cheremisin point out that such instability might occur in 0.1 micron CaAs-based HEMTs.[46] However, more experimental studies are needed in order to establish the viability of this technique. The excitation of surface plasma waves can be also used for detection, multiplication, and mixing of terahertz radiation. Weikle et al.[47] and Lu et al.[48] reported on the detection of the terahertz radiation using AlGaAs/GaAs HEMTs (see Fig. 20). Recently Ryzhii et al.[49,50] proposed to combine this approach with using resonant tunneling structure inserted between the gate and a channel of HEMT or a related NERFET-type device[51] (see Fig. 21). Their theory predicts a very large increase of the plasma wave instability increment and the plasma wave detector sensitivity.[49,50]

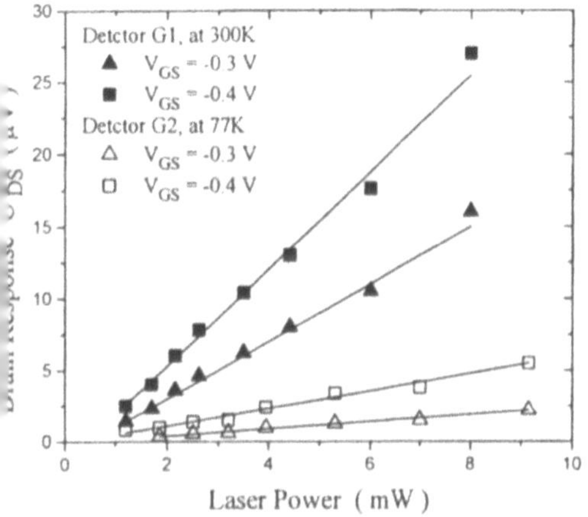

Figure 20. AlGaAs/GaAs detector response to 2.5 THz signal.[47,48]

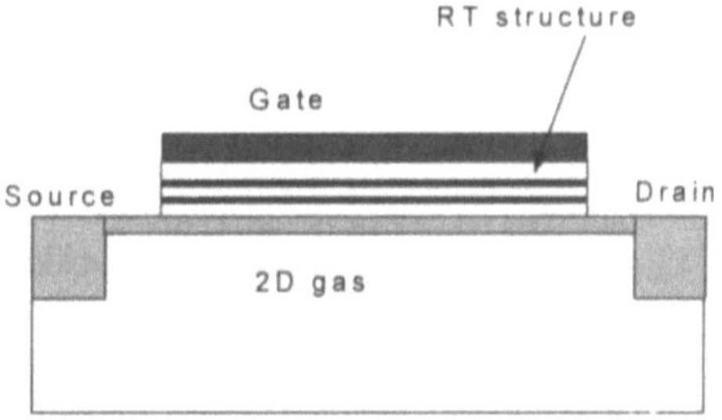

Figure 21. Resonant tunneling structures for enhancement of plasma wave instability.[49,50]

Practically all solid-state terahertz devices use compound semiconductor materials. Dyakonov and Shur discussed a possibility to use deep submicron silicon devices for plasma wave electronics applications.[52] This approach might become feasible with further scaling down of silicon device feature sizes. Also, the recent work by Dmitriev et al. showed that two-dimensional electrons in silicon should exhibit negative differential conductivity in high electric fields (see Figure 22).[53] Figure 23 shows a proposed structure that might be used for the observation of this negative differential mobility.

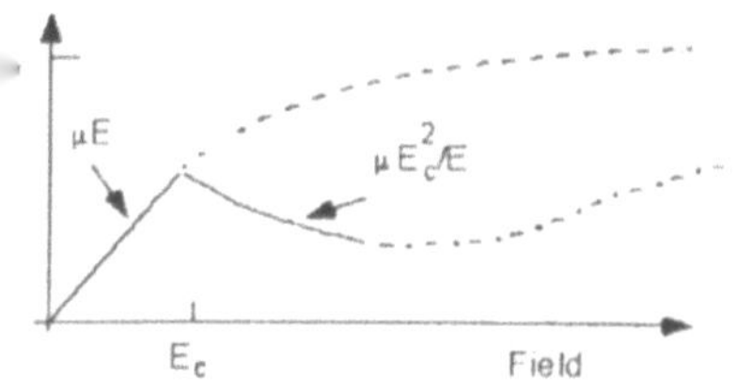

Figure 22. Predicted negative differential mobility of 2D electrons in silicon.[53]

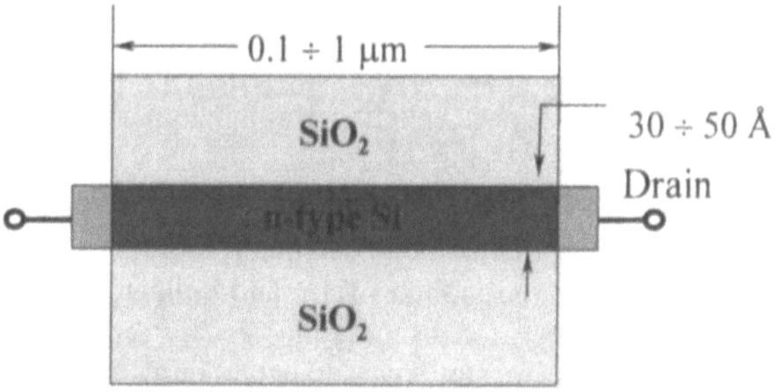

Figure 23. Proposed structure for observation of negative differential mobility of 2D electrons in silicon.

All in all, the results reviewed here show that conventional sources (Gunn diodes IMPATTs, RTDs, etc.) have difficult time approaching terahertz range. Emerging alternative technologies include plasma wave electronics devices, devices based on interband transitions and devices using hot electron instabilities. Only time will tell which, if any, of these alternatives will emerge as practical solid state CW tunable terahertz technology.

Acknowledgment.

This work has been partially supported by ARO (Project Monitor Dr. Dwight Woolard), by ONR (Project Monitor Dr. John Zolper), and by DARPA (Project Monitor, Dr. Edgar Martinez).

References.

1. http://www.tochigi-nikon.co.jp
2. Zarpette, G. (1993) Sensing Climate Change, *Spectrum*, 23.
3 Crowe, T.W., Mattauch, R.J., Weikle, R.M., and Bhapkar, U.V. (1996) Terahertz GaAs Devices and Circuits for Heterodyne Receivers, in M. S. Shur (ed.), *Compound Semiconductor Technology. The Age of Maturity*, World Scientific, pp. 209-246.
4 Gelmont, B. L. and Woolard, D. L. (1999), Polar-optical-phonon induced nonlinearity at semiconductor interfaces, *Appl. Phys. Lett.* **75**, pp. 2325-2327.
5 Ryzhii, M., Ryzhii, V., and Willander, M. (1998) Monte Carlo modeling of electron velocity overshoot effect in quantum well infrared photodetectors, *J. Appl. Phys.* **84**, pp. 3403 - 3408.
6. Ryzhii, M., and Ryzhii, V. (1998) Monte Carlo analysis of ultrafast electron transport in quantum well infrared photodetectors, *Appl. Phys. Lett.* **72**, pp. 842 - 844.
7. Brown, E.R., Smith, F.W., and McIntosh, K.A. (1993) Coherent millimeter-wave generation by heterodyne conversion in low-temperature-grown GaAs photoconductors, *J. Appl. Phys.* **73**, pp.1480 - 1484.
8. Verghese, S., McIntosh, K., and Brown, E.R. (1997) Optical and terahertz power limits in the low-temperature-grown GaAs photomixers, *Appl. Phys. Lett.* **71**, pp. 2743 - 2745.
9. Khmyrova, I., Ryzhii, V., Ryzhii, M., and Willander, M. (1998) High-frequency response of metal-semiconductor-metal photodetectors limited by dynamic and recombination effects, *Jpn. J. Appl. Phys.* **37**, pp.6352-6357.
10. Kersting, R., Unterrainer, K., Strasser, G., Kauffmann, H.F., and Gornik, E. (1997) Few Cycle THz emission from cold plasma oscillations, *Phys. Rev. Lett.* **79**, p. 3038.
11. Kersting, R., Bratschitsch, R., Strasser, G., and Unterrainer, K. (2000) Sampling a terahertz dipole transition with subcycle time resolution, *Opt. Lett.* **25**, p. 272.
12. Sollner, T.C.L.G., Goodhue, W.D., Tannenwald, P.E., Parker, C.D., and Peck, D.D. (1983) Resonant tunneling through quantum wells at frequencies up to 2.5 THz, *Appl. Phys. Lett.*, **43**, pp. 588-590.
13 Liou, W.-R., Lin, J.-C., and Yeh, M.-L. (1996) Simulation and analysis of a high-frequency resonant tunneling diode oscillator, *Solid-State Electron.*, **39**, pp. 833-839.
14. Brown, E.R., Goodhue, W.D., and Sollner, T.C.L.G. (1988) Fundamental oscillations up to 200 GHz in resonant tunneling diodes and new estimations of their maximum oscillation frequency from stationary-state tunneling theory, *J. Appl. Phys.*, **64**, pp. 1519-1529.

15 Capasso, F., Sen, S., and Beltram, F. (1990) Quantum-Effect Devices, in S. M. Sze (ed.), *High Speed Semiconductor Devices*, Wiley Interscience, John Wiley & Sons, Inc., New York.
16 Shur, M.S. (1996) *Introduction to Electronic Devices*, Wiley, New York.
17 Eisele, H. and Haddad, G. (1998) Active Microwave Diodes, in S. M. Sze (ed.), *Modern semiconductor Devices,* Wiley, New York, p.300.
18 Eisele, H. and Haddad, G. (1994) D-band InP Gunn devices with second- harmonic power extraction up to 290 GHz, *Electron. Lett.*, **30**, pp. 1950-1951.
19 Rydberg, A., (1990) High efficiency and output power from second- and third- harmonic millimeter-wave INP-TED oscillators at frequencies above 170 GHz, *IEEE Electron Device Lett.*, **1**, pp, 439-441.
20 Eisele, H. and Haddad, G.1. (1995) High-performance InP Gunn devices for fundamental-mode operation in D-band (110-170 GHz), *IEEE Microwave Guided Wave Lett.*, **5**, pp. 385-387.
21 Eisele, H., Rydberg, A., and Haddad, G. (2000) Recent advances in the performance of InP Gunn devices and GaAs TUNNETT diodes for the 100-300-GHz frequency range and above, *IEEE Transactions MTT*, **48**, pp. 626-631.
22 Jones, S.H., Tait, G.B., and Shur, M.S. (1992) Modulated-Impurity-Concentration Transferred-Electron Devices Exhibiting Large Harmonic Frequency Content, *Microwave and Optical Technology Letters*, **5**, pp. 354-359.
23 Judashke R. (2000) Comparison of Modulation-Impurity Concentration InP Transferred Electron Devices for Power Generation at Frequencies Above 130 GHz, *IEEE Trans., MTT* **48**, pp. 719-724.
24 Drexler, H., Scott, J.S., Allen, S.J., Campman, K.L., and Gossard, A.G. (1995) Photon-assisted tunneling in a resonant tunneling diode: stimulated emission and absorption in the THz range, *Appl. Phys. Lett.*, **67**, pp. 2816-2818.
25 Dellow, M.W., Cronin, N.J., and Bending, S.J. (1994) Tunneling hot electron transistor as a high power source at terahertz frequencies, *Appl. Phys. L*ett. **65**, pp. 2463-2465.
26 Rodwell, M., Lee, Q., Mensa, D., Pullela, R., Guthrie, J., Martin, S.C., Smith, R.P.; Jaganathan, S., Mathew, T., Agarwal, B., and Long, S. (1998) 48 GHz Digital ICs Using Transferred Substrate HBTs, *Proceedings of GaAs IC Symposium*, pp. 113-116.
27 Smet, J. H., Fonstad, C. G., and Hu, Q. (1996) Intra and interwell intersubband transitions in multiple quantum wells for far-infrared sources, *J. Appl. Phys.* **79**, pp. 9305-9320.
28 Gavrilenko, V.I. and Krasil'nik, Z.F.(1991) Negative mass cyclotron resonance maser, *Opt. Quant. Electron.* vol. 23, no. 2, pp. 5323-5339.
29 Starikov, E. V. and Shiktorov, P.N.(1991) Numerical simulation of heavy hole cyclotron resonance NEMAG in p-Ge, *Opt. Quant. Electron.* **23**, no. 2, pp. 5341-5349.
30 Brundermann, B., Chamberlin, D.R. , and Haller, E.E. (1985) Thermal effects in widely tunable germanium terahertz lasers, *Appl. Phys. Lett.* **73**, pp. 2757-2759.
31 Mantorov, V. V. and Sukhanov, A. A. (1985) Dynamic properties of a ballistic field - effect transistor *Sov. Phys. Semicond.* **19**, pp. 882-886.
32 Ryzhii, V. I. , Bannov, N. A., and Fedirko, V. A.(1984) Ballistic and quasiballistic transport in semiconductor structures (review) *Sov. Phys. Semicond.* **18**, pp. 481 - 491.
33 Gribnikov, Z. S., Vagidov, N. Z., Korshak, A. N., and Mitin, V. V. (2000) Negative-effective-mass ballistic field-effect transistor: Theory and modeling, *J. of Appl. Phys.* **87**, pp. 7466-7475.
34 Pozhela, Yu. K., Starikov, E.V., and Shiktorov, P.N. (1992) Dynamic NDN at transit-time resonance in InP, *Semicond. Sci. Technol.* **7**, pp. 13386-13389.

35 Waschke, C. , Roskos, H.G. , Schwedler, R. , Leo, K. , Kurz, H., and Kohler, K. (1993) Coherent submillimeter wave emission from Bloch oscillations in a semiconductor superlattice , *Phys. Rev. Lett.* **70**, pp. 3319-3322.

36 Roskos, H.G. , Waschke, C., Victor, K. , Kohler, K., and Kurz, H. (1995) Bloch oscillations in semiconductor superlattice, *Jpn. J. Appl. Phys.* **34**, pp. 1370-1375,.

37 Dyakonov, M. I. and Shur, M. S. (1996) Detection, Mixing, and Frequency Multiplication of Terahertz Radiation by Two Dimensional Electronic Fluid, *IEEE Transactions on Electron Devices* **43**, pp. 380-387.

38. Lü, J.-Q., Shur, M. S. , Hesler, J. L, Sun, L. , and Weikle, R. II, (1998) A Resonant Terahertz Detector Utilizing a High Electron Mobility Transistor, *IEDM Technical Digest* pp. 879-882, San Francisco, CA.

39. Shur, M. S. and Dyakonov, M.(1998) Two-Dimensional Electrons in Field Effect Transistors, *International Journal of High Speed Electronics and Systems* **9**, pp. 65-99.

40. Faist, J., Capasso, F., Sivco, D. L., Hutchinson, A. L., and Cho, A. Y. (1994) Quantum Cascade Laser, *Science*, **264**, pp. 553-556.

41. Kazarinov, R. R and Suris, R. A. (1971) Possibility of the amplification of electromagnetic waves in a semiconductor with a superlattice, *Soviet Phys.-Semicond.* **5**, p. 797-802.

42. Gmachl, C., Tredicucci, A., Capasso, F. , Hutchinson, A. L., Sivco, D. L., Sergent, A.M. Mentzel, T., and Cho, A. Y. (2000) High temperature (T>or=425 K) pulsed operation of quantum cascade lasers, *Electronics Letters* **36**, p. 723-725.

43. Buckle P. D., Dawson, P., Lynch M. A., Kuo C.-Y., Missous, M., and W. S. Truscott (2000) Inter-subband device with terahertz applications, *IEEE Trans. MTT*, **48**, pp. 632-638.

44. Harrison, P., Kelsall, R. W., Donovan, K., and Kinsler, P.(2000) Solid-state terahertz sources using quantum-well intersubband transitions, *IEEE Trans. MTT*, **48**, pp. 645-651.

45. Campbell, J. C. and Dapkus, P. D. (2000) Overview of the optoelectronics sessions and chapter, *in Frontiers in Electronics. From Materials to Systems*, *Selected Topics in Electronics and Systems*, **17**, World Scientific, Singapore, pp. 255-270

46. Cheremisin, M. (1999) Ph. D. Thesis, Etude d'nstabilites un liquide bidimensionnel d'electrons dans un transistor a effect de champ, University of Montpellier

47 Weikle, R. , Lu, J. , Shur, M. S., and Dyakonov, M. I. (1996) Detection of Microwave Radiation by Electronic Fluid in High Electron mobility Transistors, *Electronics Letters* **32**, pp. 2148-2149.

48 Lü, J.-Q., Shur, M. S. , Hesler, J. L., Sun, L., and Weikle, R. (1998) Terahertz Detector Utilizing Two-Dimensional Electronic Fluid, *IEEE Electron Device Letters* **19**, pp. 373-375.

49 Ryzhii, V. and Shur, M. S. (2001) Plasma Instability and nonlinear terahertz oscillations in resonant-tunneling structures, *Jpn. J. Appl. Phys.* **40**, pp. 546-550

50. Ryzhii, V. , Khmyrova, I., and Shur, M. S. (2000) Resonant detection and frequency multiplication of terahertz radiation utilizing plasma waves in resonant-tunneling transistors, *J. Appl. Phys*. **88**, pp. 2868-2871.

51 Luryi S. and Zaslavsky A. (1998) Quantum Effect and Hot-Electron Devices, in S. M. Sze (ed), *Modern semiconductor Devices*, p. 253, Wiley, New York.

52 Dyakonov, M. I. and Shur, M. S. (1995) Field Effect Transistor as Two Dimensional Electronic Flute, *Proceedings of NATO Advanced Research Workshop*, pp. 251-262 Ile de Bendor, France; (1996) Future Trends in Microelectronics. Reflection on the Road to nanotechnology, in Luryi, S., Xu, J., and Zaslavsky, A., (eds.) *NATO ASI series, Series E Applied Sciences* **323**, Kluwer Academic Publishers, Dordrecht/ Boston/ Lisbon.

53. Dmitriev, A., Kachorovskii, V., Shur, M. S., and Stroscio, M. (2000) Electron Runaway and Negative Differential Mobility in Two Dimensional Electron or Hole Gas in Elementary Semiconductors, *Solid State Comm.* **113**, pp. 565-568.

PLASMA WAVE ELECTRONICS FOR TERAHERTZ APPLICATIONS

M. DYAKONOV
Laboratoire de Physique Mathématique,
cc 070, Université Montpellier II, 34095 Montpellier, France
dyakonov@lpm.univ-montp2.fr

M. S. SHUR
Rensselaer Polytechnic Institute
CII-9017, CIEEM, Troy NY 121180 USA
shurm@rpi.edu

Abstract

A channel of a field effect transistor might act as a resonance cavity for the plasma waves. For micron or sub-micron gate lengths, the fundamental frequency of this cavity is in the terahertz range and can be easily tuned by changing the gate bias. The quality factor of this plasma wave resonator depends on the momentum relaxation time and on the plasma frequency determined by the device length. A short field effect transistor can be used as a basic device for resonant detection, mixing, multiplication, and even generation of terahertz radiation.

Introduction

The difficulty in generating and detecting terahertz radiation using solid state devices is related to the fact that on one hand the carrier transit times in semiconductor devices are typically much longer than the period of the terahertz oscillations and, on the other hand, the quanta of terahertz radiation are much smaller than room or even liquid nitrogen temperature. However, the plasma waves in a gated two dimensional electron gas (2DEG) might propagate with a much larger velocity than the electron drift velocity and their excitation is not linked to any type of intersubband or interband electronic transitions. Hence, they can be excited at room or even at elevated temperatures.

As we discuss below, a channel of a field effect transistor is, in fact, a resonant cavity for the plasma waves. For micron or sub-micron gate lengths, the fundamental frequency of this cavity is in the terahertz range and can be easily tuned by changing the

R.E. Miles et al. (eds.), Terahertz Sources and Systems, 187–207.

gate bias. If the quality factor of this plasma wave resonator is high or, at least, larger than unity (which might be achieved in High Electron Mobility Transistors as we discuss below), a FET can be used as a basic device for resonant detection, mixing, multiplication, [1] and even generation [2] of terahertz radiation.

Plasma waves are the oscillations of electron density in space and time, and their properties depend on the electron density and on the dimensions and geometry of the electronic system. Unlike in a three dimensional case, where the plasma oscillation frequency is nearly independent of the wave length, in a gated two-dimensional electron gas (2DEG), the plasma wave have a linear dispersion law similar to that of sound waves or light in vacuum. In this case, the plasma wave velocity, s, is proportional to the square root of the electron sheet density. The velocity of the plasma waves can be easily tuned by the gate bias that controls the 2DEG density.

We will start from a general discussion of plasma waves, in particular, of the plasma waves in a FET channel, and of the boundary conditions that determine the frequencies of the plasma modes in a FET. We will then consider a hydrodynamic approach to the analysis of plasma waves and invoke the shallow water wave analogy. This discussion will be followed by the review of possible devices applications, including a terahertz tunable detector, mixer, multiplier, and oscillator. In conclusion, we will consider problems and challenges to be overcome in the plasma wave electronics field.

Plasma waves in a FET channel

The dispersion relations for plasma waves in the systems of different dimensions can be derived in a simple manner by neglecting collisions and considering only the average drift velocity, **v**. In this case, the small signal equation of motion and the continuity equation are:

$$\frac{\partial \mathbf{j}}{\partial t} = \mathbf{E}\frac{e^2 n}{m} \tag{1}$$

$$\frac{\partial \rho}{\partial t} + div\mathbf{j} = 0 \tag{2}$$

where $\mathbf{j} = en\mathbf{v}$ is the current density, e is the electronic charge, n is the electron density, m is the electronic effective mass, and ρ is a small-signal charge density related to a deviation of n from its equilibrium value, **E** is the small signal electric field. Eq. (1) follows from the Newton Second Law of Motion, where the electron scattering is neglected. These equations are valid for any dimensionality of the problem. However, for the 3D case, the **j**, n, and ρ are current per unit area, electron concentration, and electric charge per unit volume, respectively, whereas in the 2D case, the **j**, n, and ρ are current per unit length, electron concentration per unit area, and the electric charge per unit area, respectively.

Differentiating Eq. (2) with respect to time and using Eq. (1), we obtain:

$$\frac{\partial^2 \rho}{\partial t^2} + \frac{e^2 n}{m} div\mathbf{E} = 0 \qquad (3)$$

In the 2D case, the electric field vector, **E**, in Eqs. (1) and (3) should be understood as the in-plane electric field having only two components, E_x and E_y, since the third component of the electric field (which is perpendicular to the plane, xy, of the 2DEG) does not contribute to the in-plane current and, hence, does not enter into Eq. (1).

The specific relation between **E** and ρ, which should augment Eq. (3) for the complete set of equations, depends on the dimensionality and/or geometry of the problem. In a three-dimensional case, this equation is obviously:

$$div\mathbf{E} = \frac{4\pi\rho}{\varepsilon}, \qquad (4)$$

where ε is the dielectric constant. Substituting Eq. (4) into Eq. (3), we obtain a harmonic oscillator equation for the charge density and the well-known expression for the bulk plasma frequency

$$\omega_p = \sqrt{\frac{4\pi e^2 n}{m\varepsilon}} \qquad (5)$$

or, in the SI system of units,

$$\omega_p = \sqrt{\frac{e^2 n}{\varepsilon\varepsilon_o m}}, \qquad (5a)$$

where ε_o is the dielectric permittivity of vacuum.)

In the case of a gated two-dimensional electron gas (i.e. of a FET), the relation between the electron concentration and electric potential is given by

$$en = CU \qquad (6)$$

where $U = U_g - U_c - U_T$, U_T is the threshold voltage, $U_g - U_c$ is the potential difference between the gate and the channel, $C = \varepsilon/4\pi d$ is the gate-to-channel capacitance per unit area, and d is the gate-to-channel separation. This equation is valid when U changes along the channel on the scale large compared to d (so-called gradual channel approximation). Hence, the in-plane electric field is given by

$$\mathbf{E} = -\frac{1}{C}\nabla\rho \qquad (7)$$

Eq. (7) replaces Eq. (4), which is valid in the 3D case. Substituting Eq. (7) into Eq. (3), we finally obtain the two-dimensional wave equation for the surface charge ρ

$$\frac{\partial^2 \rho}{\partial t^2} - s^2 \Delta \rho = 0 \tag{8}$$

where Δ is the two-dimensional Laplace operator, and

$$s = \sqrt{\frac{4\pi e^2 nd}{m\varepsilon}} \tag{9}$$

is the velocity of the surface plasma waves. The solution of Eq. (8) corresponds t waves with a linear dispersion law:

$$\omega = sk, \tag{10}$$

where ω and k are the frequency and the wave vector of the plasma waves, respectively Using Eq. (6) and (7), we can express the plasma wave velocity in terms of the gate voltage swing:

$$s = \sqrt{\frac{eU}{m}} \tag{11}$$

In a similar way, using Eq. (4) and the relation between the in-plane electric field and surface charge density for ungated 2D electron gas, one can obtain the dispersion law for the plasma waves in this system.

$$\omega = \sqrt{\frac{2\pi e^2 n}{m\varepsilon} k} \tag{12}$$

In the same way, we can obtain the dispersion law for the plasma waves propagating along a one-dimensional wire:

$$\omega = s_1 k \left(\ln \frac{1}{kr} \right)^{1/2} \tag{13}$$

Here r is the radius of the wire, s_1 is the velocity of the one-dimensional plasma wave given by

$$s_1 = \sqrt{\frac{ne^2}{m\varepsilon}} \tag{14}$$

We note that the equations for the ungated 2D gas are valid when $kd \ll 1$ (in the opposite limiting case $kd \gg 1$ the existence of the metallic gate is irrelevant, and one obtains the dispersion relation given by Eq. (12)). For the 1D case Eq. (13) holds when $kr \ll 1$.

Fig. 1 gives the summary of the plasma wave frequencies and dispersion relations for systems of different dimensions and geometry. We note the similarity between the dispersion relations for the gated 2DEG case and ungated 1DEG case. This similarity means that the results discussed below for the gated 2DEG should equally apply to ungated quantum wires.

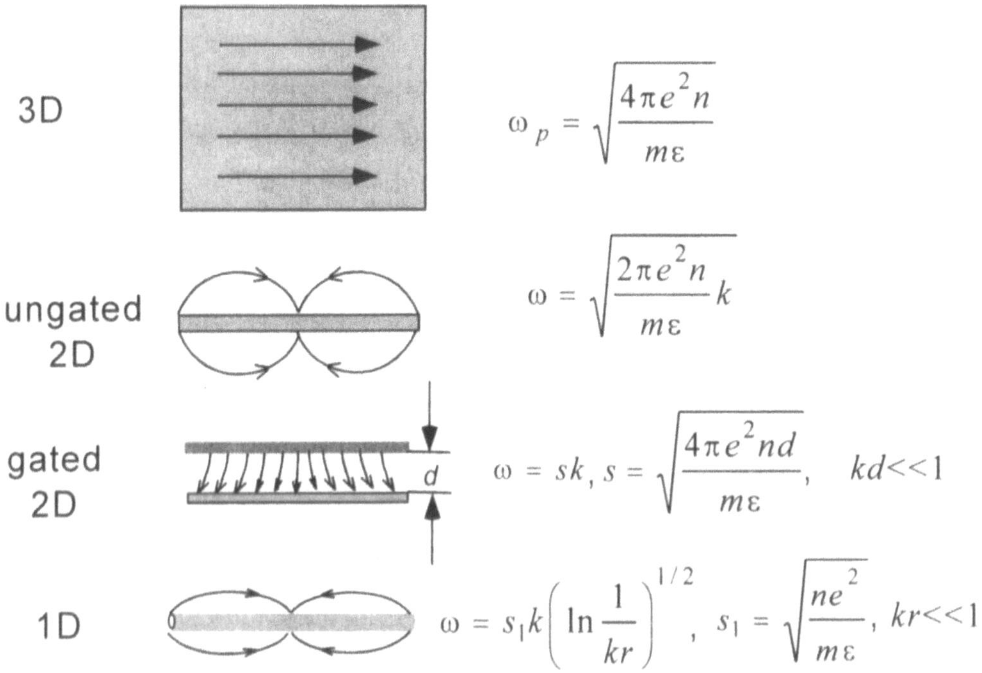

Fig. 1. Plasma wave frequencies for different sample geometries.

The fundamental frequency, f_l, of the FET plasma wave resonator is

$$f_l = const\ s/L \tag{15}$$

where *const* is a numerical constant, s is the plasma wave velocity given by Eqs. (9) or (11), and L is the gate length. The constant numerical factor, const, in Eq. (15) depends on the boundary conditions (see the following section, where we show that *const* = 1/4 for a certain type of the boundary conditions). Since s is on the order of 10^8 cm/s, we obtain and $f_l \sim 1$ THz for L=1 μm. Thus, a sub-micron FET would serve a resonator cavity for terahertz oscillations.

Plasma waves in a gated 2D electron gas were first considered by Chaplik [3] for electrons on the surface of liquid helium. For the case of a FET, they were considered by Nakayama. [4] Allen et al. [5] observed infrared absorption and Tsui et al. [6] observed weak infrared emission related to such waves in silicon inversion layers. Results of a recent study by Burke et al. [7] are presented in Figs. 2 and 3.

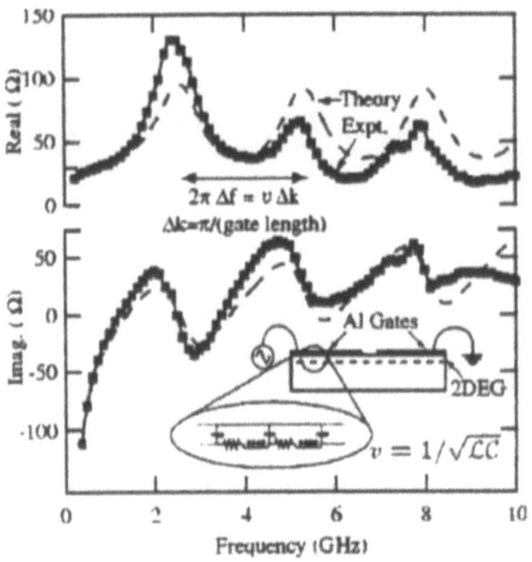

Fig. 2. Impedance of gated sample with 2D electron gas versus frequency. For this sample, $\omega_p\tau = 1$ at 1.25 GHz (where τ is the momentum relaxation time).[7]

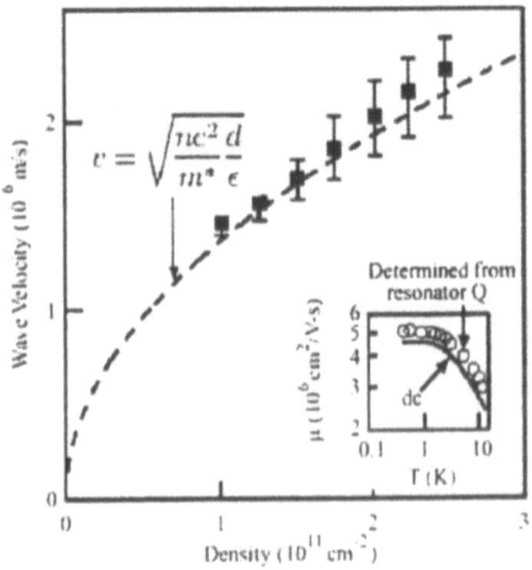

Fig. 3. Plasma wave velocity versus electron density.[7]

Burke et al. measured the impedance of a high mobility gated 2D electron gas in a gated GaAs/AlGaAs heterojunction at low temperatures. The gate length was ~ 0.4 mm, and the electron scattering time was on the order of 100 ps. In Fig. 2, one can see well-defined resonance peaks, corresponding to several plasma wave modes with frequencies in the gigahertz range. Results for the plasma wave velocity, deduced from such measurements at different gate biases, are presented in Fig. 3.

From Fig. 2 and Eq. (15) we conclude that a device with the same mobility and a gate length, L=1 μm would have the fundamental frequency of 1 THz and a quality Q=100!

Boundary conditions

The frequencies of the plasma waves depend on the boundary conditions. As discussed above, the channel of a FET works as a resonator for the plasma waves. This resonator is analogous to a segment of a transmission line. In this analogy, the role of the distributed capacitance is played by the gate-to-channel capacitance, while the equivalent inductance is related to the electron inertia. Therefore, it is natural to specify the boundary conditions in terms of the terminating impedances, Z_s and Z_d, at the source and drain side of the channel. Hence, the boundary conditions are

$$\begin{aligned} Z_s j_s &= U_s \\ Z_d j_d &= U_d \end{aligned} \quad (16)$$

where j_s, j_d, U_s, and U_d are the AC gate-to-channel current densities and potentials U at the source and drain sides of the channel, respectively (see Fig. 4).

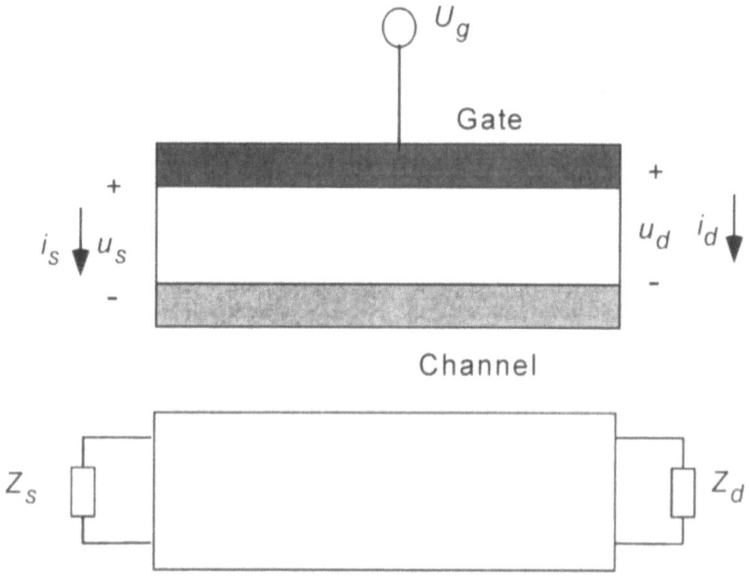

Fig. 4. Boundary conditions and transmission line analogy for plasma waves in field effect transistor.

We can understand these boundary conditions by considering the external capacitances connecting the gate contact to the source and drain contacts, respectively. A characteristic scale for these capacitances is the gate-to-channel capacitance, C, of the device. Obviously, if the terminating capacitance is much smaller then C, the terminating impedance corresponds to an open circuit at the plasma wave frequency. In the opposite limiting case, when the terminating capacitance is much larger then C, the terminating impedance corresponds to a short circuit at the plasma wave frequency. The former limiting case corresponds to the situation when the ac gate-to-channel current density is zero at the contact, while the latter limiting case corresponds to the zero ac potential at the contact.

We note that FETs are normally designed and operated in such a way that there is a natural asymmetry in these boundary conditions for several reasons. First, the high frequency transistor layout must minimize the parasitic drain-to-gate capacitance compared to the parasitic gate-to-source capacitance, since the drain-to-source capacitance is the Miller capacitance that strongly degrades the microwave transistor operation. Second, the flowing drain-to-source current caused an asymmetry in the channel potential distribution with a smaller charge in the channel toward the drain side and, as a consequence, the drain-to-gate capacitance decreases and the gate-to-source capacitance increases with the drain bias.

Having this in mind, we will adopt below the following simplified boundary conditions for the ac channel currents and gate-to-channel voltages:

$$U = 0 \quad \text{at the source,}$$
$$j = 0 \quad \text{at the drain,} \qquad (17)$$

which corresponds to a short circuit at the source and an open circuit at the drain. For such boundary conditions, the plasma wave frequencies are given by

$$f_n = sn/4L, \quad n = 1,2,3... \qquad (18)$$

Hydrodynamic approach and shallow water wave analogy

As was shown above, the plasma waves are a very general phenomenon that exists in systems with different geometry and dimensions and with different electron concentrations. However, in the special case when the concentration is large enough but the electrons are still not strongly degenerate, the electron-electron collisions are more frequent than electron collisions with impurities and/or lattice vibrations. In this case the electrons actually behave not as a gas, but rather like a fluid, and should be described using the hydrodynamic approach.[2]

As was discussed in [2], for a highly non-ideal electron gas in an AlGaAs/GaAs heterostructures with the electron surface concentration, n, on the order of 10^{12} cm^{-2} at 77 K, the thermal energy, the Fermi energy, and the Bohr energy are of the same order. Under such conditions, the mean free path for electron-electron collisions is on the order of the mean distance between electrons (~ 100 Å), which is much smaller than both the mean free path and a typical gate length, and the hydrodynamic approach should be adopted.

The basic equations describing the two dimensional electronic fluid are the relationship between the surface carrier concentration and gate voltage swing, the hydrodynamic equation of motion, and the continuity equation. The surface concentration, n, in the FET channel is related to the local gate-to-channel voltage swing, $U = U_{gc}(x) - U_T$, by Eq. (4). Here $U_{gc}(x)$ is the local gate-to-channel voltage. As discussed above, Eq. (19) represents the usual gradual channel approximation (see, for example, [8]), which is valid when the characteristic scale of the potential variation in the channel is much greater than the gate-to-channel separation.

The equation of motion (the Euler equation) is

$$\frac{\partial \mathrm{v}}{\partial t}+\mathrm{v}\frac{\partial \mathrm{v}}{\partial x}+\frac{e}{m}\frac{\partial U}{\partial x}+\frac{\mathrm{v}}{\tau}=0 \qquad (19)$$

where $\partial U/\partial x$ is the longitudinal electric field in the channel, $\mathrm{v}(x,t)$ is the local electron velocity, and m is the electron effective mass. The last term accounts for electronic collisions with phonons and/or impurities. (Here we neglect the viscosity of the electronic fluid.) Eq. (2) has to be solved together with the usual continuity equation, which (taking Eq. (1) into account) can be written as:

$$\frac{\partial U}{\partial t}+\frac{\partial (U\mathrm{v})}{\partial x}=0 \qquad (20)$$

Equations (19) and (4) coincide with the equations describing the so-called shallow water in conventional hydrodynamics [9] if U is replaced by the water level, h, and e/m is replaced by the free fall acceleration, g.

The term "shallow water" in hydrodynamics refers to the situation when the scale of spatial variations is large compared to the water depth, h. Plasma waves in the FET channel for $kd<<1$ correspond to shallow water waves with $kh<<1$. The velocity of shallow water waves is $(gh)^{1/2}$, which is analogous to the plasma wave velocity $s=(eU/m)^{1/2}$. This exact shallow water analogy [2] gives a deeper insight in the processes occurring in the FET channel under different conditions. We note that the hydrodynamic analogy goes even further, since plasma waves in the limit $kd>>1$ with the $\omega\sim k^{1/2}$ dispersion law (see Eq. (12)) exactly correspond to deep-water waves $(kh>>1)$.

An important parameter in hydrodynamics is viscosity. The viscosity of the 2D electronic fluid is of the order of $\mathrm{v}_F\lambda_{ee}$ where v_F is the Fermi velocity and λ_{ee} is the mean free path for the electron-electron collisions, which is on the order of the inter-electronic distance, $n^{-1/2}$. Then we obtain the following estimate for the viscosity of the electron fluid: $\nu = \hbar / m \approx 15$ cm^2/s for GaAs. For comparison, the viscosity of air is about 0.15 cm^2/s. For wavelengths that are short enough the electron viscosity may provide the dominant mechanism for damping of plasma oscillations.

FET as a detector of terahertz radiation

Electromagnetic radiation can excite plasma waves, and, therefore, a FET has a resonance response at the plasma wave frequency. As we discussed in [1], this effect can be used for the resonance detection and mixing of electromagnetic radiation at terahertz frequencies. The half width of the resonance curve is determined by the damping of plasma waves, i.e. by the electron momentum relaxation time and by the electron viscosity.

The detection is based on the nonlinearity of the electron fluid (the main source of non-linearity is the dependence of the plasma wave velocity on the electron concentration), which is, in fact, the usual effect of rectification of the ac radiation field. Thus, a constant drain-to-source voltage should develop in response to incoming electromagnetic radiation. Certainly, an asymmetry of the drain and source contacts should exists, in order to have this effect. Such an asymmetry can be caused by the following reasons:

a) The radiation could be fed in (by an external antenna) to one side of the transistor, e.g. between the source and the gate

b) As discussed above, the boundary conditions might be different at the source and at the drain (e.g. short circuit at the source and open circuit at the drain)

c) A dc source-drain current could be applied, depleting the drain end of the channel

As we showed in Ref. [1] at low frequencies, such that $\omega\tau << 1$, the HEMT should work as a non-resonant broad band detector with the responsivities, which are comparable to those of Schottky diode detectors. In the low frequency limit, our theory predicts that the responsivity, R, of such a HEMT detector is given by

$$R = \alpha L^4\omega^2/(6s^4\tau^2) \qquad (21)$$

The pre-factor α in Eq. (18) depends on the boundary conditions.

Recently, we fabricated a prototype non-resonant detector (operating in the microwave range) using an AlGaAs/GaAs 0.15-micron gate HEMT. [10] The measured dependencies of the detector responsivity on the gate bias and frequency were in good agreement with our theory (see Fig. 5).

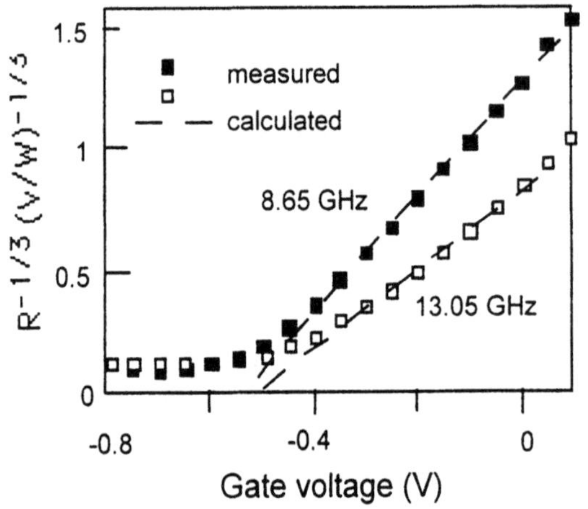

Fig. 5. Comparison of measured and predicted responsivities for an AlGaAs/GaAs HEMT. [10]

Fig. 6 shows a schematic diagram and an equivalent circuit of a HEMT subjected to terahertz radiation [11, 12]. In the case that is the most appropriate for the resonant detector, the incoming electromagnetic radiation induces an ac voltage at the source side of the channel. The drain side of the channel is an open circuit. (A coupling with a terahertz radiation might be enhanced using an appropriate antenna structure.) This ac voltage excites the plasma waves.

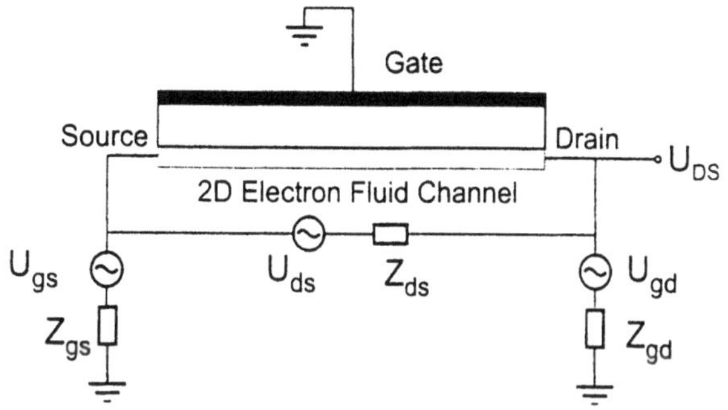

Fig. 6. Schematic diagram and an equivalent circuit of a HEMT subjected to terahertz radiation. [12]

As mentioned above, the dc gate-to-source voltage determines the velocity, s, of these waves. Because of the nonlinear properties of the electron fluid and the asymmetry in the boundary conditions, a FET biased only by the gate-to-source voltage and subjected to electromagnetic radiation develops a constant drain-to-source voltage, which has a resonant dependence on the radiation frequency with maxima at the plasma oscillation frequency and its odd harmonics. For the frequencies, such that $|\omega - n\,\omega_1| << \omega_1$, and for $s\tau/L >> 1$ where τ is the momentum relaxation time and L is the channel length, we obtain [19, 31]

$$\frac{\Delta U}{U_o} = \left(\frac{U_a}{U_o}\right)^2 \left(\frac{s\tau}{L}\right)^2 \frac{1}{4(\omega - n\omega_1)^2\tau^2 + 1} \tag{22}$$

Here where n = 1, 3, 5, 7..., ΔU is the dc voltage difference between the drain and source potentials, U_o is the gate-to-source gate voltage swing, and U_a is the amplitude of the ac source-to-gate voltage induced by the incoming radiation. The half width of the resonance curve is determined by the damping of the plasma oscillations caused by the electron momentum relaxation and/or the electron fluid viscosity. Thus, the FET acts as a tunable resonance quadratic detector of electromagnetic radiation. As can be seen from Figs. 6 and 7, the resonance responsivity of a FET can exceed typical responsivities of standard Schottky barrier detectors (which are on the order of 1000 V/W) by many orders of magnitude, if the quality factor, Q, is high enough (the responsivity is proportional to Q^2.)

Recently, we reported on the implementation of a terahertz detector utilizing 2D electronic fluid in a High Electron Mobility Transistor (HEMT). [32] To our knowledge, this is the first plasma wave terahertz detector ever demonstrated.

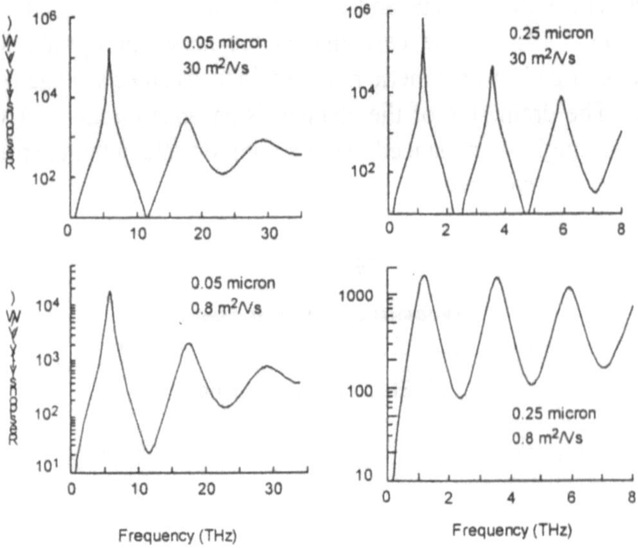

Fig. 7. Detector responsivity versus frequency for 0.05 μm and 0.25 μm gate HEMTs at 77 K and 300 K. Parameters used in the calculation: $U_o = 0.5$ V, $G = 1.5$, $\nu = 15$ cm^2/s, $m = 0.063\ m_o$, $\mu = 30$ m^2/V-s at 77 K, $\mu = 0.8$ m^2/V-s at 300 K. Here G is the antenna gain and ν is viscosity of the electronic fluid. [1, 17], © IEEE, 1996

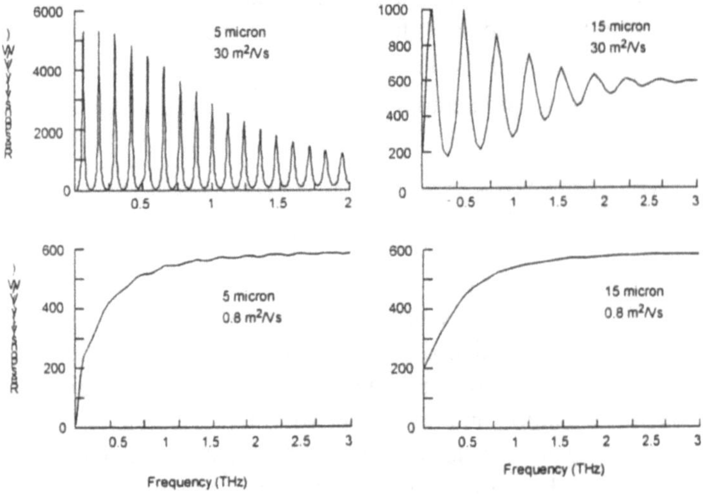

Fig. 8. Calculated responsivity of 2 μm and 15-μm gate HEMTs at 77 K and 300 K The devices behave as resonant detectors at 77 K and as broadband detectors at 300 K Parameters used in the calculation: $U_o = 0.5$ V, $G = 1.5$, $\nu = 15$ cm^2/s, $m = 0.063\ m_o$, μ = 30 m^2/V-s at 77 K, $\mu = 0.8$ m^2/V-s at 300 K. [1, 17], © IEEE, 1996

The detector was fabricated using a *Fujitsu* FHR20X HEMT mounted on a quartz substrate. The device operated at a frequency of 2.5 THz, which is about 30 times higher than the transistor cutoff frequency. A CO_2-pumped far-infrared gas laser served as a source of terahertz radiation. The laser beam was chopped and focused on the sample with the electric field polarization oriented in the drain-to-source direction. The drain was open, thus the drain current is zero. The gate current is in the order of nano-ampere for the measurement range of gate bias.

In agreement with the predictions of our terahertz detector theory [19, 31], the radiation induced a negative DC drain-to-source bias proportional to the radiation intensity. This voltage was measured using a lock-in amplifier.

The device characteristics were fitted using the HEMT model implemented in AIM-Spice. [13], [14] The parameters were extracted from measured DC characteristics and from the elements of the small-signal microwave equivalent circuit.

The measured dependencies of the detector responsivity on the gate bias are in good agreement with the gate bias dependence of the normalized responsivity predicted by the detector theory [1, 17] (see Fig. 9).

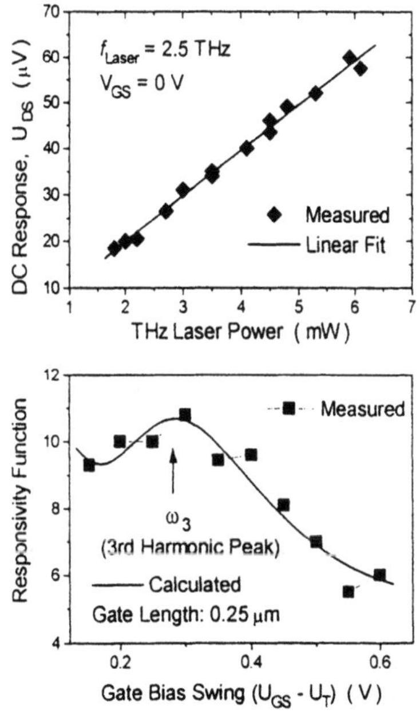

Fig. 9. Measured (symbols) response versus incident power at zero gate bias. Also shown gate bias dependence of the detector dimensionless responsivity function, which is proportional to $U = V_{gs} - V_T$, measured (symbols) at the laser frequency of 2.5 THz and the laser power of 5.7 mW. The solid line is predicted by the theory and is normalized for comparison. The peak corresponds to the third harmonic of the fundamental plasma frequency, ω_3. [11], © IEEE, 1998

The responsivity increases at smaller gate voltage swings as predicted by the theory. The dimensionless responsivity function displays a rather broad resonant peak corresponding to the third harmonic of the fundamental plasma frequency.

Since the drain current was zero, and the gate current in the measurement range was also nearly zero, the heating effect should be entirely independent of the gate bias. Hence, it could not have been responsible for the measured effect. Beside that, the wavelength of the terahertz radiation was nearly two orders of magnitude larger than the gate length. Therefore, the temperature rise, if any, would be uniform over the device, and can in no way result in the measured drain voltage response that was proportional to the terahertz radiation intensity.

More recent results showed a large increase in the detector detectivity in the regime with an applied drain-to-source bias.[15] Since the drain bias decreases the drain-to-source capacitance and increases the gate-to-source capacitance, this result is consistent with the expected impact of asymmetrical boundary conditions as discussed above.

FET as a mixer and multiplier

In practical systems, mixing of weak incoming signal, $U_a \cos(\omega t)$ with a strong local oscillator signal, $U_{loc}\cos(\omega_{loc}t)$, is often more desirable because of a much higher sensitivity. For $|\omega - n\,\omega_{loc}|\tau << 1$, the equation describing the electron fluid mixer response coincides with Eq. (`22), where U_a^2 should be replaced with U_aU_{loc}.

Nonlinear properties of the 2D electronic fluid not only lead to the rectification of the incoming electromagnetic radiation but also to the appearance of a signal at the second and higher harmonics of the incoming radiation with the resonance response at the fundamental plasma frequency (see Fig. 10).

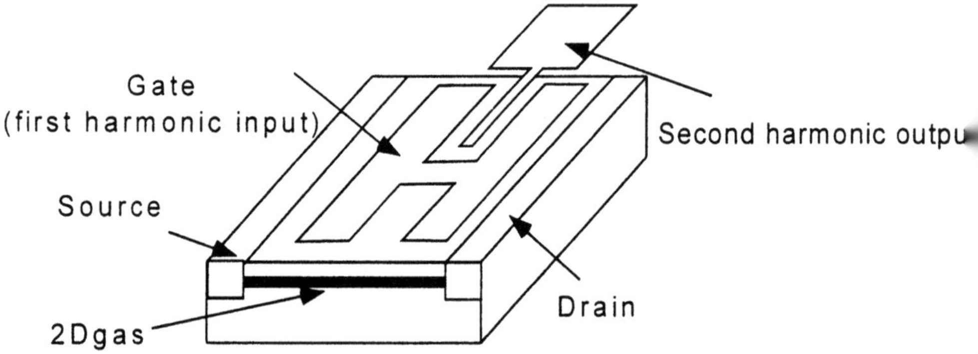

Fig. 10. Schematic design of HEMT doubler with two gates. An alternative design may utilize a single gate and an additional contact to the 2D channel at approximately 0.35 L from the source. (From [1,17], © IEEE, 1996.)

As expected, the second harmonic has a larger magnitude than higher harmonics. For the second harmonic, the voltage has a maximum closer to the source end of the device channel, which should be accounted for in the multiplier design.

The amplitude of the second harmonic is roughly of the same order as the dc voltage developed in the resonance detector.

Generation: terahertz whistle and electronic flute

In Ref. 2 we have found that under certain conditions the steady state with a dc current in the FET channel is unstable against spontaneous generation of plasma waves. This instability is the consequence of asymmetric boundary conditions at the source and the drain. The simplest asymetric conditions are the ones discussed above: short curcuit at the source and open curcuit at the drain. If, for a moment, we neglect dissipation, this instability immediately follows from Eqs. (19) and (20). These equations should be linearized with respect to small deviations of U and v from their steady state values, and the temporal evolution of these small perturbations should be studied, taking account of the boundary conditions given by Eq. (17). Such calculation yields the following expression for the growth increment [2]:

$$\omega'' = \frac{s^2 - v_o^2}{2Ls} \ln\left|\frac{s + v_o}{s - v_o}\right| \qquad (23)$$

where v_o is the steady state velocity of the electrons in the channel. The dependence of the increment on the ration v_o/s is presented in Fig.11.

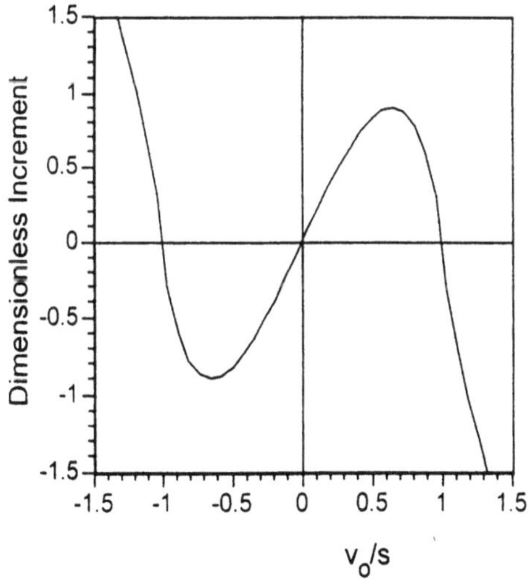

Fig. 11. Dimensionless plasma wave increment, $2\omega'' L/s$, as a function of Mach number, $M = v_o/s$. The steady electron flow is unstable when $0 < v_o < s$ and $v_o < -s$.[2]

One can see that the instability exists for any arbitrary small v_o (small current). This, of course, is the consequence of neglecting losses, caused by collisions. In reality, there would be a threshold current, at which the calculated increment exceeds the losses (see Fig. 12 which shows the net increment and threshold velocities).

The origin of this instability can be understood if one considers reflections of a plasma wave from the two boundaries. With our boundary conditions, it is easy to calculate that for a small amplitude wave of potential U the amplitude reflection coefficient at the source is equal to -1, while the reflection coefficient at the drain is equal to $(s+v_o)/(s-v_o)$, i.e. is greater than unity! Thus one could say that plasma waves are amplified during reflection at the drain contact. However, this is not quite accurate. As it was pointed out by Cheremisin [16] if one consideres the current rather than voltage wave, the reflection coefficients are reversed, but the product of the two reflection coefficients remains the same. The products of the reflection coefficients at the source and at the drain are greater than unity, if a dc electron flow exists. This is the origin of the wave amplification during a round trip, and this mplification leads to an instability of the steady state with respect to generation of plasma waves.

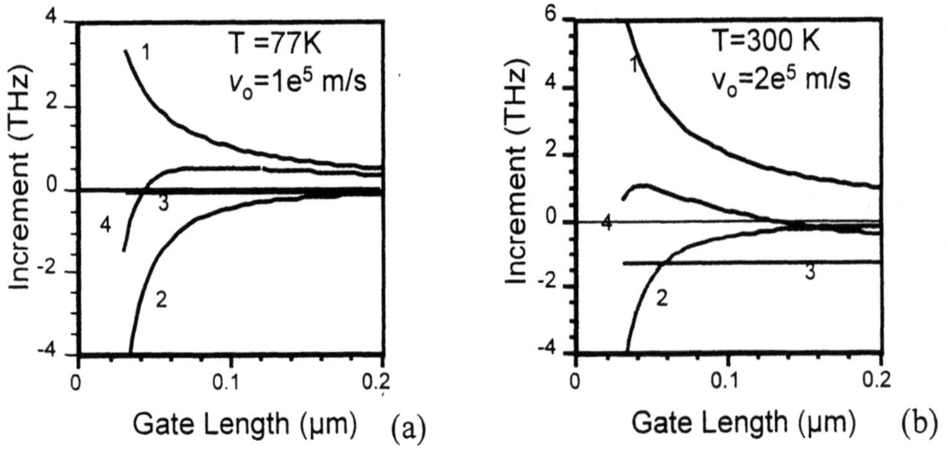

Fig. 12. Growth increment and decrements of plasma waves versus channel length for AlGaAs/GaAs HEMT at 77 K (a) and 300 K (b). 1 -growth increment, 2 - decrement caused by electron viscosity, 3 - decrement caused by electron scattering, 4 - net increment (positive values correspond to the wave growth). [17]

The precise form of the boundary conditions used above is not crucial for the instability. It can be shown, that a positive growth increment can be obtained for arbitrary terminating impedances Z_s and Z_d (see Eq. (16)), provided that the imaginary part of Z_s is greater than the imaginary part of Z_d , in other words that the gate-to-source capacitance, C_{gs}, is greater than the gate-to-drain capacitance, C_{gd}. Of course, if there are apreciable losses at the contacts, described by the real parts of Z_s and Z_d, this would increase the instability theshold or might even completely suppress the instability.

Actually, this type of instability is well known in acoustics: this is the mechanism responsible for the performance of the ordinary whistle, excitation of organ pipes and a great variety of wind musical instruments. [18] In all of these cases, if the velocity of the air flow exceeds a certain threshold (determined by dissipation), sound waves are generated in a resonant cavity. From the theoretical point of view, the problem can be devided into two: a) given the terminating (acoustic) impedances, calculate the growth increment and the threshold flow velocity, and b) calculate the terminating impedances. The latter task is difficult, since it requires consideration of the two or three dimensional flow at the pipe ends. The former problem was solved in Reference [2].

In this regard, we would like to stress that plasma waves are similar not only to shallow water waves but also to sound waves since they have a linear dispersion law. In turn, shallow water behavior is similar to the dynamics of a gas with pressure proportional to the square of the density, (see, for example, [9]). Thus, the nonlinear hydrodynamic equations for the 2D electron fluid are similar to (but not identical with) the equations for a real gas, such as air. However, the linearized equations describing small-amplitude plasma waves in a FET and sound waves in a gas are identical. Since the linearized equations determine the instability threshold for a steady flow (i. e. the wave generation threshold), the instability conditions for a real gas and for a 2D electron fluid should be similar provided that the Reynolds numbers and quality factors of resonance cavities are the same.

Hence, in principle, one can design a device, which we called an "electronic flute" [19]. As was shown in Reference [19], these dimensionless parameters for a HEMT structure might be of the same order of magnitude as for a conventional wind musical instrument (Re ~ 50 and Q ~ 1 - 100), see Table 1.

Table. 1. Parameter comparison for conventional and electronic flutes. [33]

Parameter	Conventional Flute	Electronic Flute
relevant dimension	1 cm	10^{-4} cm
flow velocity	10 cm/s	10^7 cm/s
wave velocity	3×10^4 cm/s	10^8 cm/s
viscosity	0.15 cm^2/s	15 cm^2/s
frequency	10 - 10^3 Hz	10^{11} - 10^{13} Hz
Reynolds number	60	60
Quality factor	3-100	10

Thus, a complete similarity between the plasma waves in a FET and sound waves led us to believe in the possibility of realizing an electronic flute based on the excitation of plasma waves by a direct current in gated modulation doped structures (see Fig. 13). This electronic flute should operate in the terahertz range of frequencies and emit far infrared radiation.

As a consequence of the instability, the amplitude plasma oscillations grows in time, reaching a stationary state of nonlinear oscillations with an amplitude determined by sheet carrier concentration in the channel, momentum relaxation time, and device length. [2,19] This stationary state was examined by computer simulation in Refs. [20], [21] revealing a complex behavior involving traveling shock waves. Such oscillations should result in a periodic variation of the channel charge and the mirror image charge in the gate contact, i. e. to the periodic variation of the dipole moment. This variation should lead to electromagnetic radiation. The device length is much smaller than the wavelength of the electromagnetic radiation, λ at the plasma wave frequency. (The transverse dimension, W, may be made comparable to λ.). Hence, the Ballistic FET operates as a point or linear source of electromagnetic radiation. Many such devices can be placed into a quasi-optical array for power combining.

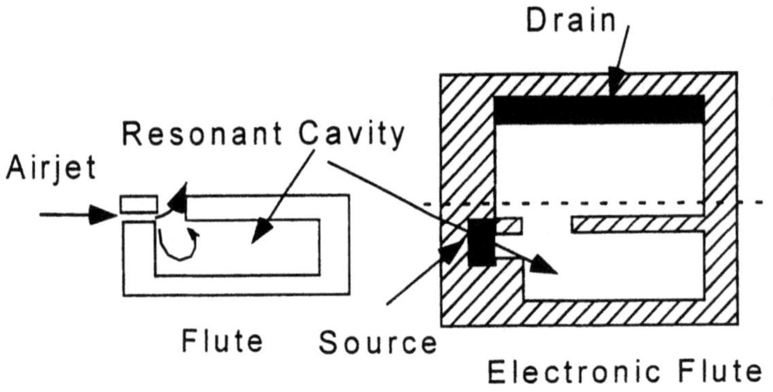

Fig. 13. Simplified diagrams of a jet driven pipe musical instrument of a flute family and of electronic flute. Arrows show the direction of airflow and streamlines of electric current. White area of the electronic flute is the gated region with the 2D-electron fluid in the channel. The part of this device shown below the dashed line is the electronic flute. The top gated portion of the device connected to the drain is similar to the outside air space for a jet driven musical instrument. The streamlines of the electron flow near the cavity opening excite the plasma waves in the cavity, just like the air jet penetrating the resonant cavity in a flute excites sound waves in the flute resonance cavity. [19]

Conclusions.

Plasma wave electronics opens up intriguing opportunities for implementing solid-state tunable generators and tunable resonant and non-resonant detectors and mixers of terahertz radiation. Such development might open up many new exciting applications. However, several theoretical and experimental problems have to be solved in order to make this dream a reality.

The theoretical challenges involve the analysis of the two-dimensional and three-dimensional problems, a more detailed understanding of the boundary conditions [22], [23], and the analysis of non-linear regimes of the plasma wave devices. [24]

Also, the predicted increment of the plasma wave instability is relatively weak, and one would need a ballistic transistor in order to obtain generation at low currents. (Some preliminary evidence of such generation was reported in [17].) We should mention, however, that recent work by Ryzhii et al. offers the solution for this problem by implementing a double barrier resonant tunneling structure between the gate an the channel. As shown in [25], [26], [27], [28], this should dramatically increase the plasma wave instability increment and make a room temperature terahertz generation possible.

Another intriguing solution of this problem (still quite speculative) might be related to the recently predicted negative differential mobility in silicon 2DEG that might be used in order to enhance the growth of plasma wave oscillations. [29]

Another problem is that of enhancing coupling between the terahertz radiation and plasma waves. This problem is common for all types of solid-state terahertz devices, and can be solved by using arrays of FETs (electronic flute) or appropriate antenna structures. Evidently, more theoretical and experimental work is required in order to solve all these problems.

Acknowledgement

This work has been partially supported by the Army Research Office (Project Monitor Dr. Dwight Woolard) and DARPA (Project Monitor Dr. Edgar Martinez), and by the Office of Naval Research (Project monitor Dr. John Zolper).

References.

[1] Dyakonov, M. I. and Shur, M. S. (1996) Detection, Mixing, and Frequency Multiplication of Terahertz Radiation by Two Dimensional Electronic Fluid, *IEEE Transactions on Electron Devices*, **43**, pp. 380-387.

[2] Dyakonov, M. I. and Shur, M. S. (1993) Shallow Water Analogy for a Ballistic Field Effect Transistor. New Mechanism of Plasma Wave Generation by DC Current, *Phys. Rev. Lett.* **71**, pp. 2465-2468.

[3] Chaplik, A. V. (1972), Zh. Eksp. Teor. Fiz. **62**, 746 (Sov. Phys. JETP, 35, 395, (1972))

[4] Nakayama, M. (1974) Theory of surface waves coupled to surface carriers, *J. Phys. Soc. Japan*, **36**, pp.393-398.

[5] Allen, S. J., Jr., Tsui, D. C., and Logan, R. A. (1997) Observation of the two-dimensional plasmon in silicon inversion layers, *Phys. Rev. Lett.* **38**, pp. 980 - 983.

[6] Tsui, D. C., Gornik, E., and Logan, R. A. (1980) Far infrared emission from plasma oscillations of Si inversion layers, *Solid State Comm.*, **35**, pp. 875- 877.

[7] Burke, P. J., Spielman, I. B., Eisenstein, J. P., Pfeiffer, L. N., and West, K. W. (2000) High frequency conductivity of the high-mobility two-dimensional electron gas, *Applied Phys. Lett.* **76**, pp. 745-747.

[8] Shur, M. S. (1996) *Introduction to Electronic Devices*, John Wiley and Sons, New York p. 384.

[9] Landau, L. D. and Lifshitz, E. M. (1966) *Fluid Mechanics,* Pergamon, New York.

[10] Weikle, R., Lu, J., Shur, M. S., and Dyakonov, M. I. (1996) Detection of Microwave Radiation by Electronic Fluid in High Electron mobility Transistors, *Electronics Letters*, **32**, pp. 2148-2149.

[11] Lü, J.-Q. , Shur, M. S. , Hesler, J. L. , Sun, L. , and Weikle, R. (1998) Terahertz Detector Utilizing Two-Dimensional Electronic Fluid, *IEEE Electron Device Letters*, **19**, pp. 373-375.

[12] Shur, M. S. and Lü, J.-Q. (2000) Terahertz Sources and Detectors Using Two Dimensional Electronic Fluid in High Electron Mobility Transistors, *IEEE Transactions on Microwave Theory and Techniques*, **48**, pp. 750-755.

[13] Fjeldly, T., Ytterdal, T., and Shur, M. S. (1998) *Introduction to Device and Circuit Modeling for VLSI*, John Wiley and Sons, New York.

[14] A student version of AIM-Spice and of FET parameter extraction software can be downloaded from http://www.aimspice.com/

[15] Lü J.-Q. and Shur, M.S. unpublished

[16] Cheremisin, M. (1999) Ph. D. Thesis, Etude d'nstabilites un liquide bidimensionnel d'electrons dans un transistor a effect de champ, University of Montpellier

[17] Dyakonov, M. I. and Shur, M. S. (1996) *Plasma Wave Electronics: Novel Terahertz Devices using Two Dimensional Electron Fluid, Special Issue on Future Directions in Device Science and Technologies*, IEEE Transactions on Electron Devices, **43**, pp. 1640-1646.

[18] Fletcher, N. H. and Rossing, T. D. (1991) *The Physics of Musical Instruments*, Springer-Verlag, New York.

[19] Dyakonov, M. I. and Shur, M. S.(1995) *Two Dimensional Electronic Flute*, Applied Physics Letters **67** , pp. 1137-1139.

[20] Dmitriev, A. P. , Furman, A. S., and Kachorovskii, V. Yu.(1996) Nonlinear theory of the current instability in a ballistic field-effect transistor, *Phys. Rev.* **B54**, pp. 14020-14025.

[21] Dmitriev, A. P., Kachorovskii, V. Yu. , Furman, A. S. ,and Samsonidze, G. G. (1997) Numerical study of the current instability in a two-dimensional electron fluid, *Phys. Rev.* **B55**, pp. 10319-10325.

[22] Crowne, F. J. (1997) Contact boundary conditions and Dyakonov-Shur instability in high electron mobility transistors, *J. Appl. Phys.* **82**, pp 1242-1254 .

[23] Crowne, F. J. (2000) Contact boundary conditions and Dyakonov-Shur instability in high electron mobility transistors, *J. Appl. Phys.* **87**, pp 8056-8063.

[24] Samsonidze, G. , Rudin, S. , and Shur, M. S. (1998) Large Signal Theory of Plasma Wave Electronics Terahertz Detectors, *1998 IEEE Sixth International Conference on Terahertz Electronics Proceedings,* University of Leeds, IEEE Catalog Number: 98EX171, P. Harrison, Editor, pp. 231-233.

[25] Ryzhii, V., Khmyrova, I., and Shur, M. S. (2000) Resonant detection and frequency multiplication of terahertz radiation utilizing plasma waves in resonant-tunneling transistors, *J. Appl. Phys.,* submitted

[26] Ryzhii V. and Shur, M. S. Terahertz radiation utilizing plasma waves in resonant-tunneling transistors, *J. Appl. Phys.*, accepted for publication

[27] Shur, M. S., and Ryzhii, V. Emerging solid state terahertz electronics in Lippens, D. (ed.) *Terahertz Systems and Sources*, Kluwer, The Netherlands

[28] Ryzhii, V., and Shur, M. S., Plasma Instability and nonlinear terahertz oscillations in resonant-tunneling structures, *unpublished*

[29] Dmitriev, A., Kachorovskii, V., Shur, M. S., and Stroscio, M. (2000) Electron Runaway and Negative Differential Mobility in Two Dimensional Electron or Hole Gas in Elementary Semiconductors, *Solid State Comm.*, **113**, pp. 565-568.

PROSPECTS FOR THE APPLICATION OF MINIATURISED VACUUM TECHNOLOGIES TO TERAHERTZ SOURCES

J.M.Chamberlain, J.J.Garcia, J.R.Fletcher, R.E.Miles, D.P.Steenson, J.R.Thorpe
Institute of Microwaves & Photonics, School of Electrical & Electronic Engineering, The University of Leeds, Leeds LS2 9JT, UK.

E. Huq,
Central Microstructure Facility, Rutherford Appleton Laboratory, Chilton, Didcot, Oxon. 0X11 0QX, UK.

C.M.Mann and M.OldField.
Space Science Department, Rutherford Appleton Laboratory, Chilton, Didcot, Oxon. 0X11 0QX, UK.

1. Introduction

It is possible, although not usual, to classify Terahertz frequency sources by the extent to which Quantum Mechanics is required to explain their operation. For example, several sources exist for which quantum mechanical explanations are irrelevant: these include devices that operate on the "bunching" principle, such as the Free Electron Lasers (FEL). The second category comprises of those devices for which Quantum Mechanics provides a background: e.g. HBVs, Gunn diodes. Finally, there are those devices for which Quantum Mechanics is essential: e.g. resonant tunnel diodes and cascade lasers.

The purpose of this Chapter is to explore the possibilities of making small scale Terahertz frequency radiation sources, lying in the first of these categories, that are based on a revived vacuum tube technology. The use of vacuum tube sources such as Klystrons and Magnetrons is common in the microwave and millimetre wave regions of the electromagnetic spectrum, although these are generally of considerable size (tens of centimetres in scale) and have substantial power requirements. In the Terahertz (or sub-Terahertz) frequency band, large sources such as Backward Wave Oscillators [1] and even Free Electron Lasers [2] are available, but the size and cost of these tubes rules them out for small-scale or mass-market applications.

Two particular technological advances over recent years may mean that sources such as Klystrons [3] can be constructed to operate at up to at least 1THz with practical power output using an essentially low-cost fabrication route. This chapter will examine the background to the proposal, indicate the fundamental physical limitations of the approach and consider in some detail the practical realisation of one type of sources (a reflex Klystron) operating up to 1THz. It is appropriate at this point to quote from a textbook, *à propos* of the Klystron:

> "As frequency increases, losses inevitably become greater, cavity sizes become smaller, difficulties of manufacture increase and the current densities required in the electron beam, whose diameter steadily decreases, become more and more difficult to obtain." [4].

R.E. Miles et al. (eds.), Terahertz Sources and Systems, 209–221.

In the three or more decades since this was written conspicuous developments have taken place within two unrelated fields: the fabrication of small and complex structures by micromachining, and the realisation of high current microfabricated silicon tips as Field Emitters. These advances suggest that the above comment is no longer correct, and in this Chapter, the possibility of making small-scale sources such as Klystrons will be considered from this fresh perspective.

This Chapter is organised as follows: the operational principles of the Klystron are first reviewed, and comment is made on the usefulness of Monte Carlo procedures in the design of these sources. The two enabling technologies needed to realise a Klystron using Field Emitter[5] sources and micromachined components[6] are then considered, and an illustrative example of a 500GHz Klystron is outlined. The effects of scaling to smaller dimensions are then summarised, with especial reference to power output, crucial dimensions, starting conditions and space charge effects. A range of practical problems is then considered, and indications given as to how these may be solved. Finally, speculations are presented concerning the applicability of these ideas to other types of Terahertz frequency source.

2. Brief review of Reflex Klystron Operation

The purpose of this short section is to briefly remind the reader of the operational principles of this simple vacuum tube device, and to indicate the issues that need to be borne in mind when designing a scaled-down structure.

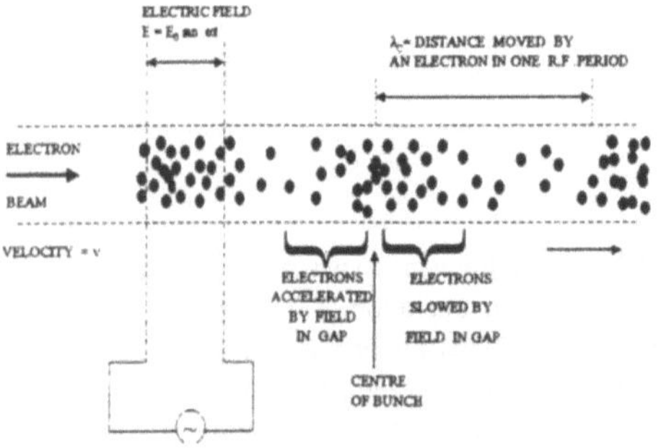

Figure 1: Density and velocity modulation produced by a gridded gap.

Figure 1 illustrates the effect of an electron beam passing through a "gridded gap" across which an alternating voltage is applied. Electrons passing through the grid will be accelerated for one half cycle, whereas for the other half cycle the result is a deceleration. For a given electron passing through the grid when the field is instantaneously zero, it is noted that the electrons immediately preceding it will have slowed down, whereas those following will go faster and gradually approach each other. The process continues until mutual repulsion effects (space charge) bring the process to a halt. Since there are no fields beyond the grids, this velocity modulation (present as the beam leaves the grids) is transformed into density modulation or "bunches". The first application of this effect was in the Two-Cavity Klystron Amplifier [3], indicated in Figure 2. Electrons are emitted from the (conventional) source (S) and pass through two such resonant cavities, finally reaching the collector (CL). DC potentials are

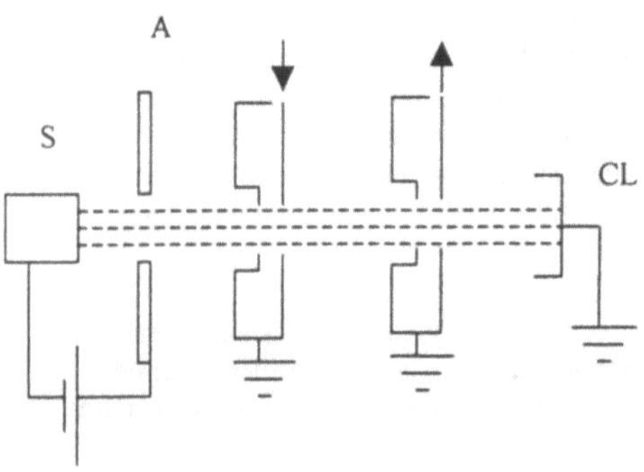

Figure 2: Two Cavity Klystron Amplifier

applied to the various components as necessary, and an RF signal applied to the first cavity. In this device, velocity modulation at the input cavity gives rise to beam current modulation at the output cavity; this beam current modulation induces a current in the output cavity. By appropriate design of the cavity structures emission of radiation occurs through the output port and the device acts as an amplifier. A simple extension of this concept is to incorporate a feedback loop, so that the amplifier becomes an oscillator. This is achieved in the Reflex Klystron, [3] which has been chosen for potential development as a Terahertz source. In the Reflex Klystron, the electron beam (from the source) passes through the resonant cavity, as before, then enters a region with a static retarding field caused by a reflector element (RFL). This field brings the beam to rest and returns it to the cavity. Bunching takes place in the beam, and by appropriate choice of cavity-reflector spacing, the bunch can arrive back in a suitable phase to give up energy to the cavity field. In this way, energy is transferred from the kinetic energy of the electron stream into electromagnetic energy in the cavity.

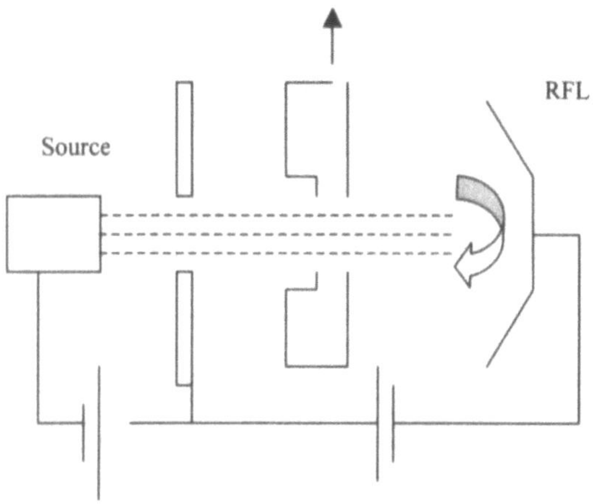

Figure3: Reflex Klystron Oscillator.

The design of vacuum tube devices of this type has, of course, reached an advanced stage over several decades of continuous development. One particularly useful design tool is Monte Carlo simulation, and there have been several reports of the use of this approach in the design of relativistic Klystrons [7]. It is possible to reproduce the bunching behaviour in a Klystron, discussed above, using a very simple Monte Carlo simulation: this has been accomplished by the present authors in a simple demonstration [8]. This simulation uses a one-dimensional ensemble Monte Carlo with 600 cells to describe the total length of the simulation. Along this length we consider three different regions: in the first one there is no applied field, in the second an oscillating field with a period T is applied, and in the last one we impose a constant field to reflect the particles. The simulation starts with a number of particles in the initial cells of the Monte Carlo, for which momentum has been chosen randomly according with a Gaussian distribution. Then, in each temporal step of the simulation ($\Delta t \approx T/100$), a constant number of particles is injected at the beginning of the device with positive momentum, which is also chosen randomly in accordance with the same distribution. With these conditions we emulate the effects of an incident electron beam in the region of the oscillating field. We have also shown that in order to take into account the power loss, it is convenient to use a Monte Carlo collisional treatment. This effect would be missing in the standard particle-in-cell (PIC) models.[9]

3. Enabling Technologies: Micromachining

As indicated in Section 1, two technologies are combined in a novel fashion in the proposed Terahertz sources. The first of these is now considered.

The use of micro-machining techniques for fabrication of Terahertz frequency components is now becoming widespread. A variety of techniques such as LIGA, laser ablation and silicon micromachining have been used for some time to fabricate waveguide cavities and related components. An alternative approach is to use UV sensitive photoresists, as originally developed in the semiconductor processing industry. The characteristic dimensions (≈ 100μm) of passive devices that can be made by such methods are entirely appropriate for Terahertz frequency use. Furthermore, the surface quality and definition is excellent, and the entire process can be accomplished at relatively low cost. Cronin *et al.* [10] developed the ideas of micromachining of waveguides using positive photoresist techniques, and other workers have demonstrated [11] an H-plane antenna for operation at 200GHz using this approach. Figure 4 shows the sequence of operations necessary to realise an antenna: a thick layer of photoresist is patterned to the desired shape, then gold or a similar metal is electroplated on the surface of the photoresist. This layer may be augmented using electroplating techniques, and the final fabrication stages involve the removal of the photoresist material in order to expose the metallic structure. Subsequently, the integration of active devices, such as Schottky diodes [12] and resonant tunnel diodes [13], into such structures has also been demonstrated.

Silicon micromachining [14] is a competitor technology to the use of photoresist based approaches. It has been successfully combined [14] with photoresist technology using a negative photoresist, SU-8 [15]. This resist has also been widely used alone (i.e.

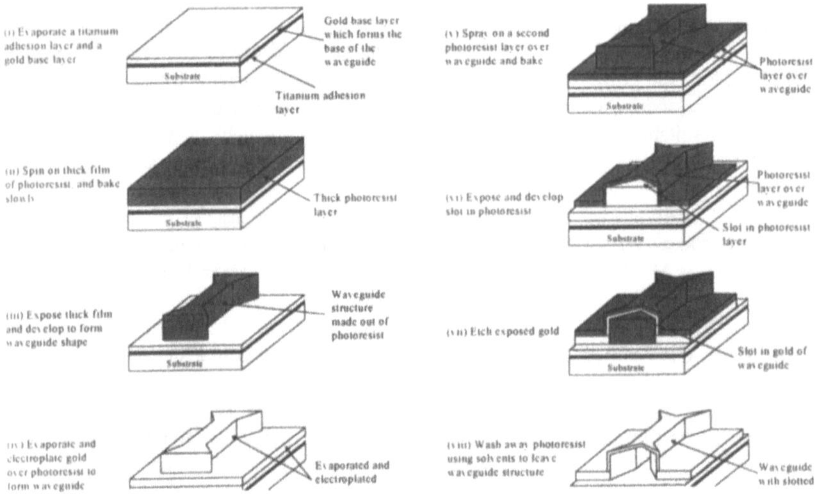

Figure 4: Micromachined Antenna structure [11].

not combined with silicon micromachining) and several workers have demonstrated that complex shapes of height up to approximately 1mm can be obtained by this method. A particularly promising development has been the realisation of "snap-together" waveguide, for use up to 450GHz, which can be used to produce complex cavities and other components [16]. The areas of the resist which conduct high frequency currents are coated with gold and a high quality surface results. There are, however, certain drawbacks with this material, especially the difficulty of removing it from inside structures, although significant progress is now being made in that area [17] . Miles et al., elsewhere in this Volume, give further discussions of the use of this resist. An interesting development, of direct relevance to the realisation of vacuum tube sources, relates to the use of an embedded mask technology [18]. Using thick positive resist structures and electro-deposition techniques, as indicated in Figure 4, it is now possible to realise the hollow metallic cavities needed for high performance integrated Terahertz waveguide components. Until now, however, such positive resist techniques have been restricted to structures of approximately 100μm in height, which in turn has limited the lowest frequency attainable using this approach. The overall limiting factor for positive resist is the depth to which Ultra-Violet radiation can penetrate and hence expose the resist. The technique described by Alderman and co-workers [18] removes this restriction by making use of cyclic exposure and development. Realignment of the mask to a pattern between exposures is avoided with the inclusion of an embedded photomask within the resist itself.

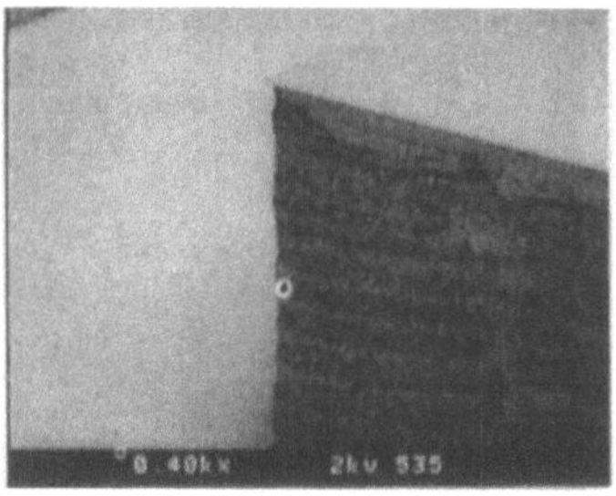

Figure 5: Edge detail of Micromachined wall, made by positive resist.

Near-vertical sidewalls and minimal undercut can be achieved using this approach with heights of up to 250μm, as shown in Figure 5.

4. Enabling Technologies: Field Emitter Sources

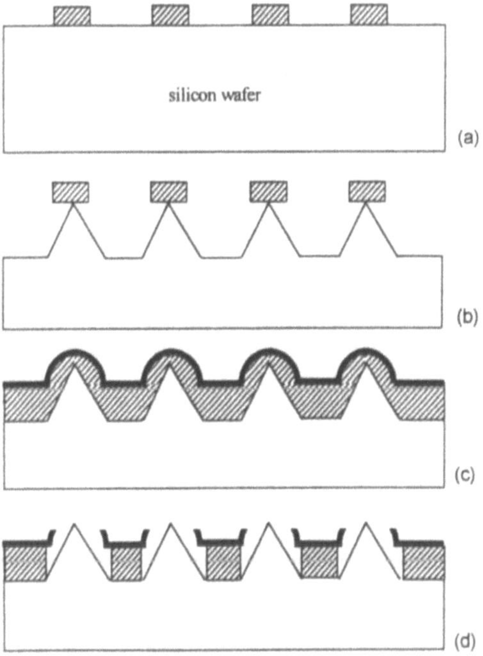

Figure 6: Schematic of Tip Production

Field Emitter (FE) Sources (otherwise known as "Cold Cathode Sources") were originally developed [19] for use in display systems . FE sources do not use thermionic cathodes, but rather rely on field emission from an array of "microtips", usually fabricated from materials such as silicon or molybdenum. Very high electric fields are present in the vicinity of the tip structure, which is less than 1 μm in size. Field emission from the surface occurs and high current densities can be achieved. A significant factor in favour of FE Emitters is that they require a relatively small voltage (of order 100V) to produce emission: furthermore, the quality of the beam structure implies that only minimal focussing arrangements are required. Since the tips are produced by a lithographic procedure, it is possible to incorporate on-chip-focussing elements of various types if these are required. The lifetime of tips was, until recently, problematic. Fortunately, recent developments in tip technology have enabled lifetimes exceeding 1000 hours of continuous operation to be achieved. Figure 6 shows a schematic of the tip production process: in (a) oxide is patterned onto the silicon substrate, followed (b) by plasma etching of the tip structures themselves. In (c), gate oxide and gate metal deposition take place, and the final stages of gate etching and aperture definition are depicted. Figures 7, 8 and 9 show: an array of single crystal silicon Field Emitters with integrated gates; a magnified

Figure 7: Array of Single Crystal Si FE tips with integrated gates.

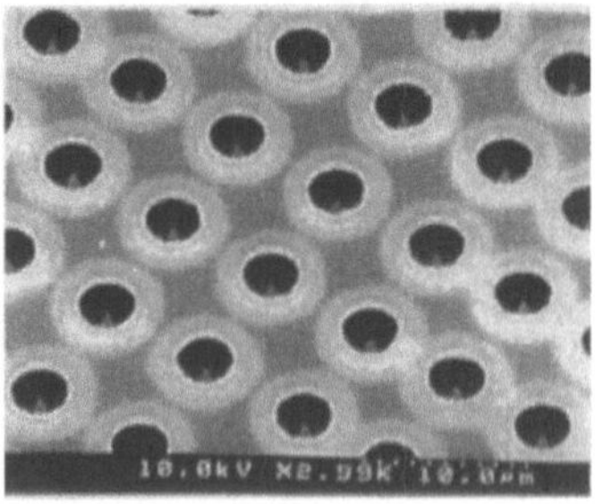

Figure 8: SEM image of the array in Fig 7.

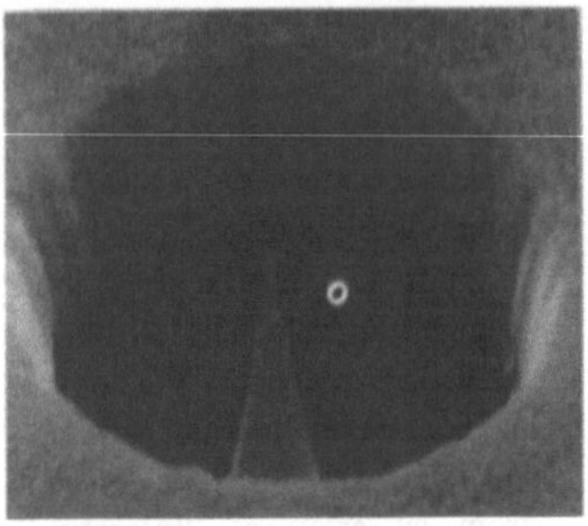

Figure 9: Individual Tip.

SEM of the array indicating the high quality Emitters (tip radius <10nm) and a single Emitter tip with a 3 micrometer diameter aperture. With current practice, a tip density of 2×10^5 mm^{-2} can be achieved for "ungated" arrays (i.e. those without additional electrode structures for focussing purposes), whereas gated arrays can be produced with a tip density of 3×10^4 mm^{-2} [20]. The current that each tip can deliver is of the order 1μA, and recent studies have demonstrated that the current noise is exceptionally low [21]. Another notable feature of the FE array is that current can be rapidly turned on and off, so that they are ideally suited for incorporation in applications requiring pulsed operation. The lifetime of such FE tip arrays can, as noted above, exceed 1000 hours: a significant factor in the achievement of this lifetime has been that they operate in extremely good vacuum conditions e.g. 10^{-6} Torr. Although proprietary vacuum encapsulation techniques are now available to enable this vacuum to be attained, it is worth noting that a very recent investigation of the suitability of FE sources (for incorporation into a mass spectrometer system) indicated that FE sources would operate satisfactorily in a much lower vacuum (10^{-3} Torr) environment.

5. Outline Design of Reflex Klystron for Terahertz Operation

The above two enabling technologies can be combined to realise a Reflex Klystron operating up to Terahertz Frequencies. The heart of the proposed design is a re-entrant resonant cavity, as shown in Figure 10. The shaded areas represent grids, through which the electron beam passes. In order to ensure that the electron beam passing through the grids traverses the gap between them in less than one-half period, δ is chosen to be approximately one-hundredth of a wavelength. Consideration of an elementary argument based on the energy stored within the cavity leads to the approximate expression for the resonant wavelengths:

$$\lambda^2 \approx 2\pi^2 (\alpha^2 \gamma / \delta) \ln. (\beta / \alpha) \quad (1)$$

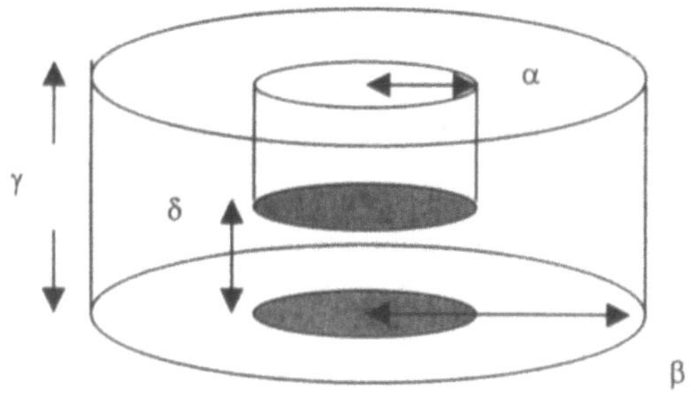

Figure10 : Re-entrant Cavity

In this expression: λ is the resonant wavelength; α, β, γ and δ are the cavity dimensions as indicated in the figure. Table 1 gives illustrative examples of the resonant dimensions (in micrometres) for operation at three chosen frequencies.

Table I: Re-entrant cavity dimensions.

Frequency	α	β	γ	δ
100 GHz	470	616	200	30
500 GHz	75	150	30	5
1 THz	47	62	20	2.2

From this table, it is evident that micromachining is an entirely appropriate technology to realise such a device. Components made by this technology have an exceptional wall quality; for the cavities proposed here, this fact will ensure a high Q factor, which is necessary for device operation.

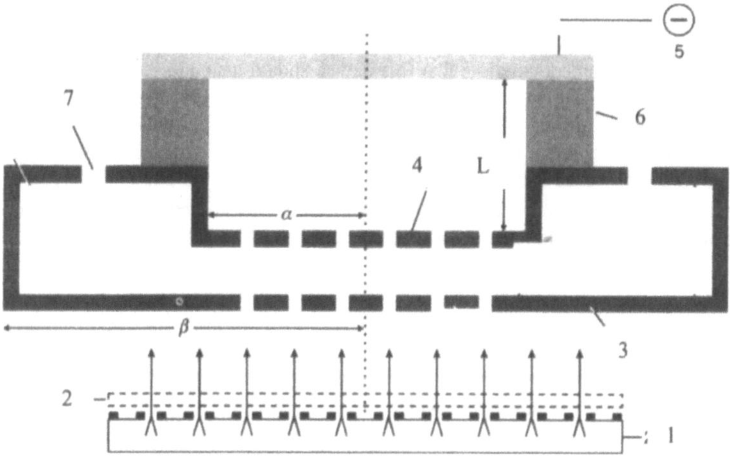

Figure 11: Terahertz frequency Klystron.

Figure 11 shows an outline of the proposed device for operation at 500GHz. A cold cathode array (1) of diamond-coated silicon or molybdenum tips, of typical size 100 μm x 100 μm, generates the electron beam. This may be focussed, if required, using (2) a conventional miniaturised electron optical lens assembly which can be processed onto the Field Emitter array during fabrication. The cavity (3) can be made by a sequence of processing steps, as discussed below, or by similar means. In this instance it consists of a grid arrangement (4) through which the electron beam passes. The beam emerges from the opposite face of the re-entrant cavity and is reflected by the reflector (5), which can be a block of gold plated photoresist. The reflector is maintained at a potential (V_R) of around 100V. The reflector is separated from the cavity, by a distance L, using an insulator block (6) also fabricated from photoresist material. Power is extracted from the slot antenna (7). The entire device is placed within a vacuum encapsulation (not shown), through which Terahertz radiation easily passes.

As an order of magnitude estimate, consider a cavity through which an electron beam of 75 μm radius passes. Present cold cathode technology can easily provide structures with a tip density of 3×10^4 tips mm^{-2} (for gated arrays) or 2×10^5 tips mm^{-2} for ungated arrays; so that approximately 500 – 3500 tips are in line with the hole in the cavity through which the beam passes. Since each tip provides about 1μA of current with a potential of 100V, a beam power of 50 – 350mW may be estimated for cw operation. The efficiency of conventional Klystrons is 5 – 10%, which indicates a THz output power of a few mW in favourable circumstances. Further consideration of scaling effects will be given below.

6. Scaling for Terahertz Operation

Let us now consider several issues related to the scaling of widely-used formulae for Klystron operation. This discussion will indicate the overall problems that are likely to arise in the design of such a source, even though more sophisticated design tools will be used in due course.

Consider, first, the roll-off of Klystron power with frequency. As indicated above, a simple numerical argument suggests that this device should generate several milliwatts of power at frequencies of up to 1 THz. This finding is broadly in agreement with state-of-the art output powers of around 100W for Klystrons [22] operating at 100GHz.

The scaling argument is now given for the two-cavity amplifier of Figure 2, but also applies to the proposed Reflex Klystron. If "well bunched", the pulsating current at the second cavity contains an AC component with amplitude ≈ average beam current. Since the beam current ∝ Grid Area i.e. $\propto \alpha^2$ (refer to Figure 10 or 11), and:

$$\lambda^2 \approx 2\pi^2 (\alpha^2 \gamma / \delta) \ln. (\beta / \alpha) \ldots\ldots\ldots\ldots (1),$$

then :

$$\text{Beam current} \propto \lambda^2 \; \ldots\ldots\ldots\ldots\ldots\ldots\ldots (2)$$

Now the RF power at the cathode $\approx I^2 R$, where:

$$R = Q / \omega C \ldots\ldots\ldots\ldots\ldots\ldots\ldots\ldots\ldots\ldots (3)$$

In this equation: R is the cavity load; C is the capacity; Q is the quality factor of the cavity and ω the angular frequency. This is somewhat of an over-simplification, but will suffice at present. Now: C ∝ (dimension) i.e. $\propto \lambda$, and $Q \propto \lambda^{1/2}$. (This latter statement comes from a consideration of the definition of Q, and the proportionality of skin depth on wavelength.) Taking all of these factors together, $R \propto \lambda^{1/2}$. Finally, combining this with (2), one has that the output power of a Klystron falls off as the inverse four and one half power of frequency, ie Power $\propto \lambda^{4 \cdot 5}$. Although this roll-off is extremely strong, the relatively large amount of power available at 100GHz indicates that 1mW at 1THz operation is still feasible.

Now consider how the reflector spacing, L, may be determined. It will be noted that it is once again appropriate to construct elements on this length scale with micromachining technology. The reflector-cavity spacing, L, is determined by a "best bunching" condition. Electrons will leave the cavity on their outward path to the reflector, and return. Returning electrons give up power to the cavity:

$$P_{in} = (g^2 I^2 V_0/T) \int_0^T \cos \omega(t + s(t))\, dt \;..........(4)$$

Where g is a loss factor relating to a single passage through the grids and V_0 is the amplitude of the oscillating voltage at the grid. The round-trip time , s(t), is determined by simple dynamics. A full treatment [23], which involves the maximisation of P_{in}, then leads to the expression:

$$m\omega\bar{u} \approx e\, V_R/ L \;....................................(5),$$

where $\bar{u}$ is the velocity modulation impressed on the beam by the field modulation and m the electron mass. Equation (5) is, essentially, the condition for optimum bunching. By manipulation of simple expressions involving: the grid capacitance; the electron velocity on leaving the source (u_0); the cavity Q and the cavity power (P); and the resonant cavity equation (1), one may finally obtain:

$$L = [\, V_R\, u_0 / \omega]\, [2PQ\, Z_0\, (\gamma/\lambda)\, \ln(\beta/\alpha)]^{-1/2} .....(6).$$

In particular, consider operation at 500GHz , assuming: P= 1mW, Q= 10^3, $\gamma/\lambda = 1/6$, V_R= 100V and $u_0 = 2 \times 10^6$ m/s. The equation (6) yields L = 30 µm . As stated above, this can easily be achieved by micromachining techniques as discussed below.

One particularly important issue is to establish that the starting conditions for the Klystron are met. In practice, this means that the power input (P_{in}) exceeds the power loss in the cavity. This latter quantity, P_{loss} may be approximately written as:

$$P_{loss} = \omega\, (E_{cav})/ Q_L \;................................(7),$$

where E_{cav} is the energy stored in the cavity with a loaded Q factor (Q_L). A full integration of-the P_{in} term leads to the expression that the starting current, I_{start}, should be:

$$I_{start} > (2CV_c) / g^2\, s_0\, Q_L \;..........................(8).$$

In this expression, V_c is the DC cavity potential with respect to the source and s_0 is the average round trip delay :

$$s_0 = [\, 2L/V_R]\, [2mV_C/ e\,]^{1/2} \;......................(9)$$

For typical operating conditions at 500GHz: $s_0 = 2\text{x}\, 10^{11}$ s, V_C = 100V, g =50%, C= 5 x 10^{-15} F, $Q_L = 10^3$. This yields the condition that $I_{start} > 150$ µ A, and since the 75 µm radius array of Emitters can supply 400 µ A, the above condition is satisfied and oscillation will take place.

The final matter for consideration is the extent of space-charge effects in the proposed structure. It is usual to characterise the effect of space charge in terms of a dimensionless parameter ξ, written in terms of the current density (J), e/m, potential across system (V), and an overall length scale (ℓ).

$$\xi := J\,[\ell^2] / [\,\varepsilon_0\,(e/m)^{1/2}\, V^{3/2}\,] \qquad \text{(10)}$$

If ξ: $\ll 1$, space charge has little effect; if $\xi \gg 1$ space charge dominates. Estimates for a cold cathode are: V = 100V, J = 10^5 A m^{-2}, ℓ = 30 μm , giving $\xi \approx 0.25$. In comparison, for a conventional cathode: V = 1000V, J = 10^4 A m^{-2}, ℓ = 10mm and ξ: ≈ 100. It may be safely concluded that space charge effects are likely to be comparatively small in structures of this type.

7. Major Components of the Proposed Structure: Cavity and Electron Beam Source

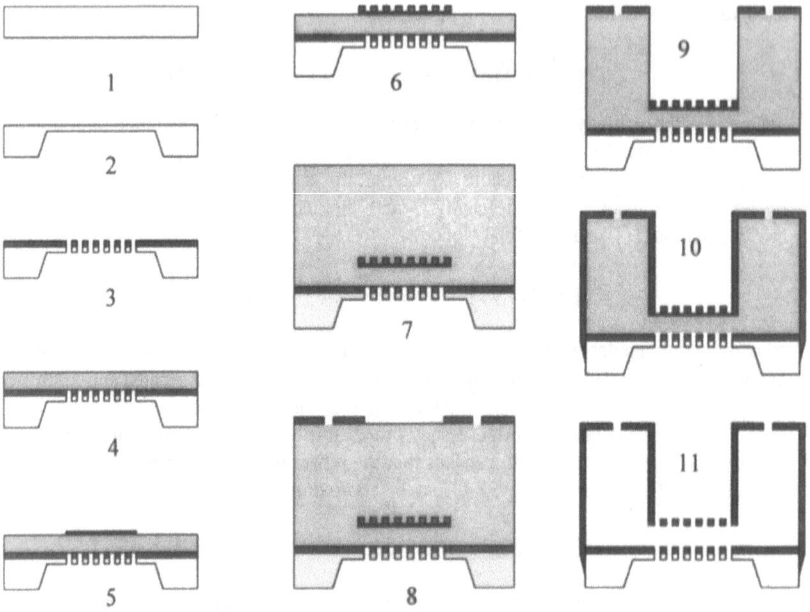

Figure 12: Fabrication steps for Klystron cavity.

Steps in the fabrication of the cavity are shown in Figure 12. It should be emphasised that this is not the only route to the manufacture of this key component, but merely indicates one possible route. Referring first to the left-hand panel: in step 1, a substrate (e.g. a silicon wafer) is taken and (2) is etched as indicated. Using conventional photoresist (3) technology involving masking, etching and evaporation (titanium/gold), the base and lower grid are formed. In (4), a resist layer is spun on to provide the desired separation (δ). In (5), a sacrificial layer of aluminium is used to define the limits of etching at a later stage. In the central panel, (6) shows the same technology steps as in (3), but re-applied to define the upper of the grids. Further strength can be given to this structure, if necessary, by electroplating. In (7), a thick layer of resist is deposited in order to define the final cavity shape: it will be noted that this technology step involves the developments introduced by Alderman *et al.* [18]. The final picture in this panel (8) shows how the top of the torus and the shapes of output antenna are formed using the technology steps of (3) and (6). In the right hand panel, the final steps towards the realisation of the cavity are presented. In (9), the cyclic expose/develop sequence [18] defines the torus walls with great accuracy. The walls are then built up with a multiple source evaporation, in (10) a, and further electroplating enhances the wall thickness. Finally, (11) shows the finished product, which results from a sequence of washing (to remove resist) and etching (to remove the sacrificial aluminium).

As discussed earlier, the source of electrons is an array of Field Emitter tips. The ability of the array to deliver the required current has been demonstrated, as has the longevity of the tips. Although the character of the beam will differ for the two cases of thermionic and Field Emitter source, it is customary to define a quantity known as the *perveance*, the ratio of current to the three-halves power of voltage. In the case of the Field Emitter beam this is approximately 1.0 micropervs, which is a very typical value for a conventional c.w. Klystron operating at microwave frequency. This means, in effect, that scaling can be carried out on all designs for Terahertz use. Although the Field Emitter array will be positioned just a few tens of microns from the bottom grid of the cavity, it is possible to include focussing elements on the array and these can be used to further concentrate the beam if necessary. It is known that beams from Field Effect emitters exhibit far less energy spread than do beams from thermionic sources, and hence the design of such focussing elements should be relatively straightforward.

8. Some practical problems

This section deals with a number of anticipated practical problems of construction. The first matter relates to alignment of the assembly of source array, cavity and reflector. The overall philosophy is to reduce the amount of manipulation involved with fabrication. One way forward may be to construct (from the *negative* photoresist, SU-8) a framework, which provides alignment and electrical isolation for the reflector. Another specific problem is that of heating: in the case considered here, the electron beam power may be up to 350mW, and heat conduction becomes a problem as much of this power is dumped as heat into the reflector structure. A simple scaling argument suggests: the beam size (and therefore the current) scales with λ^2; the temperature rise $\Delta T \propto P$ x (length/ area); hence, $\Delta T \propto \lambda$. Thus, the problem should be less at higher frequency, but is still likely to be significant due to the nature of the electrical isolation used for the reflector.

The problem of ions returning to the source, and effectively "poisoning" the tips is a familiar one in the design of Klystrons. It is usually overcome by the incorporation of some type of guard-ring assembly and it may be necessary to develop a technology of this type for the devices here proposed. The entire issue of vacuum compatibility has been raised earlier, and although the conventional wisdom is that Field Emitter tips require a very hard vacuum, recent work [21] has shown that operation at 10^{-3} Torr is possible. A particular concern centres on the use of certain photoresists (e.g. SU-8) which are known to be troublesome in vacuum conditions. Further investigation of alternative curing processes is needed to resolve this problem. It is envisaged that the entire assembly will be housed in a vacuum encapsulated unit, with sealing to a hard vacuum undertaken using a proprietary process developed for display applications. Fortunately, the encapsulation material is a plastic and Terahertz radiation easily penetrates it.

The final group of practical problems relates to stability and performance of the Klystron. As remarked earlier, the Field Emitter array current has exceptional stability, and for this reason cold cathode arrays have been used in mass-spectrometer systems. Tuning of the final output can be accomplished via adjustment of the reflector voltage and also by mechanical tuning. It is possible to envisage a piezo-electrically controlled tuning element inserted into the cavity structure to achieve such mechanical tuning. Finally, it has been demonstrated that nanotip arrays can be switched on and off in an exceptionally short time, as might be expected from their size [20]: the possibility of pulsed operation seems therefore perfectly feasible.

9. Conclusion

In this Chapter, the possibility is discussed of using a combination of micromachining and Field Emission technology to realise a Terahertz frequency source such as a Reflex Klystron. To date, no similar work has been reported, although the use of Field effect Emitters in conventional Klystron assemblies, operating at 10GHz, have now been demonstrated [24] . The question which now arises is: what is the wider application of this combination of technologies in the realisation of Terahertz sources? Initially, it would appear that crossed-field sources (e.g. magnetrons) cannot be realised, due to adverse scaling factors and the high field needed to achieve the necessary cyclotron motion at Terahertz frequency. However, some variants of the original Magnetron concept may be feasible [25], and further work on this matter continues. In general, however, the present technique does appear to be applicable to such devices as travelling wave amplifiers, where a delicate slow-wave structure is required, which could again be produced by micromachining techniques using a beam originating from a Field Emitter array.

Acknowledgement:
The authors wish to acknowledge several useful discussions with Ian Milsom , Brian Allen and Mike Wiltshire. The concepts within this article are Patented (UK Patent: 0003060.1) and are under development with assistance from the Engineering & Physical Science Research Council in the UK.

References

1. Principles of Traveling Wave Tubes, A S Gilmour, Artech House,Boston, 1994.
2. Free Electron Lasers , C.A.Brau, Academic Press, San Diego, 1990
3. Power Klystrons Today. M J Smith and G Phillips, Research Studies Press, Taunton, 1994
4. Microwave Valves, C H Dix and W H Aldous, Iliffe, London, 1966
5. *"Fed up with fat tubes"*, B.R.Chalamala and B.E.Gnade, IEEE Spectrum, 42-51, April 1998.
6. *"W-band measurements of 100 mm height micro-machined air-filled rectangular waveguides"*, C. E. Collins, J. W. Digby, R. D. Pollard, R. E. Miles, G. Parkhurst, J. M. Chamberlain, D. P. Steenson and N. Cronin (1997). IEEE International Symposium on Microwave Theory and Techniques, Denver, Colorado, USA. p1439.
7. *"Large scale parallel numerical simulations of the relativistic Klystron Oscillator"* , J J Havranek and G E Sasser, Proceedings of the IEEE 1996 International Conference on Plasma Science (ICOPS), p 166.
8. J.Garcia , Priv. Comm
9. *"3D PIC simulations of the relativistic Klystron Oscillator"*, G E Sasser, L Bowers, J W Luginsland and JJ Watrous , Proceedings of the IEEE 1999 International Conference on Plasma Science (ICOPS), p 328.
10. *"Terahertz Metal pipe Waveguides"*, A.S.Treen and N.J.Cronin, Proceedings of the 18th International Conference on Infra Red and Millimeter Waves, pp 470-471, September 1993.
11. *"Integrated micro-machined antenna for 200 GHz operation"* J. W. Digby, C. E. Collins, B. M. Towlson, L. S. Karatzas, G. M. Parkhurst, J. M. Chamberlain, J. W. Bowen, R. D. Pollard, R. E. Miles, D. P. Steenson, D. A. Brown and N. J. Cronin (1997). IEEE International Symposium on Microwave Theory and Techniques, Denver, Colorado, USA.561-564.
12. *"Active micromachined integrated Terahertz circuits."* H. Kazemi, S. T. G. Wootton, N. J. Cronin, S. R. Davies, R. E. Miles, R. D. Pollard, J. M. Chamberlain, D. P. Steenson and J. W. Bowen, *International Journal of Infrared and Millimeter Waves* **20** 967-974 , (1999)
13. *"Construction of an integrated downconverter for operation at 200GHz"* J.W.Digby, Thesis, University of Nottingham, 1999 (unpublished).
14. *"Terahertz Technology, a new era for waveguide?"*, C M Mann, Proceedings of the 23rd International Conference on Infra Red and Millimeter Waves, Colchester, September 1998. 14.-16
15. *"Micromachining applications of a high resolution ultrathick photoresist"*, K Y Lee, N LaBlanca, S A Rishton, S Zolghamain, J D Gelorme, J Shaw and T H-P Chang. J Vac Sci Technol (B) **13** 3012- 3016 (1995).

16. *"A new micro-machined millimeter-wave and terahertz snap-together rectangular waveguide technology."* C. E. Collins, R. E. Miles, J. W. Digby, G. M. Parkhurst, R. D. Pollard, J. M. Chamberlain, D. P. Steenson, N. J. Cronin, S. R. Davies and J. W. Bowen, *IEEE Microwave and Guided Wave Letters* 9 63-65 (1999).
17. C.M.Mann, Priv.Comm.
18. *"An embedded mask for application in high frequency integrated circuits using UV positive resist technology"*, B E J Alderman, C M Mann, M L Oldfield and J M Chamberlain, J.Micromech.. Microeng. **10** 334-336 (2000).
19. *"Field Effect Emitters for Display Systems"*, D Peters, I Paulus, and D Stephani, J Vac Sci Technol. B12 (2) 652 –656 (1992)
20. *"Fabrication and Characterisation of ultra sharp silicon field emitters"*, S E Huq, G H Grayer, S W Moon and P D Prewett, Materials Science & Engineering B51 150-153 (1998)
21. *"The use of microfabricated emitter arrays in a high precision mass spectrometer for the Rosetta mission"*.B J Kent, , E Huq, J N Dominey, A D Morse and N Waltham, (1999) 3rd Round Table on Micro/Nano Technologies for Space , 15-19th May 2000, ESTEC Noordwijk, The Netherlands.
22. *"Tubes: still vital after all these years"*, R S Symons, IEEE Spectrum, April 1998, 52-63.
23. J R Fletcher, J M Chamberlain and R E Miles, to be published.
24. *"Field Emitter Arrays for plasma and microwave applications"* K L Jensen, Physics of Plasmas 6 2241-2253 (1999)
25. J R Fletcher, Priv Comm.

Theme 6

Imaging

TWO-DIMENSIONAL TERAHERTZ WAVE IMAGING

Q. CHEN, ZHIPING JIANG and X.-C. ZHANG*
Physics Department
Rensselaer Polytechnic Institute, Troy, NY 12180, USA
**zhangxc@rpi.edu*

1. Introduction

The terahertz, or far-infrared, region of the electromagnetic spectrum is an important region for the spectroscopy of condensed matter, vibronic spectroscopy of polar liquids and gas phase analysis. In the past, terahertz spectroscopy was hindered by the low brightness of incoherent far-infrared sources and the limited sensitivity of bolometeric detectors. However, these difficulties are completely overcome with the advent of terahertz time-domain spectroscopy (THz-TDS). THz-TDS is based on electromagnetic transients generated opto-electrically by the femtosecond laser pulses. These terahertz transients are single-cycle bursts of electromagnetic radiation of typically less than 1-ps time duration. Their spectral range spans from below 100 GHz to more than 5 THz. The brightness of the THz transients exceed that of conventional thermal sources, and the gated detection technique adopted in THz-TDS is, in magnitude, more sensitive than conventional bolometric detection. The gated detection is coherent – both the amplitude and phase of the THz waveform can be measured. In general, many chemical compounds show very strong, highly specific frequency-dependent absorption and dispersion in the THz range, which leads to characteristic time-domain waveforms when THz wave transmits through different materials. Accordingly, highly sensitive imaging of chemical compositions in THz range is possible. The value of a practical far-infrared imaging system has been recognized in a wide range of applications for some time. A THz imaging system based on the TDS technique excludes any cryogenics or shielding for the detector; it has the potential to be portable, compact, and reliable far-infrared imaging system, which shall find its practical application in "real-world" environments.

2. Terahertz Imaging

The first two-dimensional scanning THz imaging system has been demonstrated by Hu and Nuss [1]. In this system, it is possible to obtain between 10 and 100 spectra per second with a Signal-to-Noise ratio greater than 1000:1. The data process is based on a Digital Signal Processor (DSP) card, available for a personal computer, which performs acquisitions and A/D conversions at a rate of >50k Samples/s without the use of a lock-in amplifier. Fast Fourier Transforms (FFT) or other analysis of the acquired time domain waveforms are accomplished in real-time by using the on-board DSP.

R.E. Miles et al. (eds.), Terahertz Sources and Systems, 225–239.

Figure 1 shows a schematic of the setup, which is similar to the typical THz spectrometer. It consists of a femtosecond laser, a computer-controlled fast optical scanning delay line, an optically gated terahertz photoconducting dipole emitter, a set of off-axis parabolic mirrors for collimating and focusing the terahertz beam and an optically-gated terahertz photoconducting dipole receiver. The sample can be mounted on a translation stage and scanned two dimensionally during the data acquisition. The fast scanning delay line allows for the acquisition and processing of up to 100Hz waveforms per second. The achieved spatial resolution was roughly 300 μm, close to the diffraction limit for 1 THz radiation.

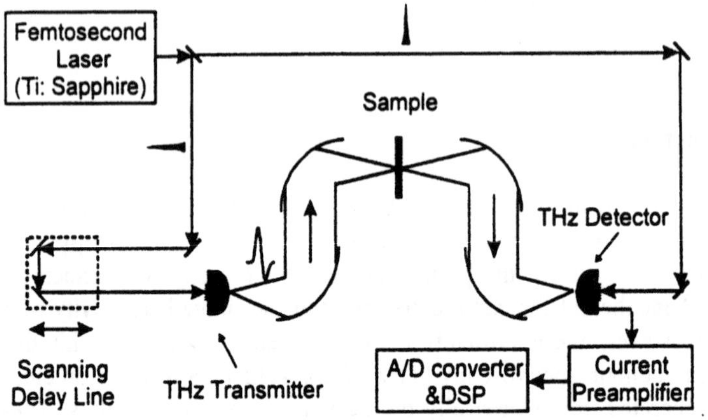

Figure 1. Schematic of experiment setup for 2D scanning THz transmission imaging.

One can place the object to be imaged at the focus of the THz beam and measure the transmitted waveform for each position of the object. A THz image can be formed by any desired aspect of the measured waveforms including amplitude, phase and any combination.

The application of the THz imaging is quite diverse [1-3]. The imaging system can detect and map out chemical compositions within an object, this has been demonstrated by the THz imaging of the water content of a leaf. The THz image of a packaged semiconductor integrated circuit shows its application in industrial inspection. Perhaps its most promising practical application lies in the area of quality control, which can be shown by the THz imaging of the voids inside solid plastic or plastic composite parts. Its potential application in the food industry can be seen by the imaging of an alien object inside the powder or flour [4]. We may also find its application in the characterization of a semiconductor wafer and the inspection of artwork. The three dimensional THz imaging (T-ray tomography) has been realized in a reflection geometry, which shows its possible application in the area of medical diagnostics as well.

3. Terahertz Imaging with EO Sampling Technique

a) An All-optical Scanning Terahertz Imaging System

In the above THz imaging system, photoconducting antennas are used as the source and detector of THz radiation. The typical peak frequency of the measured THz radiation lies at ~ 0.5 THz, which is associated with the lifetime of photocarriers and the antenna structure. The measurable THz bandwidth and spatial resolution of the above THz imaging system are limited. A desirable THz source is one that can emit THz radiation with high peak frequency and a THz sensor with a faster response that also offers a large measurable THz bandwidth. The electro-optic crystals are good candidates because of their instantaneous optical responses and THz radiation with a higher peak frequency, produced in the EO crystals by the difference-frequency mixing [5]. With the advent of the EO sampling technique, THz radiation with a frequency range from sub-THz to 37 THz has been detected [6]. Recently, this record has been extended to 70 THz [7].

An all-optical scanning THz imaging system can be implemented with electro-optic crystals as a THz source and sensor; the THz imaging system can definitely extend its application to mid-infrared range and the diffraction limited spatial resolution can be greatly improved [8-9]. With a 12 fs laser and thin EO crystals (on the order of 20 μm) as the THz source and sensor, a mid-infrared image of an array of holes of 50 μm has been realized, as shown in Fig. 2. The spatial resolution can be estimated ~ 10 μm.

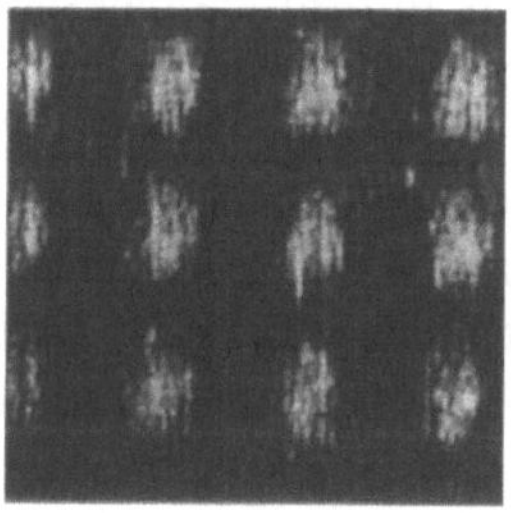

Figure 2. Mid-infrared imaging of an array of holes of 50 μm diameters.

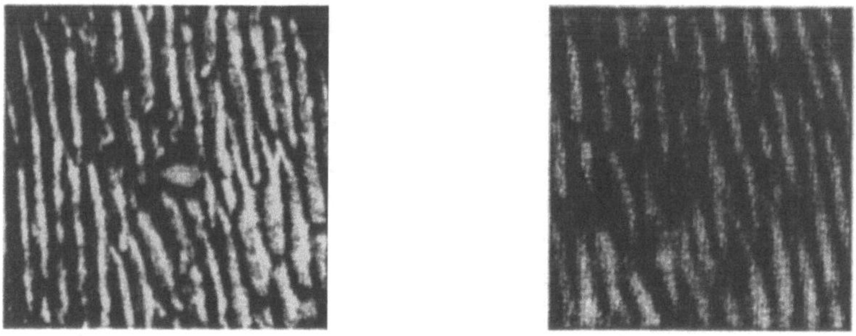

Figure 3. Images of an onion cell by optical and THz imaging.

With the increased spatial resolution, it is possible to image the onion cell using mid-infrared radiation. Fig. 3 shows the optical and THz images of an onion membrane. The contrast in these THz images is mainly the result of moisture distribution in and between the cells, leading to different amounts of absorption of the THz beam. Note that the reflected optical light forms the optical image whereas the THz image is formed by the transmitted THz radiation. These two images are similar but not identical which indicates the THz image may contain different information about the object than the optical image.

b) Near-field Terahertz Imaging with a Dynamic Aperture

According to Abbe's law, the spatial resolution that can be achieved when imaging with electromagnetic waves is limited by the wavelength of the employed radiation. In the above all-optical THz imaging system, the improvement of the system spatial resolution is only due to the involved mid-infrared radiation. However, such a diffraction limit to spatial resolution is not fundamental but rather arises from the assumption that the light source is typically many wavelengths away from the sample of interest. With the lateral scanning of a source light in close proximity to a sample, one can generate an image at a resolution functionally dependent on only the source size and the source-to-sample separation, each of which can, in principle, be made much smaller than the wavelength of the employed radiation [10]. Compared with the visible light, the above requirements are much more easily achieved in THz range. Conventionally, in near-field microscopy, the light incident upon one side of an optically opaque screen is transmitted through a subwavelength-diameter aperture to realize a tiny source. A tapered fiber could be the best choice for the near-field image in the visible light.

The concept of near-field microscopy has been adopted to improve the diffraction-limited spatial resolution in scanning THz imaging systems [11] in which the peak THz frequency is generally 0.5 THz. The tiny THz source necessary for the near-field image was realized by a tapered metal tube with nearly circular aperture; the diameter of the aperture was smaller than 100 μm. The achieved sub-wavelength spatial resolution was demonstrated to be better than $\lambda/4$. The main disadvantage of this near-field imaging is the high-pass filtering of the THz signal, due to the waveguide effect of the metal tube, which decreases the THz signal and limits the measurable THz bandwidth.

Another THz near-field imaging technique that does not require the introduction of a physical aperture has been recently demonstrated [12]. In this technique, the near-field THz aperture is realized by a transient photocarrier layer induced by an optical gating beam whose size is decided by the focus of the optical beam and can be easily adjusted. Since the thickness of this photo-carrier layer is determined by the absorption depth of the optical beam on the gating material, that is generally at the order of 1 μm, the high-pass filtering of the THz signal due to the waveguide effect can be avoided, the measurable THz bandwidth can be maintained.

Figure 4 shows the schematic setup of this near-field THz imaging. The laser source is a Ti: sapphire laser (Coherent RegA 9000) with a 250-fs pulse duration at 800 nm and a repetition rate of 250 kHz. A THz beam with its spectrum peak at 0.9 THz is generated by a pump beam through optical rectification in a 2-mm-thick <110>-cut ZnTe. The main difference between this setup and the conventional THz imaging setup is the introduction of an optical gating beam. This gating beam, which is split off from the same laser beam, is focused by a lens with f = 30 cm and illuminates on a GaAs wafer along with the THz beam. The average power of the gating beam is ~ 65 mW and the timing can be independently controlled by a mechanical delay stage.

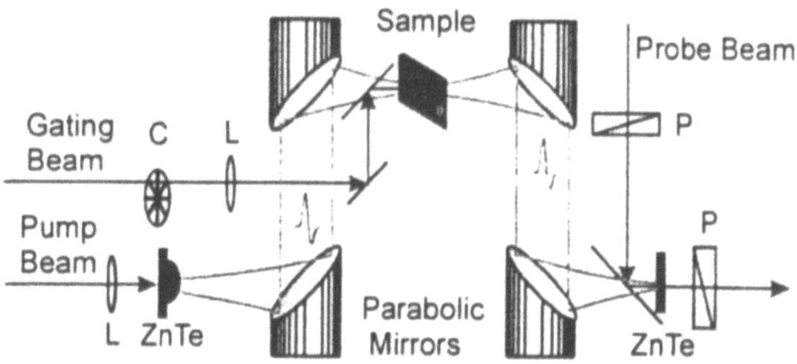

Figure 4. Schematic illustration of the near-field THz imaging with a dynamic aperture.

In a typical optical pump and THz probe experiment, when the THz beam is chopped, the transmission of the THz beam through the sample will show a step-function-like variation with a time delay between the THz and optical pump beam. Such variation of the THz transmission is not quite discernible when the optical beam is tightly focused on the semiconductor wafer because the effective modulation area is too small to have significant modulation of the THz transmission. By chopping the optical gating beam instead of the THz beam, one can obtain a step-function-like variation of the THz transmission with a greatly improved sensitivity.

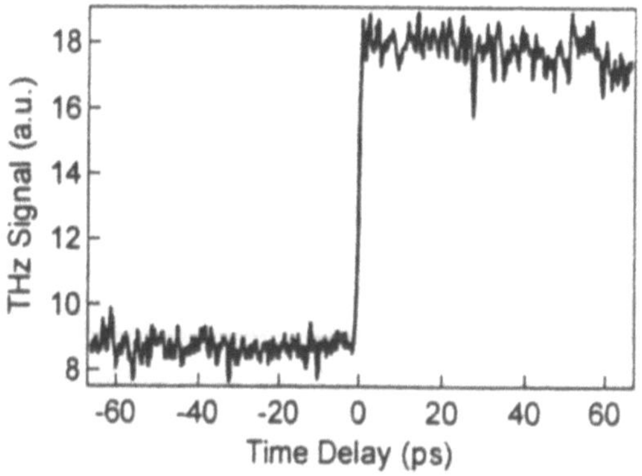

Figure 5. Variation of the THz signal with a time delay between the THz beam and optical gating beam; negative timing means the optical gating pulse arrives later than the THz pulse.

Fig. 5 shows the change of the THz signal with the time delay between the THz pulse and the gating pulse. The measured THz signal is proportional to the fluctuation of a THz transmission during the on and off states of the optical pump beam that is passing the mechanical chopper. When the optical gating pulse arrives at the GaAs wafer earlier than the THz pulse the increased local GaAs wafer conductivity, induced by the photocarriers, decreases the THz transmission in the area decided by the focus size of the gating beam. The increase in conductivity is due to the electron intervalley scattering within the first several picoseconds, which can last for a time scale of 100 ps until the electrons and the holes diffuse into the sample and eventually recombine [13].

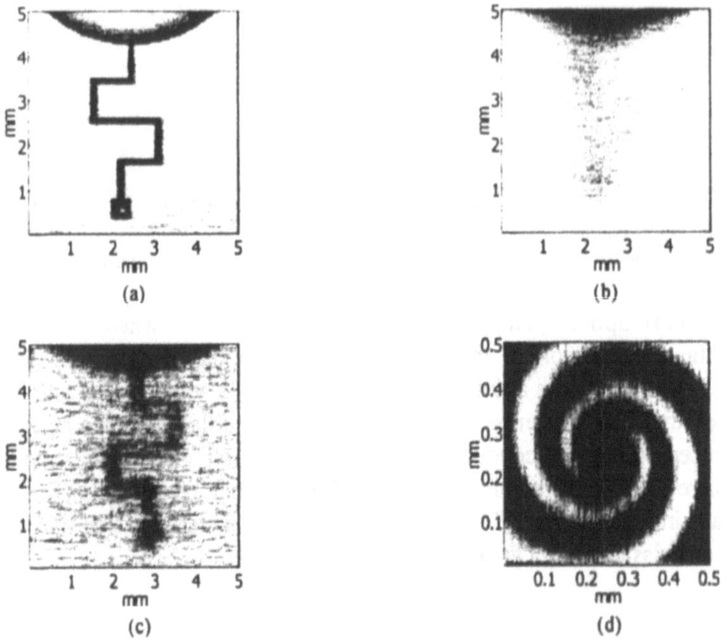

Figure 6. THz images of some metal circuits deposited on a GaAs wafer by: (a) near-field technique with a dynamic aperture, (b) the conventional THz imaging technique, (c) near-field technique with a dynamic aperture (the sample was flipped), and (d) near-field technique with a dynamic aperture (image of a THz antenna, image size: $0.5 \times 0.5\ mm^2$).

When the optical gating beam is incident on the metal line deposited on a semiconductor wafer, the THz signal detected in the above described way will not exist because of the disappearance of the modulation to the THz transmission. Fig. 6(a) shows a THz image of a simple metal circuit, deposited on a semi-insulating GaAs wafer, obtained in the scanning imaging setup with a dynamic aperture. The optical gating pulse was focused by a lens with f = 15 cm and the pulse's timing was set so that it arrived at the sample about 50 ps earlier than the THz pulse. In comparison, this circuit was also imaged by the use of the conventional THz imaging technique in which the optical gating beam was blocked and the THz beam was modulated, as shown in Fig. 6(b). From Figs. 6(a) and 6(b) it is clear that the spatial resolution and the image contrast have been greatly improved by the introduction of a dynamic aperture.

The transient photocarrier layer, excited by the optical gating beam, serves as a near-field aperture or, more precisely, a filter, which partially blocks the THz transmission in an area much smaller than that of the THz beam focal spot. Since the dynamic aperture and the metal circuits are directly on the sample surface, THz propagation is not involved and the spatial resolution of the THz image is thus solely decided by the focus size of the optical gating beam. To test the effect of THz propagation, we flipped the circuit sample and obtained a THz image of the circuit on the backside of the wafer as shown in Fig. 6(c). Since the wafer thickness is ~ 0.4 mm, the circuit pattern was then 0.4 mm from the dynamic aperture. The THz beam propagated this distance to reach the pattern, and the spatial resolution and contrast were relatively reduced, clearly showing the diffraction effect. As shown in Fig. 6(d), we also imaged the metal pattern of a THz antenna, whose size is $0.5\,\text{mm} \times 0.5\,\text{mm}$. The result again indicates the great improvement on the spatial resolution.

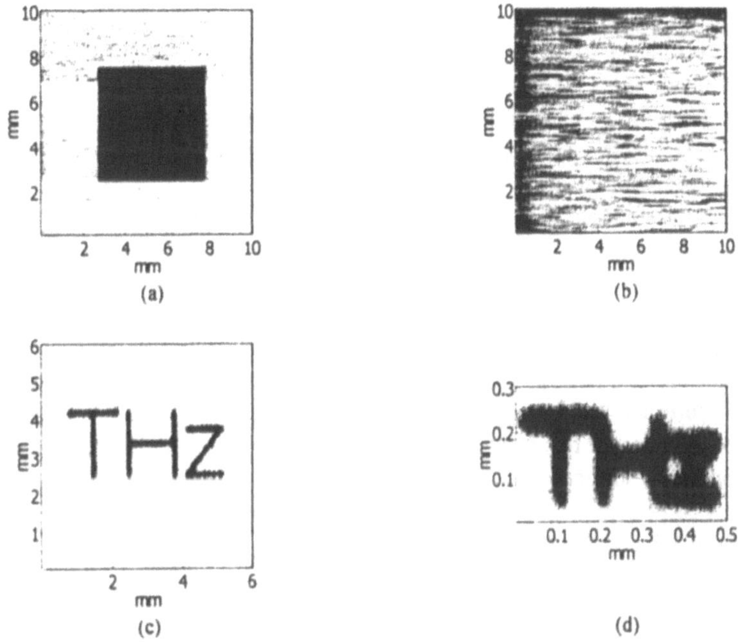

Figure 7. THz images of some optical damaged area on a Silicon wafer (a) by near-field technique with a dynamic aperture. (b) by the conventional THz imaging technique.(c) by near-field technique with a dynamic.(d) by near-field technique with a dynamic aperture. The spatial resolution is better than 50 μm.

In addition to the image the metal patterns deposited on a semiconductor wafer, the near-field THz imaging with a dynamic aperture is able to image the optically damaged area on a semiconductor surface, which is not readily achieved by the conventional THz imaging and the near-field THz image with a physical aperture. As shown in Fig. 7(a), an optically damaged area on a silicon wafer has been successfully imaged by the introduction of a dynamic aperture. The same area was also imaged by the conventional way (Fig. 7(b)), which could not show any difference in the THz transmission between the damaged and undamaged areas. The conventional way is not sensitive enough to pick up such a small change on surface quality. However, the difference can be shown by the introduction of a dynamic aperture on the surface because the modulation depth of the dynamic aperture to the THz transmission changes from the damaged to the undamaged areas, which is mainly because of the photocarrier lifetime's difference in these two areas. Therefore, this technique has the potential to image the surface quality of a semiconductor wafer, such as the measurement of doping levels and types, with a spatial resolution limited by the near-infrared wave and the near-field effect.

To estimate the aperture-size-defined spatial resolution, a **THz** word was carved on the surface of a silicon wafer by optically damaging it. The total size of these words were within an area of $6\ \text{mm} \times 6\ \text{mm}$ and $0.3\ \text{mm} \times 0.5\ \text{mm}$ respectively. Fig. 7(c) and (d) show the THz image of the word **THz**. The estimated spatial resolution is better than 50 μm. The spatial resolution is limited by the signal-to-noise ratio of the system and the spot size of the gating beam on the sample. The decrease in the spot size of the gating beam on the sample can increase the spatial resolution, but it can also decrease the modulation of the THz beam, leading to a worse signal-to-noise ratio. There is a trade-off between the image resolution and the signal-to-noise ratio. Since the measured quantity is THz electric field, the THz signal is proportional to the diameter of the gating beam instead of the area.

As for another adopted near-field technique in the research reported in Ref. 14, the THz generation is limited to the mechanism of optical rectification and the spatial resolution is determined by the focal size of the optical pump beam as well as the THz diffraction through the electro-optic crystal. The spot size of the optical beam cannot be too small, otherwise two-photon absorption will limit the generation efficiency through the saturation effect [14-15]. The dynamic aperture technique excludes the use of a physical aperture and can be applied to a THz imaging system with an arbitrary THz source. The limitation of this technique is that the dynamic aperture can be induced only on a semiconductor surface.

(c) Real-time 2D Terahertz Imaging System

In all the above implementation of THz imaging, the sample, rather than the THz beam, is scanned. The imaging process is time consuming, and for example, will limit the application of THz imaging *in situ* study of dynamical processes in biological samples and plasma. Although in principle it is possible to fabricate an antenna array to realize a focal plane imaging and speed up the imaging process, but such an attempt would be hindered by some technical issues such as high optical power consumption and complexities in packaging. With the advent of the electro-optic measurement of THz pulses, the scheme of 2f-2f two-dimensional THz imaging can be implemented to realize real-time imaging [16].

The schematic of the experiment setup shown in Fig. 8 consists of a femtosecond laser, an optically gated terahertz larger aperture antenna, the object to be imaged, a high-density polyethylene lens and a high speed CCD camera. In this setup, instead of scanning the sample, the distribution of THz electrical field of the sample is first mapped on the EO crystal by the polyethylene lens in a 2f-2f imaging scheme and is upconverted to the optical region by an expanded probe beam via the Pockels effect. The corresponding optical image is then recorded by the CCD camera. The data acquisition rate for this imaging technique is thus only limited by the speed of the CCD camera, which could be as high as 70 frames/s. Uniquely, this imaging system has the capability of making single-shot 2D THz imaging. The image of peak field distribution of a THz beam at its focal point has been realized in this scheme at the speed of the CCD camera.

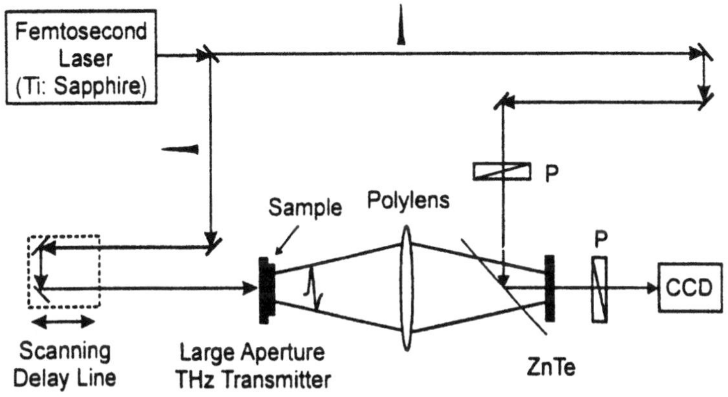

Figure 8. Schematic of the experiment setup for the 2D THz imaging with a CCD camera.

In the above all-optical THz imaging system, the EO modulation in the EO sampling of the THz beam usually works at the linear optical bias point and gives the largest dynamic range. However, in the case of parallel measurement, such as real-time 2D imaging with a CCD camera, the crossed polarizer geometry has to be used to achieve a high signal-to-noise ratio. The EO modulation has to be realized at the near-zero optical bias point. The following analysis will show a practical way of finding the optimum working point at this crossed polarizer geometry [17].

In the crossed polarizer geometry, the intensity of the transmitted probe beam light can be written as

$$I = I_0\left[\eta + (\Gamma_0 + \Gamma)^2\right] \tag{1}$$

where I_0 is the input probe light intensity, η counts the scattering contribution, Γ_0 is the optical bias and Γ counts the THz electric field induced birefringence contribution, and $|\Gamma_0| << 1$ and $|\Gamma| << 1$ holds at a near-zero optical bias point.

The modulation depth γ defined as signal-to-background ratio can be readily obtained as

$$\gamma \equiv \frac{I_{\Gamma\neq 0} - I_{\Gamma=0}}{I_{\Gamma\neq 0} + I_{\Gamma=0}} = \frac{2\Gamma_0\Gamma + \Gamma^2}{2\eta + \Gamma_0^2 + (\Gamma_0 + \Gamma)^2} \tag{2}$$

The optimal working point that gives the maximum modulation depth or signal-to-noise ratio is given by

$$\Gamma_0^m = -\frac{\Gamma}{2} \pm \sqrt{\left(\frac{\Gamma}{2}\right)^2 + \eta} \approx \pm\sqrt{\eta} \tag{3}$$

and the maximum of modulation depth is

$$\gamma_{max} \approx \frac{\Gamma}{2\sqrt{\eta}} \tag{4}$$

The transmitted background light at this optimal optical bias point is given by

$$I_b \approx 2\eta I_0 \tag{5}$$

Fig. 9 shows the experimental and calculation results of the modulation depth versus the optical bias. The strong agreement between the experiment and calculation proves that the model Eq. (1) is valid.

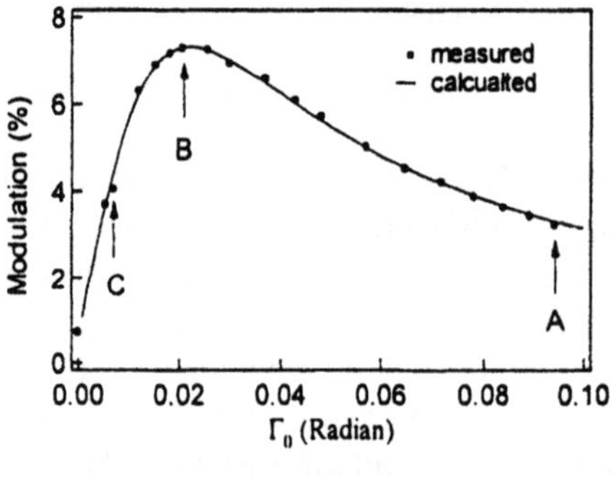

Figure 9. Modulation depth versus the optical bias. B is the optimal working point.

The importance of the above analysis is that it actually indicates a practical procedure to find the optimal working point. One can first adjust the compensator, which is set after the second polarizer, to get the minimum background light intensity and then adjust the compensator to make the background light intensity twice its minimum (as seen by Eq. (5)), one thus finds the optimal working point which offers the largest modulation depth or signal-to-noise ratio.

However, since the lock-in amplifier, which is generally used in the scanning imaging system and can suppress measurement noise dramatically, cannot be used with the CCD camera, the real-time 2D imaging system still suffers a poor signal-to-noise ratio. To overcome this disadvantage, the phase-sensitive detection principle of the lock-in amplifier has been adopted by using a dynamic subtraction technique [18]. The operation principle of the dynamic subtraction technique is shown in Fig. 10. The optical pump beam is modulated by an EO modulator synchronized to the CCD camera with a modulation frequency equaling half of the inverse of the CCD frame rate. This is in order for the THz signal to be alternately be detected by the probe beam and the CCD camera. The difference between two consecutive frames gives the THz image an improved signal-to-noise ratio. More frame accumulation can show results with a higher signal-to-noise ratio since the in-phase THz signal does not change during the averaging process, but the out-phase noise can be averaged out dramatically. The low part of Fig. 10 shows the timing between the CCD camera and EO modulator. Experiments with Coherent RegA-9000 shows that the signal-to-noise ratio improves 2 orders by this dynamic subtraction technique. The detectable fraction modulation can be as small as 5×10^{-5}.

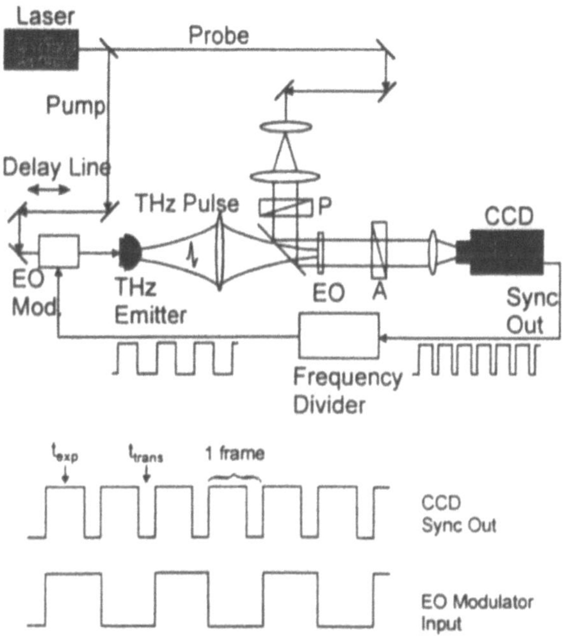

Figure 10. A 2D real-time THz imaging system with a dynamic subtraction technique.

Fig. 11 shows THz signals obtained by the above CCD based imaging system with the technique of dynamic subtraction. The emitter was a 2-mm-thick <110> ZnTe. In comparison to a large, biased aperture photoconductive antenna emitter, the THz radiation from the optical rectification emitter ZnTe is much weaker and the modulation depth on the probe beam is generally several percentages, which is in the same order of the laser fluctuation. Without the dynamic subtraction technique, it would not be possible to detect a THz signal with a good signal-to-noise ratio. Figs. 11(a) ,and (b) show the 2D peak field distribution of THz pulses and the temporal THz waveform, indicating the good signal-to-noise ratio. Figs. 11(c) and (d) are the spatio-temporal distribution of the THz field, when the EO sensor is located at the THz focal point and 2 cm away from the focal point, respectively. The curved wavefront in Fig. 11(d) tells that the THz beam is divergent.

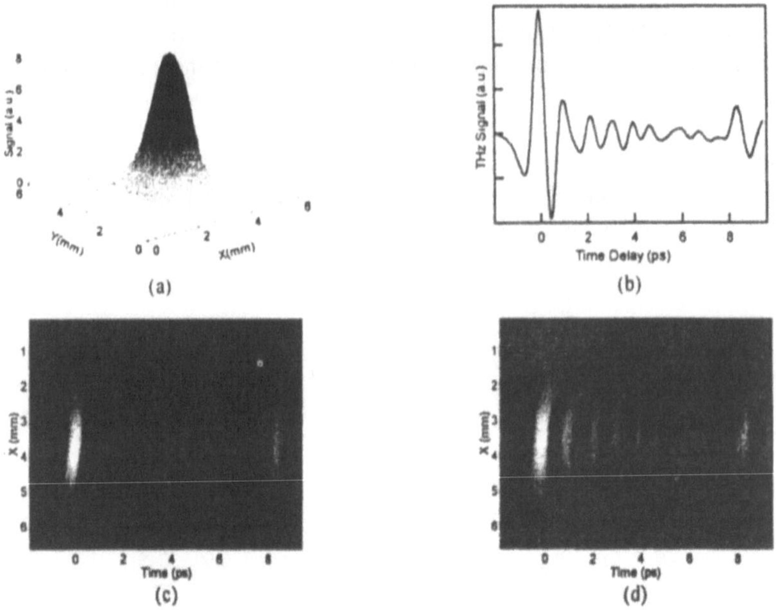

Figure 11. Electric field distribution measured with dynamic subtraction from an optical rectification emitter. (a) distribution of the peak signal. (b) Temporal waveform of the THz beam (c) 1D spatial and 1D temporal distribution of the THz beam at it focal point. (d) 1D spatial and 1D temporal distribution of the THz beam at a position of 2 cm away from its focal point.

The CCD based THz imaging system is a powerful tool in visualizing the spatial profile of the THz beam. The beam quality can be monitored on the computer screen in real-time, which helps to optimize the optical alignment.

(d) Near-field Imaging in Real-time 2D Terahertz Imaging System

The system spatial resolution of this 2f-2f 2D imaging system is also limited by the diffraction effect of the THz beam, because the EO sensor is set at the imaging point of the object that corresponds to the far field of the THz beam. To improve the system spatial resolution, a counter-propagating geometry of the THz and probe beam at the EO sensor has been applied to realize 2D THz near-field imaging [17], as shown in Fig. 12. In this setup, the object to be imaged is set directly in front of the EO sensor. The expanded probe beam reads out the near-field THz electric field distribution. Since the electric field of the THz beam is measured before it diffracts out, the spatial resolution of THz imaging can be improved.

To estimate the possible improvement of spatial resolution by counter-propagating EO sampling geometry in the conventional scanning THz imaging setup, the near-field THz beam size was measured by scanning a razor blade, which was almost in contact with the ZnTe sensor in the counter-propagating EO sampling geometry. The similar measurement was also carried out by scanning the razor blade through the THz focal point. In this case, the measured THz beam size was larger, showing the diffraction effect. The result is shown in Fig. 13. From the distance between the 90% and the 10% of the maximum signal, the spatial resolution is estimated from the near-field and focal point measurements and can be quantified as 0.4 mm(curve a) and 1.2 mm(curve b), respectively. It is seen that the spatial resolution was improved by 3 times.

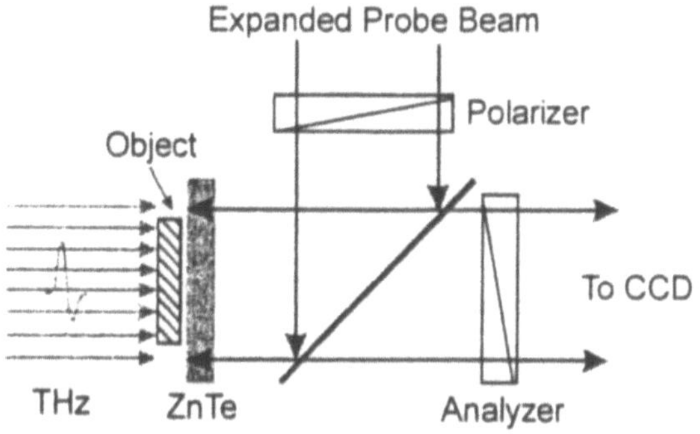

Figure 12. Schematic illustration of 2D THz near-field imaging with a CCD camera.

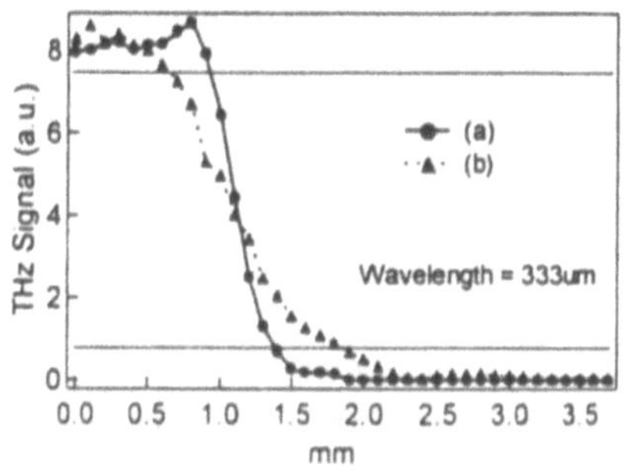

Figure 13. The scan of a razor blade through the THz beam. (a) Razor blade scans through the THz near-field. (b) Razor blade scans through the THz focal point.

It should be emphasized that the above result is from the scanning imaging system with focused THz and probe beams on the EO sensor and a single detector. In principle, the spatial resolution of the scanning imaging system and CCD imaging system should be the same if THz sources and system numerical apertures are the same. However, by taking the aberration and the difference of the SNR into account, the scanning system generally has better spatial resolution. Compared with the 2f-2f 2D imaging system, the spatial resolution of near-field imaging setup (Fig. 12) can be improved to a greater degree, as shown by following experimental results. The metal word, **THz**, was imaged by positioning it directly in front of the ZnTe sensor in the above setup. The image can be displayed on the computer monitor in real-time, as shown in Fig. 14. The metal letters were formed by a metal line with a width of 0.5mm, the size of the word was

$1\text{cm} \times 0.5\text{cm}$. The system spatial resolution can thus be estimated to be at least 0.5 mm. This spatial resolution cannot be realized in the 2f-2f THz imaging system. In the above

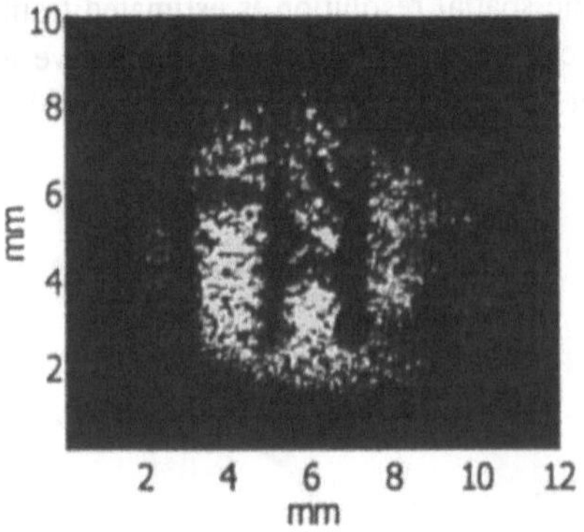

Figure 14. The real-time 2D THz near-field imaging of a word "THz".

experiment, a larger aperture antenna was used as a THz emitter whose peak frequency is 0.3 THz. If such an emitter was used in the 2f-2f imaging system with a 0.5 system numerical aperture, the best system spatial resolution can be estimated as ~ 2.5mm in paraxial condition. Considering the system aberration, it can be reasonably concluded that the spatial resolution is improved by about one order in this near-field CCD imaging setup. The result also shows the capability of this system to image moving target or living objects.

4. Conclusion

We have shown state of the art techniques in all THz imaging systems with the EO sampling technique. The scanning mid-infrared THz imaging system and the near-field THz imaging with a dynamic aperture have greatly improved the system spatial resolution and extended the potential practical applications for THz imaging. With the implementation of the near-field imaging and the dynamic subtraction technique, the performance of real-time 2D THz imaging has been improved, showing its potential in the study of THz beam optics and the realization of real-time THz imaging with better quality.

Acknowledgement

The work is supported in part by the U.S. Army Research Office and the U.S. National Science Foundation.

References

1. Hu, B. B. and Nuss, M. C. (1995) Imaging with terahertz waves, *Opt. Lett.*, **20**, 1716-1718.
2. Mittleman, D., Jacobsen, R. , and Nuss, M. C. (1996) T-Ray imaging, *IEEE J. Sel. Top. Quantum Electron.*, **2**, 679-692

3. Mittleman, D., Gupta, M., Neelamani, R., Baraniuk, R., Rudd, J. and Koch, M. (1999) Recent advances in terahertz imaging", *Appl. Phys. B*, **68**, 1085-1094, and references therein.
4. Herrmann, M., and Sakai, K. (2000) Objects in powders detected and imaged with THz radiation, in *Conference on Lasers and Electro-Optics* (Optical Society of America, Washington, D.C., 2000), 479.
5. Nahata, A., Weling, A., and Heinz, T. (1996) A wideband coherent terahertz spectroscopy system using optical rectification and electro-optic sampling, *Appl. Phys. Lett.*, **69**, 2321-2323.
6. Wu, Q. and Zhang, X.-C. (1997) Free-space electro-optic sampling of mid-infrared pulses, *Appl. Phys. Lett.*, **71**, 1285-1287.
7. Leitenstorfer, A., Hunsche, S., Shah, J., and Nuss, M. (1999) Detectors and sources for ultrabroadband electro-optic sampling: experiment and theory, *Appl. Phys. Lett.*, **74**, 1516-1518.
8. Chen, Q., Jiang, Zhiping, and Zhang, X.-C. (1999) All-optical THz image, *Proceedings of SPIE*, **3617**, 98-103.
9. Han, P., Cho, G. and Zhang, X.-C. (2000) Time-domain transillumination of biological tissues with terahertz pulses, *Opt. Lett.*, **25**, 242-244.
10. Betzig, E. and Trautmann, J. (1992) Near-field optics: microscopy, spectroscopy, and surface modification beyond the diffraction limit, *Science*, **257**, 189-192, and references therein.
11. Hunsche, S., Koch, M., Brener, I., and Nuss, M. C. (1998) THz near-field imaging, *Opt. Comm.*, **150**, 22-26.
12. Chen, Q., Jiang, Z., Xu, G. and Zhang, X.-C. (2000) Near-field THz imaging with a dynamic aperture, *Opt. Lett.* (in press).
13. Greene, B., Sateta, P., Dykaar, D., Schmitt-Rink, S. and Chuang, S. L. (1992) Far-infrared light generation at semiconductor surface and its spectroscopic applications, *IEEE J. Quantum Electron.*, **28**, 2302-2312.
14. Wynne, K. and Jaroszynski, D. (1999) Superluminal terahertz pulses, *Opt. Lett.*, **24**, 25-27.
15. Sun, F.G., Zhang, X.-C., and Ji, W. (2000) Two-photo absorption induced saturation of THz radiation in ZnTe, in *Conference on Lasers and Electro-Optics* (Optical Society of America, Washington, D.C., 2000), 479.
16. Wu, Q., Hewitt, T., and Zhang, X.-C. (1996) Two-dimensional electro-optic imaging of THz beams, *Appl. Phys. Lett.*, **69**, 1026-1028.
17. Jiang, Zhiping, Sun, F. G., Chen, Q. and Zhang, X.-C. (1999) Electro-optic sampling near zero optical transmission point, *Appl. Phys. Lett.*, **74**, , 1191-1993.
18. Jiang, Z., Xu. G., and Zhang, X.-C. (2000) Improvement of terahertz imaging with a dynamic subtraction technique, *Appl. Opt.*, **39**, 2982-2987.
19. Chen, Q., Jiang, Z. and Zhang, X.-C. (2000) Electro-optical based two-dimensional THz near-field imaging, in *Conference on Lasers and Electro-Optics* (Optical Society of America, Washington, D.C., 2000), 527.

THz IMAGING OF BIOLOGICAL SAMPLES

M. Koch[1], K. Schmalstieg[1], P. Knobloch[1], S. Hunsche[2], M.C. Nuss[2], E. Rehberg[3], I. Sautter[4], J. Fromm[4], M. Hempel[5], R. Haferkorn[5], I. Libon[5], N.E. Hecker[5], and J. Feldmann[5]

1) Institut für Hochfrequenztechnik, Technische Universität Braunschweig, Schleinitzstr.22, 38106 Braunschweig, Germany
2) Bell Laboratories - Lucent Technologies, 101 Crawfords Corner Rd., Holmdel, NJ 07733, U.S.A.
3) Klinik für Hals-Nasen-Ohrenheilkunde der Universität Witten/Herdecke an der Katholischen Krankenhaus gGmbH, Bergstr. 56, 58095 Hagen, Germany
4) Fachgebiet Angewandte Holzbiologie, Wissenschaftszentrum Weihenstephan, Technische Universität München, Winzererstr. 45, 80797 München, Germany
5) Lehrstuhl für Photonik und Optoelektronik, Ludwig-Maximilians Universität, Amalienstr. 54, 80799 München, Germany

Abstract:

With the development of fast and efficient THz imaging using THz-time domain spectroscopy (THz-TDS), it has become possible to image biological samples in this frequency range. Such imaging studies are complementary to studies in other parts of the electromagnetic spectrum, because they convey different information due to the fact that materials exhibit different properties in the THz frequency range than, for example, in the optical. There are many advantages of using THz-TDS. Some of these include that this technique is not damaging to tissue since the THz energy and intensity are very low. In addition THz-TDS can be used for *in-situ* measurements, a complete THz-system can be made portable, and the technique is very sensitive to water content as well as density and chemical fluctuations in a biological sample. In this paper, we will discuss the following topics. A typical THz-TDS experimental setup for generation and detection of THz waveforms and configured for THz-imaging is presented. This section contains a brief discussion of both the generation and detection process using photoconducting (also known as Auston) switches. Then, we give three examples in which imaging on biological samples can be useful and present our experimental results on these materials using THz-TDS. First, we examine the wood density variations which occur over the period of a tree's life and which are manifest by imaging the cross sections of tree trunks in the THz frequency range. Second, we measure the increase in the water content of a leaf of drought stressed plants after rewatering. Third we

R.E. Miles et al. (eds.), Terahertz Sources and Systems, 241–258.

investigate the fast water transport that occurs in mimosa after mechanical stimulus. Forth, we explore the potential of THz imaging for medical diagnostics by investigating the larynx of a pig which has been dehydrated and embedded in wax.

1. Introduction:

Today innumerable modern imaging techniques provide a wealth of information for biological, medical and material science applications. While some of these techniques are more or less mechanical, including those which belong to the field of scanning probe microscopy, most imaging techniques use electromagnetic radiation. In addition to x-ray imaging which exposes samples to undesirable high energy ionizing radiation, visible and near-infrared radiation is used to obtain spatially-resolved information on an object. Typical examples for these optical techniques include confocal and laser scanning microscopy, optical tomography, thermography and infrared imaging. At the longer wavelength end of the electromagnetic spectrum, microwaves are also widely used for imaging purposes although they cannot provide good spatial resolution. Nevertheless they are still useful for many applications. Typical examples include ground penetrating radar and synthetic aperture radar.

Until recently however the far-infrared region located between the infrared and the microwave region of the electromagnetic spectrum was not accessible to imaging, because of the difficulty of producing a high power THz source combined with a simple detection scheme. In the past 10 years however, the gap ranging from 0.1 to about 10 terahertz (THz) became accessible because of the developments in THz time-domain spectroscopy (THz-TDS) [1-3]. This innovative technique makes use of femtosecond optical pulses produced from standard commercial laser systems which were simultaneously being developed in the past two decades. These ultrashort optical pulses can be used to "drive" small photoconductive dipole antennas much faster than that possible with the fastest electronic ac-devices available to date. The nearly instantaneous excitation of free carriers into the antenna structure and their subsequent acceleration in an applied electric field results in the emission of few-cycle THz bursts with a typical bandwidth of several THz. These THz pulses can be detected with a second antenna of similar design which is gated by ultrashort laser pulses and which is extremely sensitive to the impinging electric field of a THz-pulse.

Nuss and coworkers were the first to further develop this photoconducting antenna emitter-receiver scheme to the point where two-dimensional images of objects a few centimeters in diameter could be accumulated [4]. Since this first demonstration in 1995, more advanced procedures such as THz tomography [5,6] and THz near-field imaging [7-9] have been explored.

2. Description of the THz spectrometer

The THz spectrometer is depicted in Fig. 1. A Ti:sapphire laser is optically pumped by an argon ion laser or by a frequency doubled Nd:YVO_4 laser, which has lately become more popular, to produce 100 fs pulses with a repetition rate of around 80

MHz. These laser pulses are used to gate both the generation and detection photoconductive dipole antennas. These antennas consist of a semiconductor substrate and a pair of coplanar metal striplines which have been evaporated several 10 μm apart from one another onto the substrate surface. The laser pulse with a photon energy above

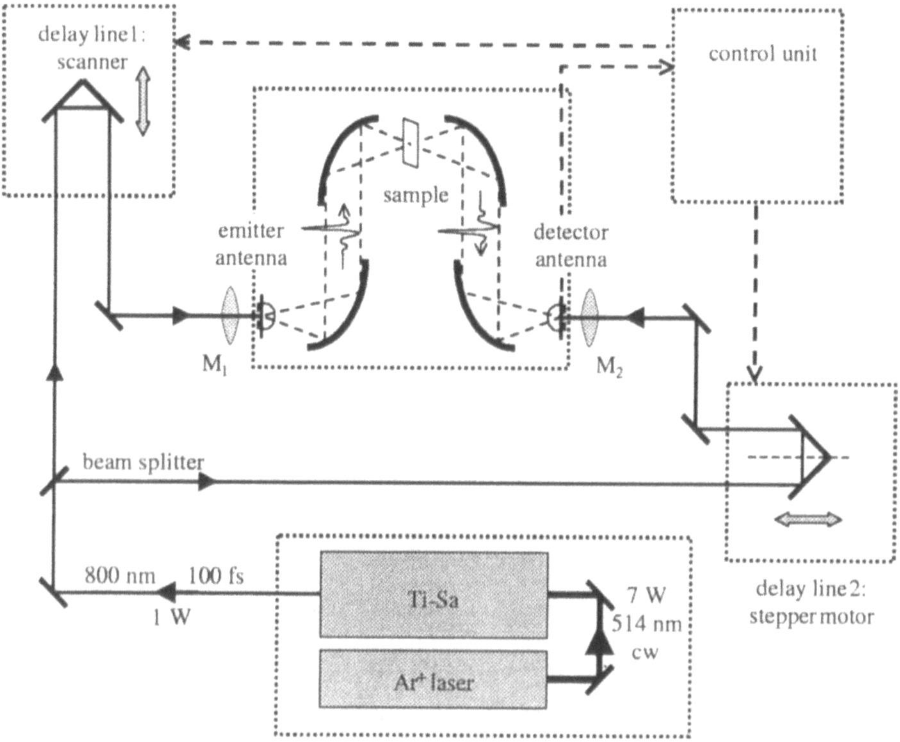

Figure 1: Schematic drawing of the THz imaging setup.

the semiconductor's energy gap is focussed between the two striplines in order to introduce electrons and holes into the substrate. Because in the generation antenna a bias voltage of several 10s of volts has been applied across the striplines which leads to a typical field strength around 5 kV/cm, the optically excited carriers are accelerated. The resulting current pulse acts as a radiating dipole and emits THz-radiation. This current surge typically persists only for a picosecond or less. This is due to the fact that the carriers are accelerated only for a short time since the carriers themselves screen the applied bias field. Hence after a few hundred femtoseconds, the carriers are decelerated. Fig. 2a shows such a short current spike calculated using a Drude-Lorentz model which incorporates dynamical screening of the bias field [10,11]. According to Maxwell's law, the emitted THz pulse shown in Fig. 2b is given by the negative temporal derivative of the current spike. After passing some optical components such as silicon hyperspherical lenses, off-axis parabolic mirrors, and the sample, the transmitted THz radiation is focussed onto the receiver antenna. The propagation through the spectrometer and the dispersion in the emitter and receiver antenna leads to a reshaping of the THz pulse. The electric field of the THz pulse which actually arrives at the receiver antenna is plotted in Fig. 2c (dashed line). The detector antenna which also consists of two striplines and a semiconductor material is gated by another 100 fs laser pulse. The semiconductor

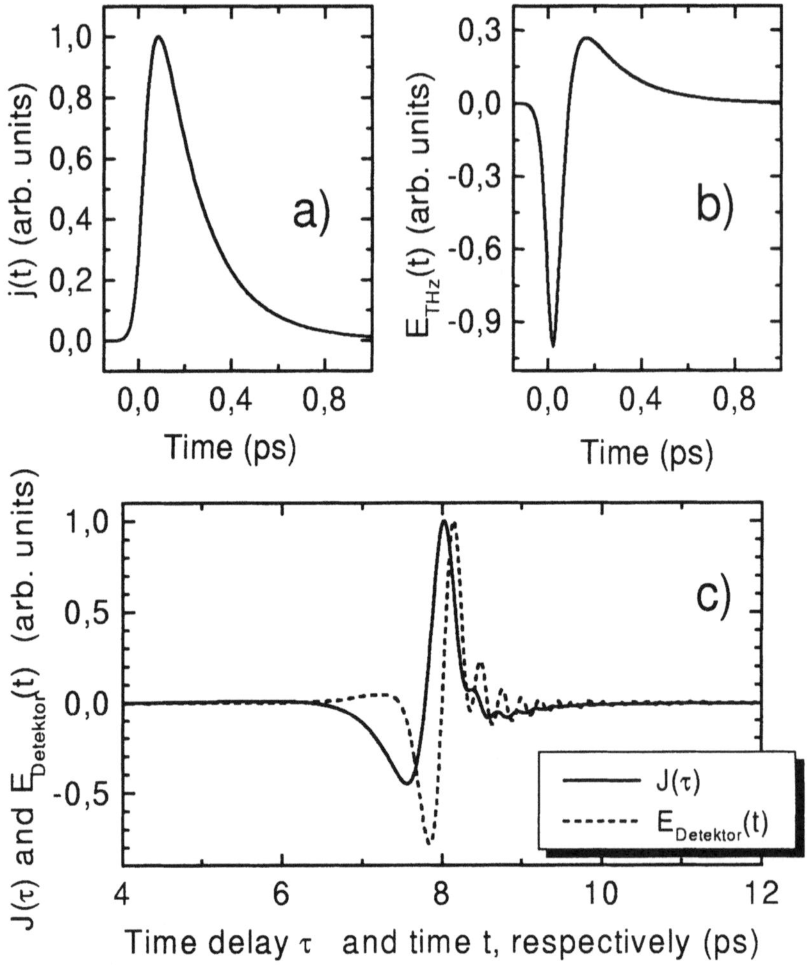

Figure 2: a) photocurrent in the emitter antenna. b) emitted THz pulse, c) THz pulse after passing through the spectrometer (dashed) and the THz waveform that is finally measured (solid).

material chosen for this antenna must have a very short carrier life time (typically radiation damaged silicon-on-sapphire (SOS) or low-temperature grown GaAs is used) in order to resolve the time-dependent THz-field strength. In contrast to the antenna used for THz generation, the optically excited carriers are not accelerated by an applied external field but instead by the field of the incident THz pulse. Hence the THz field biases this antenna which gives rise to a net current. Generally this current must be amplified (since it is in the nanoampere range), detected using lock-in techniques, and finally monitored by a computer. By varying the delay between the gating laser pulse and the incoming THz pulse, one aims to map out the THz field via a cross correlation process (see e.g. Ref. 10 for details). If the carriers had a δ-function-like lifetime than a measurement of the current would be a direct determination of the THz field strength. However since this is not the case, the measured waveform (solid line in Fig. 2c) differs from the actual THz pulse that impinges onto the antenna.

In order to accumulate an image of the THz transmission though an extended object, the sample is simply translated through the focus of the THz beam and a transmitted THz waveform is collected at each position of the object . An image is therefore acquired pixel by pixel. The final processed image may be used to represent any desired aspect of the measured waveforms, including amplitude and phase of the THz waveform or any combination thereof [12]. Mostly, however, the waveform is Fourier transformed to acquire the transmitted spectrum and the integrated signal intensity within a given frequency range (typically 0.7 to 1.2 THz) is translated into a gray value to obtain a pixel. Using this scheme the spatial resolution (typically 0.2-0.4 mm) depends critically on the spectral range over which the integration is performed.

In order to accumulate images on a relatively short time scale such as minutes, the waveforms must be digitized and the desired information stored in electronic form. This can be accomplished using a commercial digital signal processor in a standard personal computer. The translation of the object through the THz focus and the waveform acquisition must also be synchronized and controlled by a computer [12].

An example of THz imaging demonstrates the capability of this technique to reveal the internal structure of objects. In Fig. 3, we show the THz transmission images of the handle of a typical kitchen knife. The gray scale shown in (a) corresponds to the

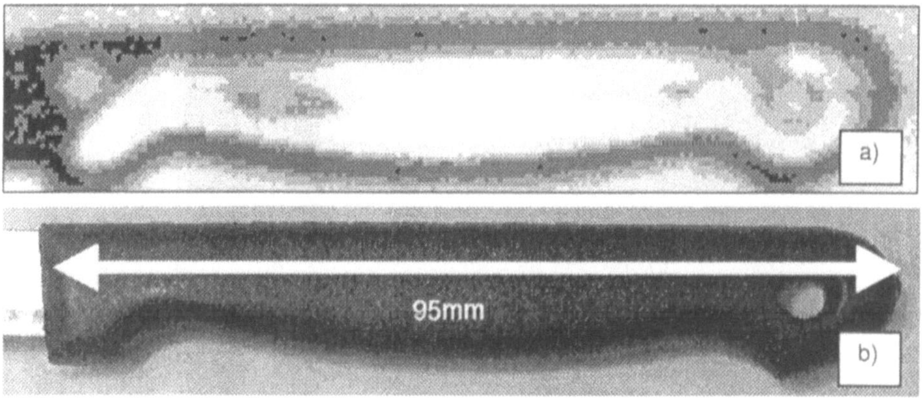

Figure 3: a) THz transmission image of the handle of a kitchen knife, b) fotograph.

integrated signal amplitudes from 1.2 to 1.75 THz. Because metal is opaque to THz radiation whereas plastic is transparent, the THz picture clearly reveals the structure of the metal inside the knife's plastic handle. In the following sections we will show that THz imaging is sensitive to density fluctuations and variations in water content.

Furthermore it allows us to distinguish between different types of biological tissue and may therefore also hold considerable potential for medical diagnosis.

3. Density determination of wood:

The density of wood is not only important for the wood and paper industry, but also for basic research on plant growth and dendrochronology. Dendrochronology allows researchers to extract information about the environmental conditions effecting plant growth via investigations of its year rings. Fluctuations in wood density also effects its mechanical properties and are tested for at wood processing plants. In order to determine the local density of wood, various methods have been employed over the years. In addition to simple graviometric methods [13], density determination techniques include the 'hardness feeler' [14] which determines the pressure involved in boring a needle into a section of wood and the acid corrosion method [15] in which wood is subjected to acid erosion and the resulting surface is investigated using an optical light microscope. More advanced methods include radiodensitometry with x-rays [16,17] and sandblasting [18] in which the resulting surface height profile after damage is directly related to the density.

Because THz radiation is strongly absorbed by water and we are interested in density not water profiling, all investigations discussed in this paper were performed on heat-dried wood slices. However after storage in ambient conditions, humidity from the air is reabsorbed by the wood so that the final water concentration in the measured samples is approximately 12 % in weight. Three samples were used for the

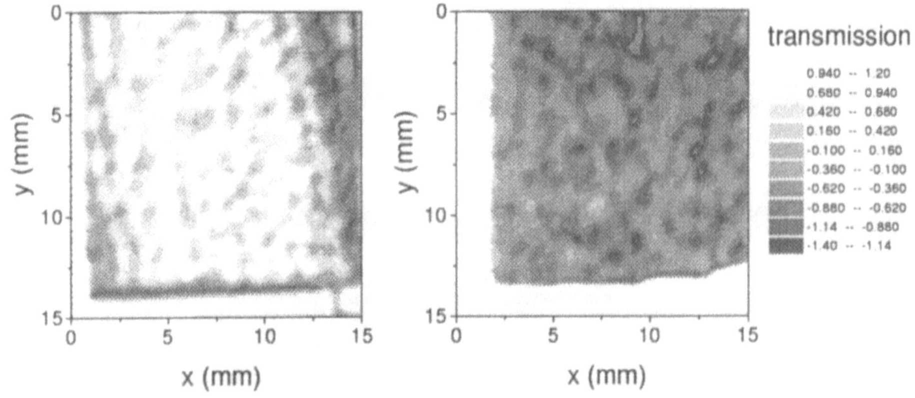

Figure 4: THz images of a piece of balsa (left) and a piece of bongossi (right).

investigations discussed here: balsa (Ochroa pyramidale), bongossi (Lophira alata) a west African wood, and beech (Fagus sylvatica). All samples were cut to the same size (14 x 14 x 1.7 mm^3) to allow for direct comparison between THz transmission experiments. The first two samples were compared, because their densities are dramatically different and this should be evident in their THz transmission intensities. Balsa has a density of 0.1 - 0.25 g/cm^3 whereas bongossi is 1 g/cm^3. Figure 4 shows the two resulting THz transmission images obtained with an integration range from 0.75 to

1.2 THz. It is evident from the images that the structures are seemingly homogeneous (which is to be expected since year rings are not visible in these two wood samples) and that balsa transmits more THz radiation than bongossi. This is a first indication that the optical density in the THz range scales with the wood density.

A sample from beech which exhibits strong visible year rings indicating large density variation over the course of growth was also studied. A THz-image taken of a cross section of this plant is shown in Fig. 5a. Annual rings show themselves as vertical

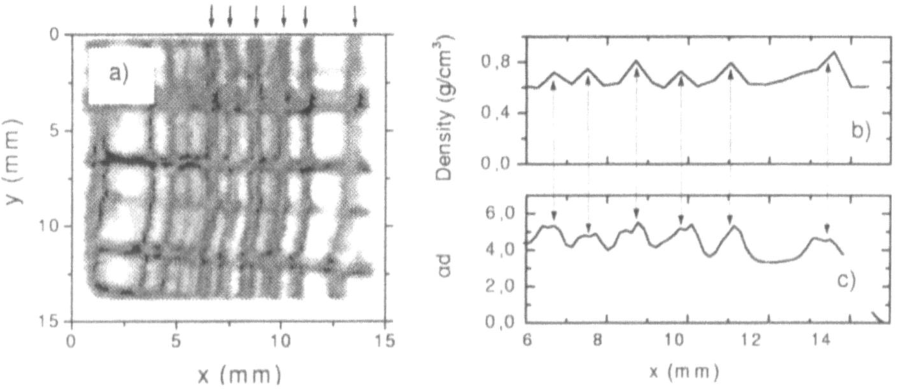

Figure 5: a) THz image of a piece of beech wood. b) Density profile obtained with the gravitometric-volumetric method. c) Absorption profile extracted from a defined area of the THz image in the left part (see text).

structures. The darker areas represent latewood which is less transparent due to its greater density. Earlywood, on the other hand, has larger cell lumina and thinner cell walls and consequently is more transparent and appears grey to white. In addition, radial rays can be observed as horizontal structures.

In order to relate this THz transmission image to a true wood density, the transmission data must be calibrated to a density obtained by the gravitometric-volumetric method. This was achieved by cutting part of the sample into slices along the y-direction. Subsequently the volume and weight of these pieces were then determined. Fig. 5b shows the determined density for the range from x=6 to 15 mm and y=0 to 4.5 mm. Clearly, several peaks can be observed. They correspond to regions of latewood which are marked by the set of arrows in Fig. 5a. Since reflection losses at the two interfaces of the sample are estimated to be below 3 % (we determine an average refractive index of 1.4 for the THz range) they are neglected here and the transmission data can be transformed into the optical thickness αd according to Beer's law:

$$\int_{0.75THz}^{1.2THz} d\nu\alpha(\nu)d = -2\ln\left(\int_{0.75THz}^{1.2THz} d\nu E(\nu) \,/ \int_{0.75THz}^{1.2THz} d\nu E_0(\nu)\right)$$

where $\alpha(\nu)$ is the absorption coefficient at frequency ν, d the sample thickness of 1.7 mm, and E and E_0 represent the transmitted and incident THz field amplitude, respectively. The absorption data extracted for the sample area investigated by the gravitometric-volumetric method are integrated over y to be comparable to the density data obtained for the slices cut along the y-direction. The absorption data are plotted in Fig. 5c. A reliable correlation between the density data and the THz data is apparent. A quantitative calibration between absorption and density can be obtained by extracting several data points from each plot and plotting the values for the density n versus the corresponding absorption data (triangles in Fig. 6a). The data points in Fig. 6a can be approximated by a line (dashed line) obeying:

$$n = 0.145 \frac{g}{cm^3} \cdot \alpha d$$

This function can be used to directly correlate the transmission image of Fig. 5b into a density map of this sample which is shown in Fig. 6b. While white and light gray regions represent low density domains of early wood, the darker areas

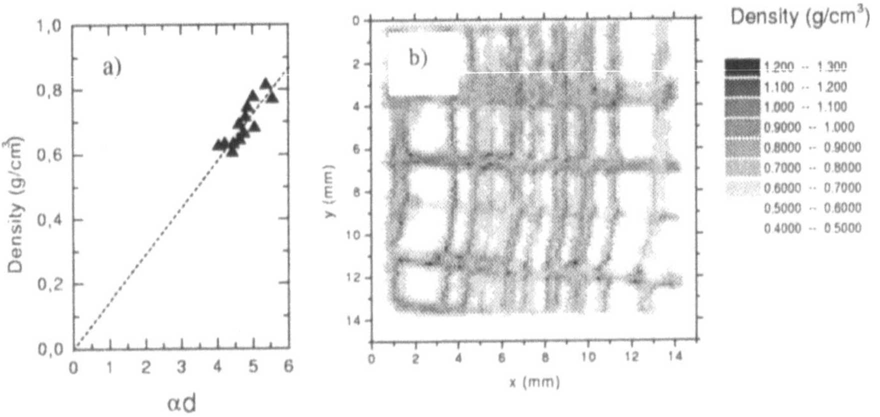

Figure 6: a) Density plotted versus absorption. b) Density map of the beech sample.

represent the high density regions of late wood (see Ref. 19 for a colored reproduction). Overall, the density in this particular sample varies from 0.4 to 1.3 g/cm^3.

The above results demonstrate that THz imaging represents an interesting alternative for mapping out density profiles of wood specimen with sub-tree ring resolution. In contrast to the sandblasting method the section of wood can be reused for further studies since it has not been destroyed. In addition, because THz radiation is not harmful in the field strengths used in these measurements, it has advantages over direct scanning with x-rays. Concomitantly, THz radiation, since it is strongly absorbed by water, could be furthermore used to reveal water depots in wood and other biological specimens: a task which can not be performed by radiodensitometry or the sandblasting method. In the next section we will show that THz imaging allows us to monitor water transport in living plants.

4. Water Flow in Plants

The mechanisms involved in long-distance water transport in higher plants are still controversial. A simply estimation shows that capillary forces alone cannot be responsible for the water transport from roots to the leaves in tall trees with heights up to 120 m. The diameter of the xylem vessels, the "arteries" through which water flows, have typical diameters between 7 and 200 μm, depending on the plant type. In a capillary with a radius r=20 μm, however, water rises only 75 cm. For nearly one century the so called cohesion theory, which assumes that water transpiration form tree leaves effectively pulls water from the ground up, was widely accepted. This theory was introduced by Dixon and Joly [20] in 1895 and postulates a continuous water column with a pressure gradient between the roots and the leaves of the trees. It is based on a variety of experiments which indicated that transpiration is very important for water transport in plants [21-23]. Recent experimental evidence has led to a reexamination of the cohesion theory due to the incompatibility of data from two competing techniques which use either pressure chambers [24] or xylem pressure probes [25]. Current controversies of the cohesion theory were discussed recently by Milburn [26] and Tyree [27].

To test the validity of the transpiration theory, more accurate experimental data on water flow rates, preferably on individual xylem vessels, needs to be gathered. One technique that has proven to be useful in recent experiments is nuclear magnetic resonance (NMR) [28,29]. Kuchenbrod and coworkers [30], for example, investigated maize plants under controlled transpiration conditions. They could show that the flow velocity of water in the large vascular bundles reached 2 mm/sec and was considerably lower in the smaller vessels. NMR is however an extremely expensive technique which needs to be supplemented by other methods which are portable and can be performed on large plants.

Since THz radiation is strongly absorbed by water, THz imaging represents an interesting, alternative method for obtaining information on water transport in plants. Fig. 7 shows the absorption spectrum of water around 1 THz. Although it is widely accepted that translational motions of the water molecules are at the origin of this featureless absorption, the detailed mechanisms involved are still controversially discussed [31-35]. The absorption coefficient at 1 THz is 235 cm^{-1}. Using this value we can calculate the attenuation of the THz signal by a water film (losses due to reflection are neglected for simplicity). The inset in Fig. 7 shows the transmitted field amplitude at 1 THz as a function of the film thickness. From this plot we can conclude that a biological sample should ideally contain an amount of water that is equivalent to a water film of between 100 to 200 μm thickness. This criteria is nicely fulfilled by most leaves. In fact the first demonstration of THz imaging was performed on a leaf cut from a tree [4].

A THz investigation of a plant is non-invasive, yet in most cases it will be limited to a single leaf. Therefore one does not obtain information on flow rates in the different sections of the plant. However, one can investigate the dynamics of the water uptake and, thus, answer the question of how long does the rehydration process take in drought stressed plants. In addition, THz measurements should give an estimate of how much the water uptake is.

A first THz investigation of a living plant demonstrating that THz imaging is an alternative useful method to obtain information on plant water transport was conducted by Mittleman and coworkers [12]. They studied water uptake in a house plant (coleus). In this proof of principle experiment, they found that the overall transmission

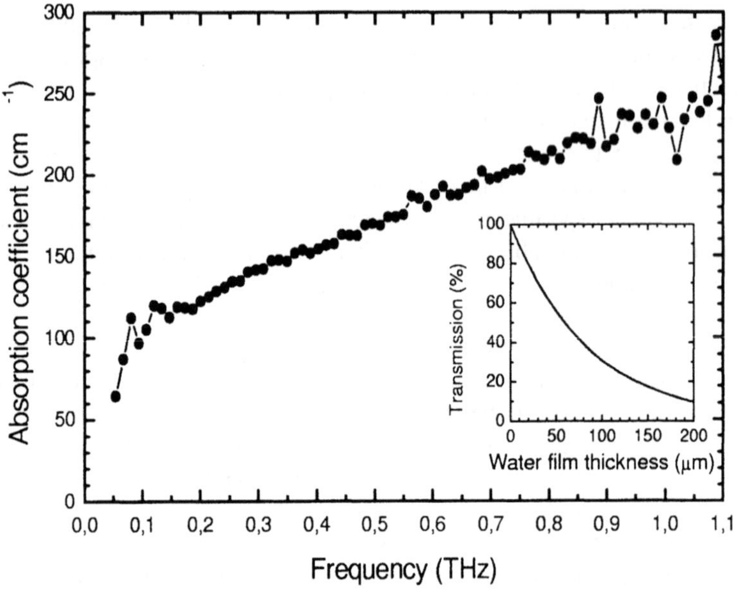

Figure 7: Absorption spectrum of water. Inset: Attenuation of the THz signal by a water film as a function of the film thickness.

through a leaf drops by more then ten percent within eight hours after watering [36]. This drastic decrease of the transmitted signal is attributed to the water which is taken up during the rehydration process. A close inspection reveals that the rehydration process divides into two regimes [36]: the THz transmission shows an initial fast drop within the first hour (in the following referred to as fast rehydration region (FRR)) and a subsequent nearly linear decrease (in the following referred to as slow rehydration region (SRR)) which was still not completed at the end of the measurement, i.e. after nearly eight hours.

In the following we present THz transmission data on the rehydration process in clematis. In agreement with the data by Mittleman *et al.*, we observe two time regimes. Yet, the initial fast rehydration regime is completed in our measurements after 15 minutes, i.e. four times as fast as that observed in the first Mittleman experiments. We also estimate the amount of water that is taken up by the leaf.

In a second experiment we investigate the change in the water content of leaf joints in mimosa after an external stimulus to the plant. Our experiments show a fast decrease of the THz signal transmitted through this section of the plant. We can therefore conclude that the net amount of water in the leaf joint increases. Important information which is not simply obtained otherwise.

4.1. Clematis

Prior to the experiments our clematis had not been watered for a period of about a week. In order to scan a living leaf attached to the plant through the THz focus, we have developed a special holder which allows us to clamp the leaf softly. We found after many attempts at measurements that leaves often exhibited signs of stress if the leaf was not gently held. Since normal lab illumination is not intense enough to trigger adequate transpiration, the plants were illuminated with white light from mercury halide lamps during the measurements. To monitor the water uptake in a leaf, we do not take a complete image but restrict ourselves to a line scan in order to decrease the data aquisition time. By increasing the number of waveforms averaged per data point, we can further reduce experimental noise. A complete line scan thus takes approximately five minutes to obtain.

Fig. 8 shows the spatially resolved transmission through a clematis leaf directly before watering and 144 minutes after watering. The integration range is 0.2 to 0.9 THz. The data are normalized by assuming that transmission through air (from x=0 mm to 10 mm, data points not shown in this expanded view) is 100 %. The data are not

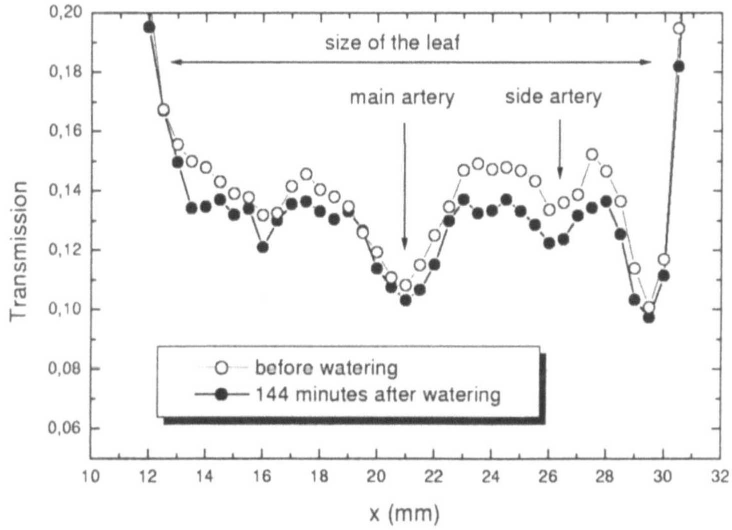

Figure 8 : Spatially resolved transmission through a Clematis leaf.

corrected for reflection losses. The two pronounced minima visible in Fig. 8 at x=21 mm and 26 mm correspond to two of the vascular bundles containing an aggregate of xylem vessels. After watering, the overall transmission drops as water flows into the leaf. This can be seen most clearly at the positions of the minima discussed previously.

Fig. 9 shows the decrease in the overall transmission through the leaf as a function of time after watering (here the data taken from Fig. 8 have been averaged from x=14 mm to x=28 mm). Within the first 15 minutes we observe a fast drop of the transmitted THz signal (FRR). Subsequently, the THz transmission further decreases, but with a reduced rate (SRR). The data are hence in qualitative agreement with that of Mittleman *et al* [36]. Yet, for clematis the duration of the FRR is much shorter than for

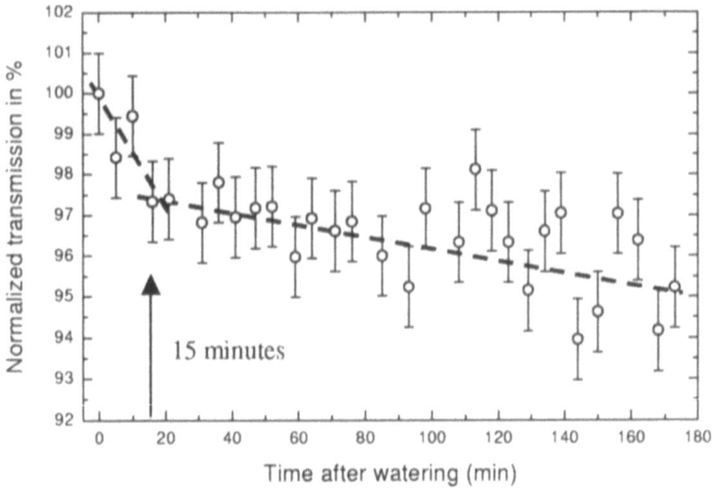

Figure 9: Overall transmission through the leaf as a function of time after watering.

coleus. From these observations several interesting questions arise. Why does the rehydration process divide into two regions? Is this behavior a typical effect that can be observed in every plant? Is the duration of the FRR characteristic of a particular species or does it vary within a single plant, e.g. does it depend on the distance of the leaf under investigation form the ground? How does the rehydration dynamics depend on environmental conditions? To answer all these questions much more data and extensive measurements are needed

We can, however, draw further information from the change in THz transmission. We can crudely estimate the amount of water that is taken up by the leaf. Before watering, the averaged transmitted THz intensity is 13.43 %. With an average

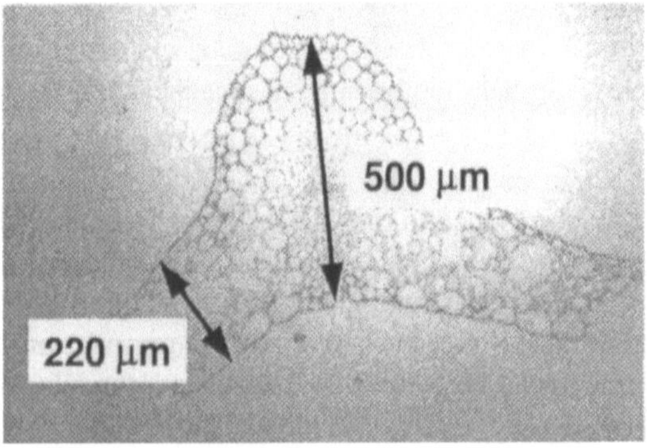

Fig. 10: Cross section through a leaf of clematis.

absorption coeffcient of 165 cm^{-1} (this is the value for 0.5 THz (see Fig. 7), this attenuation would correspond to a water film with a thickness of 121.6 μm (ignoring losses due to reflection). This is a reasonable number since it is on the order of the leaf's physical thickness which typically varies between 180 and 220 μm (in the main arteries it reaches 500 μm). Fig. 10 shows a magnified optical image of the cross section of a leaf in the region of an artery. After three hours the transmission has dropped to 12.8 %. With the above assumptions this corresponds to a water film of 124.6 μm thickness. This is an increase in the water amount of 2.5 %. Even though the average absorption coefficient is somewhat uncertain (since we are averaging over the measured THz spectral range) we can conclude that the amount of water in the leaf has increased by 2 to 3 %. We would like to stress, that it is extremely difficult to obtain this information in living plants from any technique used today.

4.2. Mimosa

The rehydration process studied above involves water transport that occurs on the time scale of hours. There are plants which show a much faster water transport. One example is mimosa (mimosa pudica) which originates from the tropical part of the American continent. Fig. 11 shows a schematic drawing of this sensitive plant. When mechanically stimulated, the plant responds with rapid leaf movements. While primary pulvini respond with downward motions, tertiary pulvini respond with upward movements of the leaflets so that the leaf folds up (right part in Fig. 11). Anatomically

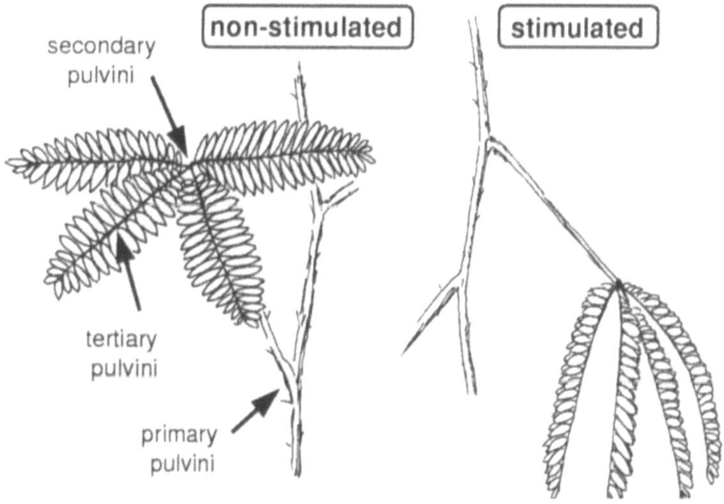

Fig. 11: Schematic drawing of Mimosa pudica before and after mechanical stimultation.

all pulvini have a core of vascular tissue surrounded by a bulky cortex of thin-walled parenchyma cells. It is agreed that these cortex cells cause movements by either losing or gaining turgor [37,38]. Yet, what is unknown is whether there is a net increase or decrease of pulvinar water during the movement cycle.

To answer this question we measured the THz transmission through a single tertiary pulvini pair. We clamped one leaf gently between two transparent plastic sheets in order to prevent the leaf from closing in response to the stimulus. The leaf was mounted with this configuration so that the joint remains in place and we can avoid any changes in signal amplitude due to a variation in sample geometry. To prevent multiple reflections of the THz pulses between the plastic sheets, we cut holes of a few mm in diameter in the plastic exactly at the position of the THz focus.

Fig. 12 shows the THz transmission integrated from 0.1 to 1.0 THz through tertiary pulvini as a function of time. At time zero the plant is mechanically stimulated. Instantly the transmission drops by six percent. It rapidly recovers within a few seconds to 98 percent. This fast recovery is followed by a slow recovery on the time scale of

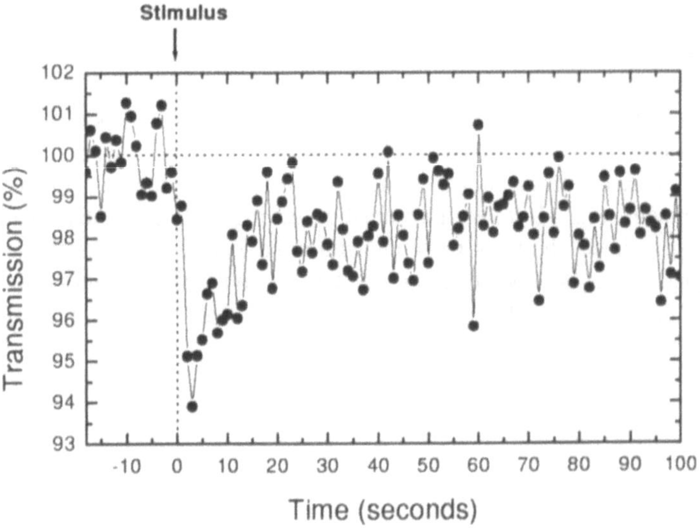

Fig. 12: THz transmission through tertiary pulvini after mechanical stimulus of the plant.

many minutes. This slow recovery is typically completed after 30 minutes, and the leaf appears non-stimulated again.

From the drop in THz transmission, we can conclude that there must be a net increase of turgor in the tertiary pulvini. Yet, it is difficult to obtain quantitative values since the spot size of the THz radiation is larger than the entire leaflet joint, and a portion of the THz beam passes the joint and does not interact with it. This is particular true for the low frequency components which form a larger focus.

For the science community working on water flow in plants, THz imaging shows great promise as a noninvasive tool for probing water concentration in leaves. In addition with the further development of ultrashort pulsed lasers, the cost of such a system is becoming a viable method for general plant researchers. Finally in the not so distant future it seems that THz spectrometers will become available which will run from portable battery operated systems.

5. Medical imaging

Since the first demonstration of THz imaging, only a few publications have focussed on the future use of THz radiation in medical diagnosis. Both the groups of X.-C. Zhang and of D. Arnone could identify different types of tissue in "fresh" untreated samples [39,40]. Mittleman *et al.* showed that THz imaging may be useful for burn diagnosis [8,12]. Caries detection has been demonstrated by Arnone *et al.* [40]. Yet, much more work is needed to fully explore the potential of THz imaging for medical diagnosis.

Here we investigate a sample that has been treated by the standard procedure for histo-pathological examination. We chose the larynx of pig since it contains cartilage and soft tissue.

Fig 13a shows an optical image of the sample. The larynx has been cut in slices several mm in thickness. The slices were fixed in 4 % formalin solution for one week.

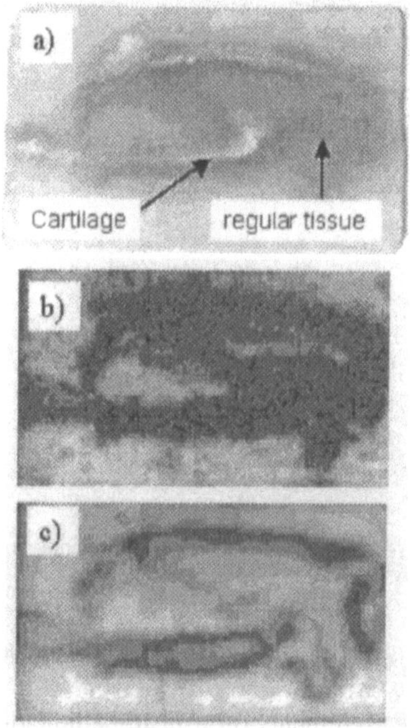

Fig. 13: a) Optical image of a larynx sample taken from a pig. b) THz transmission image of this sample for the frequency window from 1.0 to 1.5 THz, c) Image for 0.2 to 0.5 THz,

Subsequently, they were treated by a series of alcohol solutions to remove water from the tissue. Finally, they were embedded in paraffin. For histo-pathological examination such paraffin blocks are normally cut into thin slices which are then scrutinized with an optical microscope. Since this type of sample preparation leads to dehydration, THz radiation can penetrate several mm of tissue whereas generally these frequencies are strongly absorbed by water. We therefore can investigate the entire block with dimensions of 7.3 x 4.0 x 0.8 cm including the regions of cartilage and soft tissue that are hard to identify in Fig. 13a.

Fig. 13b shows an image of the THz transmission integrated from 1.0 to 1.5 THz. The dark areas correspond to regions of low transmission through the biological tissue. All biological tissue in the sample seems to absorb the THz radiation equally well. Hence a distinction between different types of tissue cannot been made in this frequency window. The non-polar paraffin that surrounds the larynx has negligible absorption and appears bright.

The situation is different if the spectral integration is performed from 0.2 to 0.5 THz (Fig. 13c). In this case the features that can be identified in the THz image correspond to the regions of cartilage. Obviously the absorption of cartilage is much stronger than that of soft tissue in this frequency window. This is confirmed by Fig. 14 which shows the absorption spectra of cartilage and soft tissue obtained by an

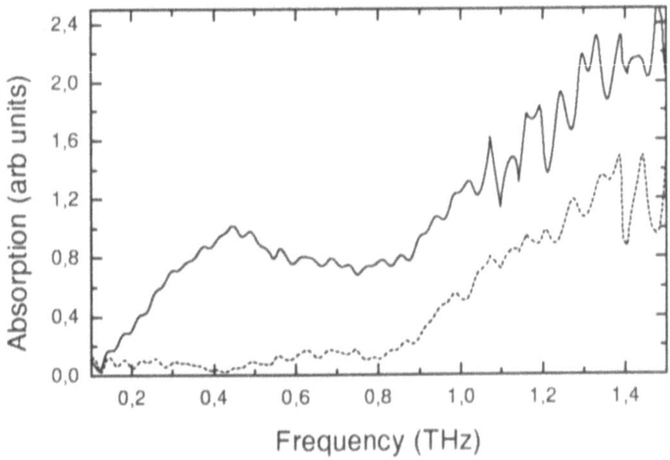

Fig. 14: Absorption of cartilage (solid line) and soft tissue (dashed line).

independent measurement. At 0.45 THz the cartilage absorption exceeds that of soft tissue by a factor of 20.

These first measurements on a histo-pathological sample demonstrate that in these kind of samples distinct types of tissue can be identified from a comparison of images obtained using different integrated frequency windows. Yet, it should also be noted that images taken for different spectral windows have, by nature, a different spatial resolution. For a fair comparision of images obtained for different frequency ranges a numerical treatment may prove necessary.

6. Conclusion

In this paper we have discussed a few experiments using THz-imaging which lead to the conclusion that imaging in this frequency range can be complementary to other conventional imaging techniques. For instance in wood samples, THz imaging may prove to be useful for determining its density and perhaps even its water concentration, both of which are useful for the wood industry. In addition we have observed water flow in plants which has been a subject of controversy for over a hundred years and for which THz imaging may be able to play a significant role in helping to answer this debate. Finally in biomedical imaging of a histopathological sample, it appears that THz imaging may help pinpoint areas of differing tissues even though the diffraction limited resolution of this technique is on the order of 100s of microns. We would like to stress with our studies shown here that although THz imaging is still in its infancy it shows great promise for use in obtaining information on biological samples.

Acknowledgements:

N.E.H. would like to thank the Alexander von Humboldt Stiftung for support of this work.

References:

1) P. R. Smith, D. H. Auston, and M. C. Nuss, IEEE J. Quant. Elec., **24**, 255 (1988).
2) M. van Exter and D. Grischkowsky, IEEE Trans. Microwave Th. Tech., **38**, 1684 (1990).
3) L. Xu, X.-C. Zhang, and D. H. Auston, Appl. Phys. Lett., **61**, 1784 (1992).
4) B.B. Hu, and M.C. Nuss, Opt. Lett. **20**, 1716 (1995).
5) D.M. Mittleman, S. Hunsche, L. Boivin, and M.C. Nuss, Opt. Lett. **22**, 904 (1997).
6) R.A. Cheville, and D. Grischkowsky, Appl. Phys. Lett. **67**, 1960 (1995).
7) S. Hunsche, D.M. Mittleman, M. Koch, and M.C. Nuss, IEICE Trans. Electron. **E81-C**, 269 (1998).
8) D.M. Mittleman, M. Gupta, R. Neelamani, R.G. Baraniuk, J.V. Rudd, and M. Koch, Appl. Phys. B **68**, 1085 (1999).
9) Q. Chen, Z. Jiang, G.X. Xu, and X.-C. Zhang, Opt. Lett. **25**, 1122 (2000).
10) P. Uhd Jepsen, R. H. Jacobsen, and S. R. Keiding, J. Opt. Soc. Am. B, **13**, 2424 (1996).
11) M.Koch, R. Haferkorn, J. Feldmann, postdeadline paper at the THz '99 conference, Nara, Japan (1999).
12) D.M. Mittleman, R.H. Jacobsen, M.C. Nuss, IEEE Journal of Selected Topics in Quantum Electronics **2**, 679 (1996)
13) O. Lenz, E. Schär, and F.H. Schweingruber, Holzforschung **30** 114 (1976).
14) H. Mayer-Wegelin, Allg. Forst- und Jadgzeitung **122** (1950).
15) J. Kisser, and I. Lehnert, Mitt. Österr. Ges. Holzforschung, Wien (1951).
16) H. Polge, Wood Science and Technology **12**, 187 (1978).
17) D. J. Cown, and B.C. Clement, Wood Science and Technology **17**,91 (1983).
18) G. Lesnino, Wood Science and Technology **28**, 159 (1994).
19) M. Koch, S. Hunsche, P. Schumacher, M.C. Nuss, J. Feldmann, and J. Fromm, Wood Science and Technology **32**, 421 (1998).

20) H.H. Dixon and J. Joly, Phil. Trans. Roy. Soc. Lond. B **186**, 563 (1895).
21) G. Huber, in "Sap flow in Plants", Springer, Berlin (1956).
22) J. Böhm, Ber. Dtsch. Bot. Ges. **11**, 203 (1893).
23) M.H. Zimmermann, "Xylem Structure and the Ascent of Sap", Springer, Berlin (1963).
24) P.F. Scholander, H.T. Hammel, E.D. Bradstreet, and E.A. Hemmingsen, Science **148**, 339 (1965).
25) A. Balling, U. Zimmermann, K.-H. Büchner, and O.L. Lange, Naturwissenschaften 75, 409 (1988).
26) John A. Milburn, Annals of Botany **78**, 399 (1996).
27) Melvin T. Tyree, J. of Exp. Bot. 48, 1753 (1997).
28) H. Van As, and T.J. Schaafsma, Biophysical Journal, **45**, 469 (1984).
29) G.A. Johnson, J. Brown, and P.J. Kramer, Proc. Nat. Acad. Sci. **84**, 2752 (1987).
30) E. Kuchenbrod, M. Landeck, F. Thürmer, A. Haase, and U. Zimmermann, Bot. Acta **109**, 184 (1996).
31) L. Thrane, R.H. Jacobsen, P. Uhd Jepsen, and S.R. Keiding, Chem. Phys. Lett. **240**, 330 (1995).
32) J.T. Kindt, and C.A. Schmuttenmaer, J. Phys. Chem. **100**, 10373 (1996).
33) P.L. Silvestrelli, M. Bernasconi, and M. Parrinello, Chem. Phys. Lett **277**, 478 (1997).
34) C. Rønne, P.-O. Åstrand and S.R. Keiding, Phys. Rev. Lett. **82**, 2888 (1999).
35) M. Koch, I.H. Libon, M. Hempel, S. Seitz, N.E. Hecker, J. Feldmann, D. Mittleman, F. Bartl, A. Hayd and G. Zundel, Proceedings of the LASER'98 conference, Eds. V.J. Corcoran and T.A. Goldman, (STS Press, McLean, VA 1999) p.225-231.
36) D.M. Mittleman, private communication.
37) E. Brücke, Müllers Arch. Anat. Physiol. 443 (1848).
38) J. Fromm, and W. Eschrich, in: The pulvinus: motor organ for leaf movement, R.L. Satter, H.L. Gorton, T.C. Vogelmann (eds), The American Society of Plant Physiologists, Rockville, USA (1990).
39) P.Y. Han, G.C. Cho, and X.-C. Zhang, Opt. Lett. **25**, 242 (2000).
40) D.D. Arnone, C.M. Ciesla, A. Corchia, S. Egusa, M.Pepper, J.M. Chamberlain, C. Bezant, E.H. Linfield, SPIE Proc."Terahertz spectroscopy and applications II", Munich meeting (1999).

Theme 7

Towards Communication Systems

TOWARDS TERAHERTZ COMMUNICATIONS SYSTEMS

C.M. MANN
Rutherford Appleton Laboratory, Chilton, Didcot, Oxfordshire, UK
Flann Microwave Ltd, Dunmere Rd, Bodmin, Cornwall, UK

1. Introduction

Within the wireless communications industry the pressure is on to provide ever more services such as e-mail, fast access to the World Wide Web and ultimately videophones. More and more people are requiring the ability to send and receive large documents and files from wherever they are in the world. As this demand increases it is likely that bandwidth is going to become an even more valuable commodity. There will be therefore an increase in the frequency used for communication and data transferral. This trend is already taking place with the point to point links required for LMDS moving to 40GHz. Whether there will ever be a need for even higher frequencies is unclear but this paper briefly investigates the feasibility of implementing terahertz wireless communication.

It will be demonstrated that there are indeed some benefits offered by wireless links operating in the terahertz region, particularly for high bandwidth, short path, line of sight wireless links. However, what will undoubtedly determine if and when such systems are used will be the commercial viability rather than technological issues.

Mobile communications has now become an everyday requirement for literally billions of people worldwide. The basic reason for this is the convenience of not requiring any physical connection to the network. It can be safely stated therefore that wireless communication is here to stay and whilst the number of users will eventually reach saturation the demand for new and evermore data hungry services is always going to rise. Coupled with this growth

R.E. Miles et al. (eds.), Terahertz Sources and Systems, 261–267.

there will be an ever-increasing need for bandwidth as the volume of information and services evolves.

The first section if this paper investigates the role of the atmosphere in determining whether there is any potential for terahertz wavelengths. Potential technical advantages are then considered by comparing a hypothetical terahertz front-end based on current state of the art components with a commercially available system at 60GHz. Finally the commercial aspects are considered and conclusions drawn.

2. Atmospheric Behaviour in the Terahertz Region

To investigate the feasibility of a terahertz wireless link the absorption presented by the atmosphere and precipitation must be considered. The plot in figure 1 shows the absorption of the atmosphere in dB/kilometre up to 6 THz.

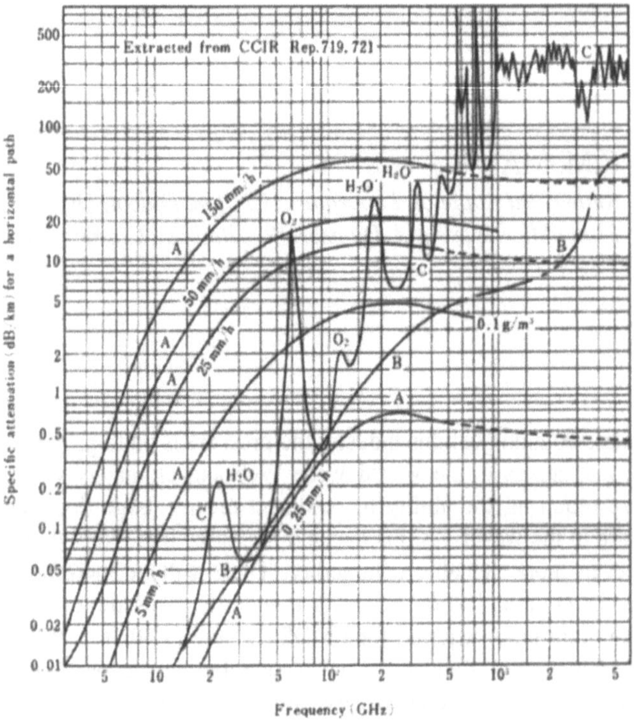

Figure 1. Absorption due to the atmospheric fog and rain

Curve A gives the attenuation as a function of rain, curve B the attenuation as a function of fog and curve C the attenuation due to the atmosphere. The main points to note are that fog (and smoke, not shown) has little or no effect up to 1 THz. This is in sharp contrast to the infrared region where fog presents a serious challenge to any wireless link. In addition the effect of rain becomes more or less flat above 50GHz. Finally it is clear that the atmospheric attenuation displays a number of strong molecular absorption lines. The level of each absorption line is mainly dependent on the relative pressure and abundance of the gas and can therefore be considered fixed for a given altitude.

It is apparent that the atmospheric window around 400GHz is favourable for the evaluation excercise. This wavelength is sufficiently short to qualify as being terahertz yet provide sufficient attenuation to gauge its effect.

3. Physical Considerations

As is usually the case, to evaluate a new systems potential it is best to return to basics. For this we consider the Friss antenna equation given below and describing the configuration shown in figure 2

$$P_r = P_t(A_{et}\, A_{er}) \,/\, r^{\,2} \lambda^2$$

Where:

A_{et} = *Effective Aperture of the transmitting Antenna*
A_{er} = *Effective Aperture of the receiving Antenna*
P_r = *Power received*
P_t = *Power transmitted*
r = *Distance between antennae*

λ = *The wavelength*

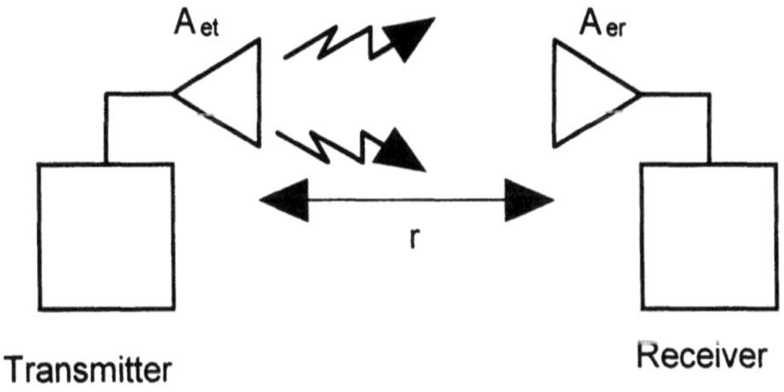

Figure 2. Absorption due to the atmospheric fog and rain

What is clear that is the gain of the system is inversely proportional to λ^2. This means for a given aperture dimension the required transmit power drops dramatically as the frequency increases. The data below in table 1 compares a representative 400GHz system with a commercially available 60GHz system. The numbers used for the transmit power and receiver sensitivity represent conservative estimates of what can now be achieved using an all solid state source and heterodyne mixer. Shown underneath each system is a breakdown of the approximate atmospheric absorptions in dB/Km. Figure 3 shows a comparison of each system's performance in terms of distance that can be transmitted over for the same conditions and same area antenna. It should be noted that the atmospheric conditions would be considered severe, however higher rainfall levels can occur.

*Table 1:*Comparison of a 60GHz and 400GHz system

60GHz commercial wireless duplex systems

Frequency	*-> 60GHz*
Baseband	*-> 0-100MHz*
Transmit Power	*-> 20dBm*
Receiver Sensitivity	*-> 10^{-12} W*

At 60GHz for Fog + Rain (25mm/h) + Gaseous □ 0.15+ 10+17 □ <u>27dB/Km</u>

Hypothetical Duplex System at 400GHz might have the following performance:-

Frequency	*-> 400GHz*
Baseband	*-> 0-?MHz*
Transmit Power	*<u>-> 6dBm</u>*
Receiver Sensitivity	*-> 10^{-12} W*

At 400GHz for Fog + Rain (25mm/h) + Gaseous □ 4+ 12+10 □ <u>26dB/Km</u>

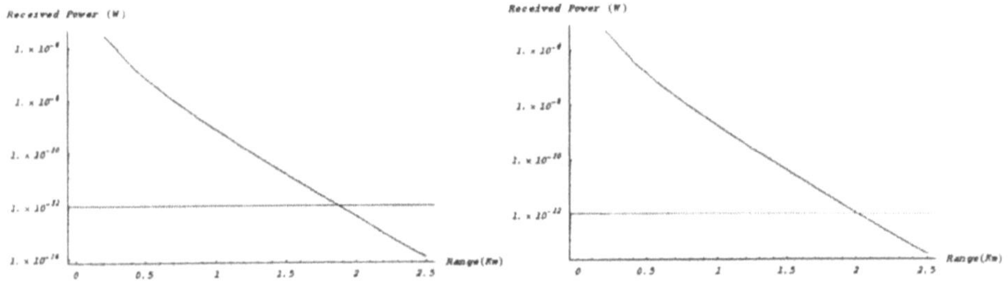

Figure 3. Transmission range for the 60GHz and 400GHz duplex systems described in Table 1

Assuming a 0.01m^2 aperture for both 60 and 400GHz systems the range at the 10^{-12} (W) break point is 1.85km and 2.0km respectively. Clearly this is an interesting result, whilst there is far less power available at 400GHz it is possible to transmit similar distances. This is entirely due to the increased gain of the antennae at the shorter wavelength.

One of the additional benefits of moving to higher frequencies is that even for very small antenna reasonable transmit distances can be achieved. If we now take the case of a 1cm^2 antenna for a 400GHz system operating in the same atmospheric conditions we can transmit 400 metres or so. The addition of a lens to the aperture increases this distance to 1350 metres as shown in figure 4.

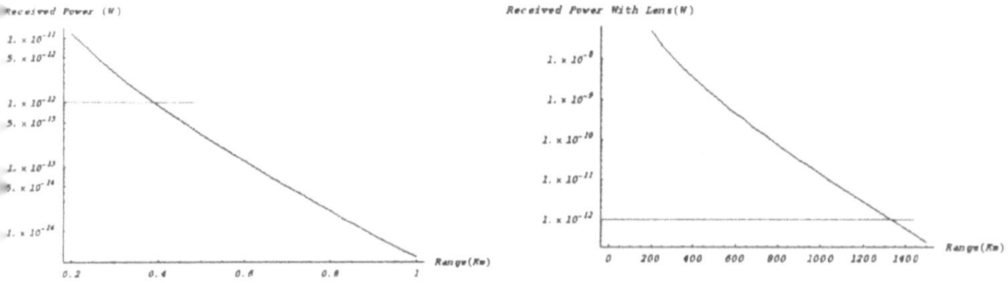

Figure 4. Transmission range for a 400GHz duplex system fitted without and with a 1cm^2 lensehorn

The conclusion of this section is that at first sight atmospheric absorption in the terahertz region does not rule out applications of relatively short paths. At higher frequencies, for selected bands the total atmospheric absorption can be comparable to those currently being used in the millimetre wave region. The lower available transmit power can be compensated for by the resulting increased gain of the antennae. Power received is inversely proportional to the square of the wavelength.

Additional advantages appear to be low transmit power requirement and compact systems. Finally, it would be possible to provide enough attenuation to permit frequency re-use for relatively short distances again increasing effective bandwidth.

4. Commercial Considerations

If terahertz systems are to be adopted by the wireless industry then there needs to be a strong commercial incentive to do so. Because of the highly directional nature of a terahertz system and terahertz radiation's inability to pass through buildings etc, if it is to find a niche then it will be an application where very high data rates are required over sort distances in a multi-point to point/multipoint basis. One such an application may be the provision of fast-dedicated Internet access for users in cities where the cost of laying optical cables is prohibitively high. The compact nature of a terahertz system would allow it to comply with the strict planning regulations imposed in most cities.

Another requirement for commercial applications is that the cost of implementation must be low and that the technology adopted is mature and reliable. At present this is not the case, conventional terahertz electronic components are prohibitively expensive. Reliability is not such an issue as many systems have shown themselves to be very reliable and have successfully undergone space qualified. Recent advances in fabrication technology have started to tackle the cost issue. This has largely been due to the adoption of micromachining techniques to manufacture the intricate waveguide circuits and active devices used[1,2]. An example of a 500GHz micromachined subharmonic mixer is shown below in figure 4. This image shows one half of the waveguide circuit with the active components incorporated on a quartz microstrip circuit. The cost associated with such a fabrication approach can be greatly reduced when compared with conventional technology.

Figure 5. Scanning electron micrograph of a micromachined 500GHz subharmonic mixer.

The use of micromachining also offers the potential for full system integration. This however appears to be some way of.

5. Conclusions

Present state of the art solid state electronic terahertz components possess sufficient performance to realise a terahertz wireless link of reasonable transmission range. Such a system would have a number of advantages over its lower frequency counterpart, namely reduced mass and compact size, low transmit power and increased bandwidth capability.

Atmospheric attenuation can be adjusted by choosing the appropriate transmission frequency to tune in or out of the molecular absorption lines. By selecting a highly absorbing frequency the repeat range for the re-use of frequency can be adjusted to suit the application again leading to higher effect bandwidth.

In order that useful transmit ranges are achieved it is necessary that the gain of the antennae are increased thus making the system highly directional. Likely applications therefore are line of sight multi-point-point/multi-point in crowded environments such as dedicated Internet access in cities

The commercial incentive for the adoption of such high frequencies is not obvious although the impact of recent technological advances is likely to reduce the high cost associated with such systems.

6. References

[1] 'Millimetre and Submillimetre Wave Components for Future Diagnostic Instrumentation', C. M. Mann, MRS April 2000, San Francisco, California USA

[2] '*2.5-THz GaAs Monolithic Membrane-Diode Mixer*', Peter H. Siegel, IEEE, R. Peter Smith, Michael C. Gaidis, Suzanne C. Martin, 596 IEEE Transactions On Microwave Theory And Techniques, Vol. 47, No. 5, May 1999

Towards Terahertz Communications - Systems Requirements

J.W. BOWEN
Department of Cybernetics, The University of Reading, Whiteknights, Reading, RG6 6AY, UK

1. INTRODUCTION

This chapter concentrates on identifying the subsystems and techniques that need to be developed if we are to fully exploit the communications potential of the large bandwidths available to us in the terahertz range. We will pay little attention to current device and component performance, concentrating more on the underlying information and communication theory. The chapter looks at the links between bandwidth, signal-to-noise ratio and the rate at which information can be transmitted before considering the modulation and multiplexing techniques that could be employed.

2. BANDWIDTH AND CHANNEL CAPACITY

A fundamental limit on the bandwidth of any communications system is imposed by the range of frequencies over which it is possible to obtain low-loss, dispersion-free propagation in the wave guiding components from which the system is constructed. For rectangular metal waveguide, the lower frequency limit is usually set a little higher than the cut-off frequency in order to avoid the high losses that occur in its vicinity. The upper frequency limit is set by the onset of multi-mode propagation, at which point the propagation becomes dispersive. This results in a bandwidth of a little less than an octave for full-height rectangular metal waveguide. However, as the centre frequency increases, and the waveguide is reduced in size

R.E. Miles et al. (eds.), Terahertz Sources and Systems, 269–283.

accordingly, its bandwidth increases commensurately. The usable bandwidth for full-height rectangular metal waveguide is plotted against centre frequency in Figure 1. It is clear that the bandwidth increases linearly with centre frequency as the waveguide dimensions are reduced and that a significantly larger bandwidth can be achieved if the operating frequency is increased.

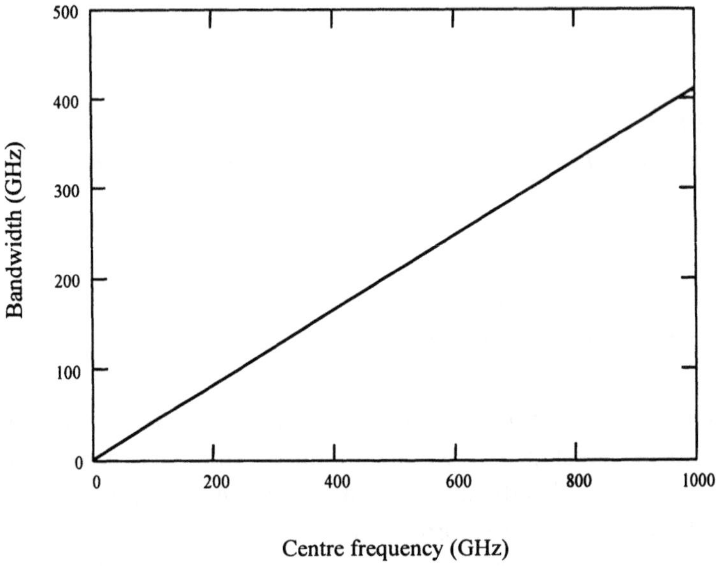

Figure 1. Usable bandwidth of full-height rectangular waveguide

In order to relate the bandwidth to the rate at which information can be conveyed we need to consider some fundamental results from information theory. Ignoring noise induced errors for the moment; the quantity of information in a message depends on both the length of the message and on the probability each symbol in the message has of being transmitted. For example, if only one symbol is ever transmitted, the message received at the receiver is completely predictable and conveys no information. The less likely a symbol is of being sent, the more information it conveys when it is actually received. So that the quantity of information in a series of symbols forming a message adds correctly, the quantity of information I in a symbol of probability p is defined as [1]

$$I = -\log_2 p \text{ bits} \tag{1}$$

For example, if we assume that the letters of the English alphabet are equiprobable, each has a probability of 1/26 and thus contains 4.7 bits of information. If the symbols have different probabilities of occurring, we can calculate the average information, which is termed the entropy. The maximum entropy occurs when the symbols are equiprobable. For example, in a binary system, each binary digit can carry a maximum of 1 bit of information.

In order to quantify the maximum rate at which information can be conveyed by a continuous analogue signal it is convenient to consider digitising the signal by sampling it at regular intervals. The information per second will then be given by the maximum number of independent samples per second multiplied by the information per sample. The Nyquist sampling rate gives the maximum number of independent samples per second as $2B$, where B is the signal bandwidth. In order to determine the information per sample, we can recognise that the smallest detectable signal amplitude is given by the root mean square noise in the system σ. If s represents the maximum signal amplitude, the number of distinguishable amplitude levels must be s/σ. Hence, if all levels are equiprobable, the information rate is

$$C = 2B\log_2\left(\frac{s}{\sigma}\right) \text{ bps} \tag{2}$$

The units are bits per second, usually abbreviated as bps. Alternatively, C may be expressed in terms of the signal-to-noise power ratio

$$C = 2B\log_2\left(\frac{S}{N}\right)^{\frac{1}{2}} = B\log_2\left(\frac{S}{N}\right) \text{ bps} \tag{3}$$

In reality, the approach we have taken in deriving equation (3) is an approximation. A rigorous analysis yields

$$C = B\log_2\left(1+\frac{S}{N}\right) \text{ bps} \tag{4}$$

This is known as Shannon's Ideal Communication Theorem. C is called the channel capacity and represents the maximum error-free communication rate for a communications channel of bandwidth B and signal-to-noise power ratio S/N. Hence, in theory, it is possible to convey information at a rate up to C with no errors in the received message at all. However, it can be shown that to achieve this it would be necessary to encode each symbol to be sent as a noise-like waveform; this is currently not practicable.

In practice, the information rate in real communications systems falls somewhat short of C and the inevitable presence of noise will cause some errors in the received message. Nonetheless, the channel capacity is a useful figure for comparing communications systems and the following useful deductions can be made from equation (4).

Firstly, equation (4) indicates that the same amount of information can be transmitted in a shorter time if the bandwidth is increased.

Figure 2 shows the channel capacity for full-height rectangular metal waveguides plotted against their centre frequency for a range of different signal-to-noise values. From the plot, it is clear that a higher signal-to-noise ratio gives a higher channel capacity.

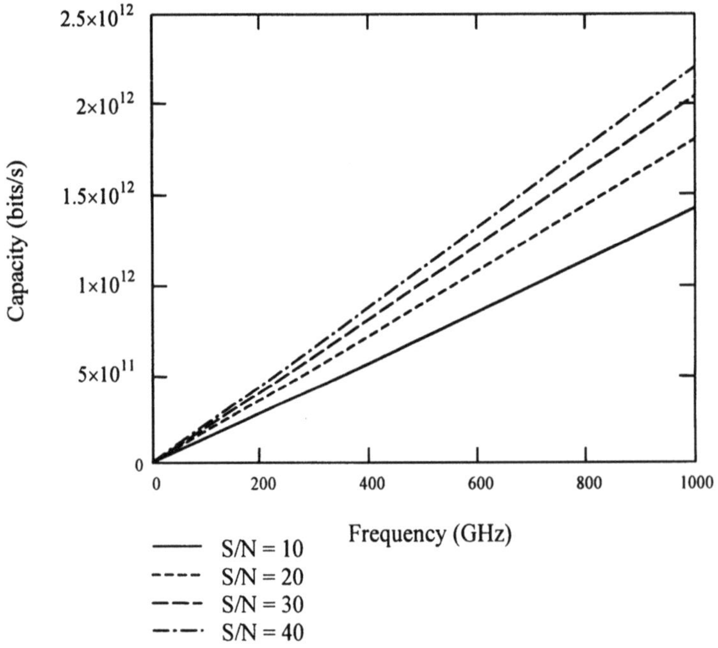

Figure 2. Channel capacity for full-height rectangular waveguide

Figure 3 shows the bandwidth plotted against signal-to-noise for a fixed channel capacity of 500 Gbps. This shows how signal-to-noise ratio can be traded off against bandwidth. For example, a bandwidth of 100 GHz, corresponding to a waveguide with a centre frequency of about 250 GHz, requires a signal-to-noise power ratio of 31 for a channel capacity of 500 Gbps. However, if the bandwidth is increased to 200 GHz, corresponding to a waveguide with a centre frequency of about 490 GHz, a signal-to-noise power ratio of only about 4.5 is necessary for the same channel capacity. So, in this case, for a doubling of the bandwidth, the same information rate can be obtained with almost a factor of 7 reduction in signal-to-noise ratio. In practice, there is a lower threshold for signal-to-noise below which the error rate increases rapidly. This threshold level varies depending on the modulation scheme employed, but for frequency modulation is at about 5.

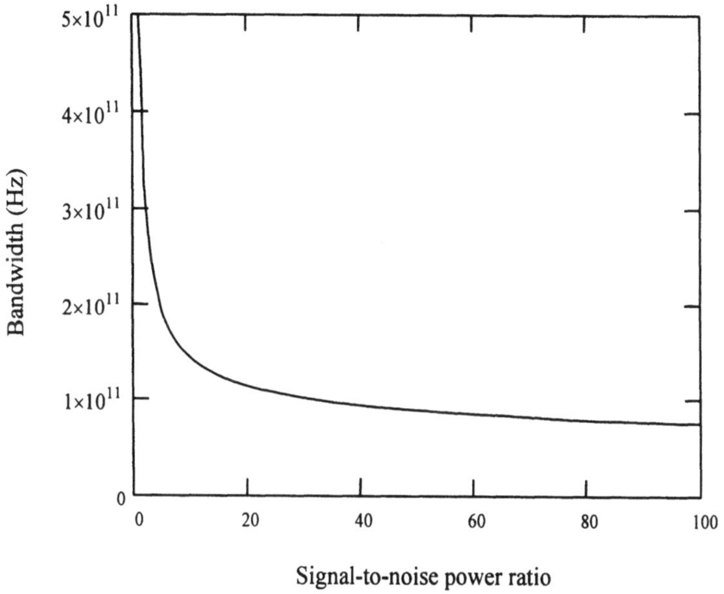

Figure 3. Trade off between bandwidth and signal-to-noise ratio for a fixed channel capacity of 500 Gbps

For comparison, the information rate requirements for current applications varies from about 6.4 kbps for a digital telephone channel up to 600 Mbps for broadband ISDN (Integrated Services Digital Network).

3. MODULATION

In order to transmit information on a THz carrier wave, the carrier has to be modulated in some manner. There are three main types of modulation employed in communications systems: amplitude modulation (AM), frequency modulation (FM) and pulse modulation. Each is discussed below.

3.1 Amplitude Modulation

Figure 4 shows a carrier wave, a signal represented by a lower frequency sine wave, and the results when the signal amplitude modulates and frequency modulates the carrier wave. It is clear that the envelope of the AM waveform follows the original signal.

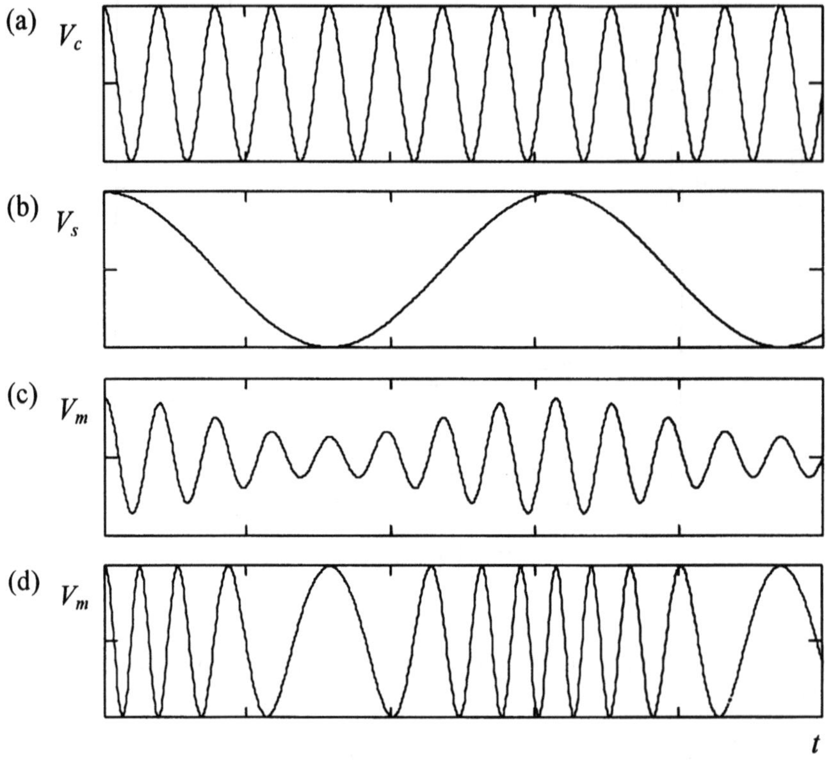

Figure 4. (a) Carrier, (b) signal, (c) amplitude modulated waveform, (d) frequency modulated waveform

Mathematically, we can represent the carrier as $V_c = a_c \cos\omega_c t$ and the signal as $V_s = a_s \cos\omega_s t$. The amplitude modulated waveform has the form

$$V_m = a_c(1 + m\cos\omega_s t)\cos\omega_c t \tag{5}$$

where $m = a_s / a_c$ is called the modulation index.

Rewriting equation (5) we find

$$V_m = a_c \cos\omega_c t + \frac{ma_c}{2}\cos(\omega_c + \omega_s)t + \frac{ma_c}{2}\cos(\omega_c - \omega_s)t \tag{6}$$

Hence, the AM waveform contains the unchanged carrier plus components at the sum and difference of the carrier and signal frequencies.

For a more general signal $f(t)$, the AM waveform is

$$V_m = a_c(1 + mf(t))\cos\omega_c t = a_c \cos\omega_c t + a_c mf(t)\cos\omega_c t \tag{7}$$

Hence, its spectrum is the unchanged carrier plus the convolution of the signal and carrier frequencies. This is illustrated in Figure 5. All of the information in the original signal is contained in any one of the two sidebands and so it is only strictly necessary to transmit one sideband. Three types of AM can be defined, dependent on how much of this spectrum is transmitted: double-sideband, large carrier (DSB-LC) – in which the whole AM waveform is transmitted; double-sideband, suppressed carrier (DSB-SC) – in which the unchanged carrier is not transmitted; and single sideband (SSB) – in which one sideband and the unchanged carrier are not transmitted.

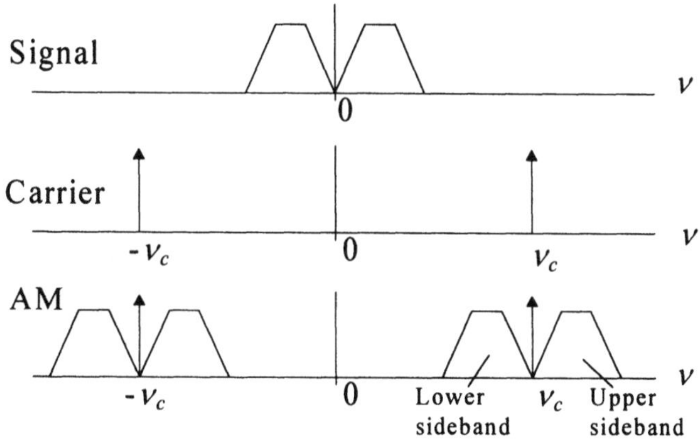

Figure 5. AM spectrum

3.1.1 Double-Sideband, Large Carrier

Generation of DSB-LC can be carried out using a diode modulator, a non-linear device similar to a mixer. If the device has a characteristic of the form

$$i = a + bV + cV^2 \tag{8}$$

where i is the current, a, b and c are constants and the input V is

$$V = a_c \cos\omega_c t + a_s \cos\omega_s t \tag{9}$$

the output will be of the form

$$\begin{aligned} i &= a + ba_c \cos\omega_c t + ba_s \cos\omega_s t \\ &+ ca_c a_s \cos(\omega_c - \omega_s)t + ca_c a_s \cos(\omega_c + \omega_s)t \\ &+ ca_c^2 \cos^2 \omega_c t + ca_s^2 \cos^2 \omega_s t \end{aligned} \tag{10}$$

This contains the necessary carrier, sum and difference frequency terms. Appropriate tuning of the embedding impedance should allow power to be optimised at the desired frequencies.

As the whole AM waveform is transmitted, the signal can be extracted by simply detecting its envelope. This is easy to achieve using direct video detection with a diode detector and smoothing filter.

Although DSB-LC is relatively easy to modulate and demodulate, it is enormously wasteful of power. Typically 90% of the transmitted power is in the carrier, which contains no information and is removed by the demodulator.

3.1.2 Double-Sideband, Suppressed Carrier

DSB-SC is much more efficient than DSB-LC because no power is wasted in transmitting the carrier. Generation of DSB-SC is conventionally achieved using a balanced modulator based on two non-linear devices. A block diagram of a typical example is shown in Figure 6. The modulating signal is applied in anti-phase to the two non-linear devices whereas the carrier is in-phase. The output from each non-linear device will be of the form of a DSB-LC waveform:

$$i_1 = a_c(1 + m\cos\omega_s t)\cos\omega_c t \tag{11}$$

$$i_2 = a_c(1 - m\cos\omega_s t)\cos\omega_c t \tag{12}$$

Taking the difference gives DSB-SC modulation:

$$i_1 - i_2 = 2a_c m\cos\omega_c t \cdot \cos\omega_s t = a_c m[\cos(\omega_c - \omega_s)t + \cos(\omega_c + \omega_s)t] \tag{13}$$

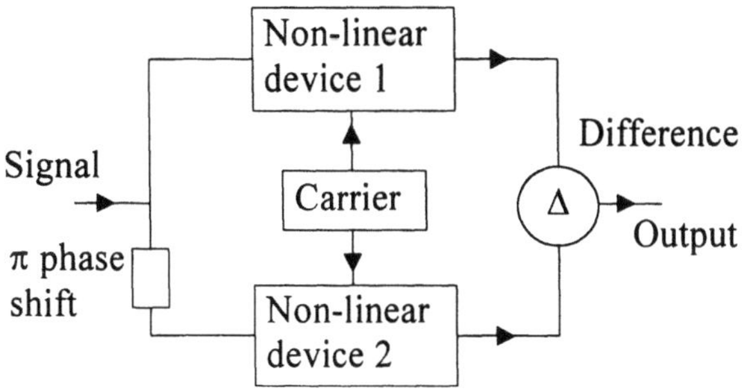

Figure 6. Balanced modulator

Because the carrier is not transmitted, demodulation of DSB-SC requires the carrier to be effectively supplied at the receiver. Hence, coherent detection must be used, a local oscillator supplying a local carrier. In general, the demodulated output is

$$V = a_c \cos \omega_s t \cdot \cos(\Delta\omega t + \phi) \qquad (14)$$

where $\Delta\omega$ and ϕ are the frequency and phase errors in the local carrier. These must be zero to avoid distortion of the signal. A Costas loop demodulator [2] can be used to ensure frequency stability of the local carrier.

3.1.3 Single Sideband

SSB modulation can be generated in two ways. A DSB-SC signal with a low (e.g. radio) frequency carrier can be generated, the unwanted sideband removed with a single sideband filter and then the signal upconverted using a mixer. Alternatively, the arrangement in Figure 7 could be used, although the need to provide exactly π/2 radian phase shift to all frequencies present in the modulating signal would add considerably to the complexity.

Coherent detection has to be used for demodulation, as with DSB-SC. SSB requires only half the transmitter power and half the bandwidth of a DSB-SC system.

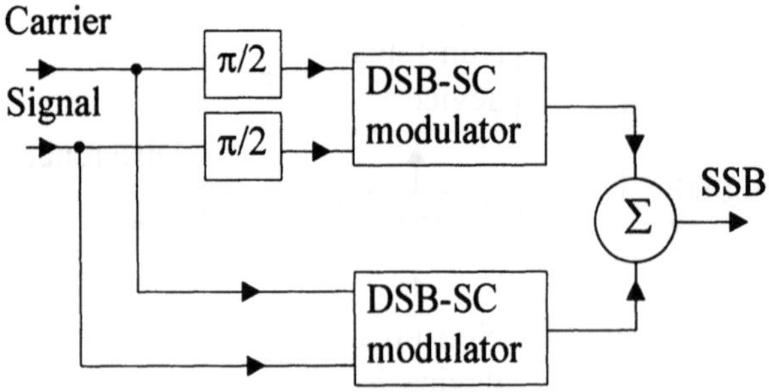

Figure 7. Single sideband modulator

3.2 Frequency Modulation

In an FM waveform, the instantaneous frequency depends on the signal. Figure 4 shows the example of a sinusoidal signal frequency modulating a carrier. The equation for an FM wave is

$$V_m = a_c \cos(\omega_c t + \beta \sin \omega_s t) \tag{15}$$

where

$$\beta = \frac{\Delta \nu_c}{\nu_s} \tag{16}$$

$\Delta \nu_c$ is the peak frequency deviation of the carrier wave, and ν_s is the frequency of modulating wave.

The resulting spectrum in Figure 8 can be seen to consist of a series of lines at $\nu_c \pm n\nu_s$ with amplitudes given by Bessel functions of order n i.e. $J_n(\beta)$. The amplitudes become negligible for $n > \beta + 1$. Hence, FM automatically spreads the signal energy over a wide bandwidth. This also holds true for a general modulating signal, although the spectrum would then be much more complicated, frequency modulation being a non-linear process. The channel bandwidth required for a signal bandwidth of B_s is given to a good approximation by Carson's rule:

$$B_c \approx 2(\beta + 1)B_s \tag{17}$$

From our earlier discussion of bandwidth and channel capacity it is clear that FM needs a lower signal power than an AM system to achieve the same information rate. Furthermore, FM has a higher immunity to noise than AM

– frequency is less affected by noise than amplitude, and the signal is spread over many separate sidebands in FM but only two in AM.

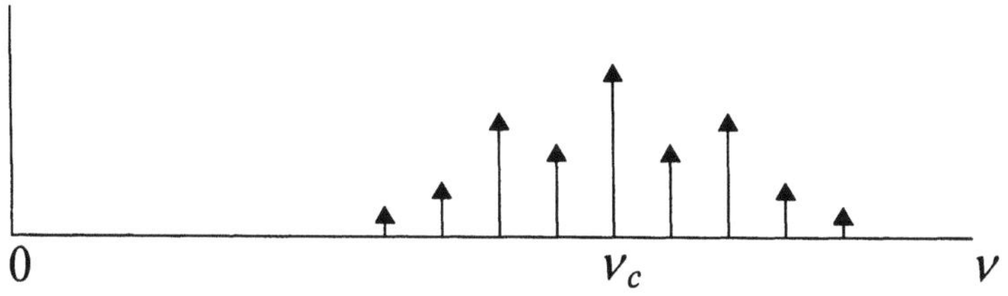

Figure 8. FM spectrum

FM can be produced directly by voltage tuning of an oscillator, e.g. tuning the bias on a Gunn oscillator. Alternatively, varactor tuning of the oscillator embedding impedance could be employed. The Armstrong modulator [3] can be used to generate FM indirectly. Here, narrow-band FM is generated by integrating the signal before feeding it to a phase modulator formed from a balanced modulator and $\pi/2$ phase shifter. Frequency multiplication and frequency conversion stages convert this output to wide-band FM with the desired β.

FM demodulation can be achieved using a frequency discriminator, the action of which is to produce a voltage proportional to the instantaneous frequency of the modulated wave. This is usually accomplished by using a differentiator, which can be a relatively simple tuned circuit, followed by an envelope detector. Alternatively, demodulation can be by a phase-locked loop. Here a voltage controlled oscillator in a feedback loop follows the instantaneous frequency (by means of a phase comparitor), the controlling voltage being proportional to the frequency of the modulated wave and providing the demodulated output.

3.3 Pulse modulation

Instead of using AM or FM, information may be transmitted using various forms of pulse modulation. The commonest form of modulation for communications purposes is pulse code modulation (PCM), in which the signal to be transmitted is digitised and sent as a sequence of pulses representing the resulting binary code. The signal is impressed on the carrier wave using schemes such as amplitude, frequency or phase shift keying (ASK, FSK or PSK). In ASK the amplitude of the carrier wave is switched between two levels (sometimes on and off) to represent binary 0's and 1's. In FSK the carrier wave is switched between two frequencies. In PSK the carrier wave is switched between two different phase delays. More elaborate schemes can be used in which, for example, more than two different phase delays are used, so that a group of binary digits can be represented by one pulse, thereby increasing the information rate. At the expense of transmitting extra binary check digits, error correction coding techniques can be used to further increase the information rate.

As an example, if we have a system with a bandwidth of 150 GHz, we can use simple binary ASK or PSK to transmit at an information rate of up to 150 Gbps. In reality, if the signal-to-noise power ratio were 5, using this sort of PCM at this information rate would result in an unacceptably high bit error rate (about 0.01 for ASK). A much more acceptable bit error rate of 10^{-5} could be achieved at the same signal-to-noise ratio if the information rate were reduced to about 40 Gbps (PSK, which has better noise immunity, could cope with twice this rate). However, if we wish to transmit at even this rate, it requires us to be able to controllably switch a terahertz transmitter between its two states at up to 40 GHz – this is well beyond the capabilities of current technology!

4. MULTIPLEXING

Multiplexing enables several signals to be transmitted, more or less in parallel, along the same communications channel. As the information rate required for typical signals may be rather less than the channel capacity available at terahertz frequencies, multiplexing will be important to make the best use of the available bandwidth.

4.1 Time division multiplexing

In time division multiplexing (TDM) the pulses from several pulse modulated signals are interleaved and transmitted sequentially as indicated in Figure 9. In this simple example, there are three signals, designated A, B and C; the first, fourth and eighth pulses relate to signal A, and so on. A practical implementation of TDM requires a switch at the transmitter to switch between the signals and a similar switch at the receiver to separate the pulses into the separate signals.

In the section on pulse modulation above we stated that a system with a 150 GHz bandwidth could cope with an information rate of at least 40 Gbps. If we wanted to build a TDM system operating at this rate, we would need to be able to switch between signal channels at a rate of 40×10^9 signal channels per second. This would, however, relax the need to switch each signal channel at this rate. For example, if we were to multiplex 100 channels, each would need a clock rate of only 400 MHz, transmitting 400 Mbps. While the

TDM switch has to switch at a rate of 40 GHz, all it has to do is repeatedly switch between the signal channels – this should be easier to realise than having to produce a PCM signal with a 40 GHz pulse rate.

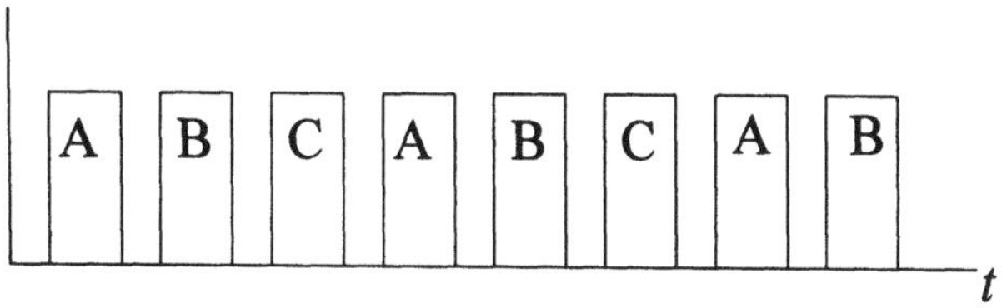

Figure 9. Time division multiplexing

4.2 Frequency division multiplexing

In frequency division multiplexing (FDM), several signals are transmitted simultaneously using different frequency ranges within the available channel bandwidth. Figure 10 is a block diagram of an FDM transmitter that multiplexes 3 signals. Each signal originally occupies the same bandwidth, but modulates carriers at 3 different frequencies ν_{c1}, ν_{c2} and ν_{c3}. The resulting modulator outputs are superimposed to give the baseband signal. This then modulates the high frequency carrier ν_c to give the transmitted signal $f_T(t)$. In the receiver shown in Figure 11, the incoming signal is first demodulated to recover the baseband. The original signals are recovered by selection with a filter bank and demodulation. The two stages of modulation employed here could be both AM, both FM, or any combination of AM and FM.

5. CONCLUSIONS

The broad bandwidths of terahertz systems offer the potential of high information rates for wireless communications. However, increased bandwidths mean that the total noise power in the receiver will increase. We have seen that this can be offset to some extent by trading off some of the increased information capacity to allow communications with a lower signal-to-noise ratio. Furthermore, as the frequency is increased, antenna gain increases (for fixed antenna dimensions), and the signal power at a receiving antenna increases as the square of the frequency. The Friis transmission formula (see the Chapter by C. M. Mann) can be used to calculate how much transmitter power is required to cover a given distance, bearing in mind the required signal-to-noise ratio at the receiver, worse case atmospheric attenuation and other losses. Modest transmitter powers of a few tens of milliwatt can cover distances of the order of a kilometre at 400 GHz with a channel capacity of about 380 Gbps. At this frequency, increasing the transmitter power by a factor of 10^3 increases the range by about 1km. Transmission over greater distances would require the development of low noise amplifiers or repeaters (these would receive the signal and then retransmit it onwards) to boost the signal power.

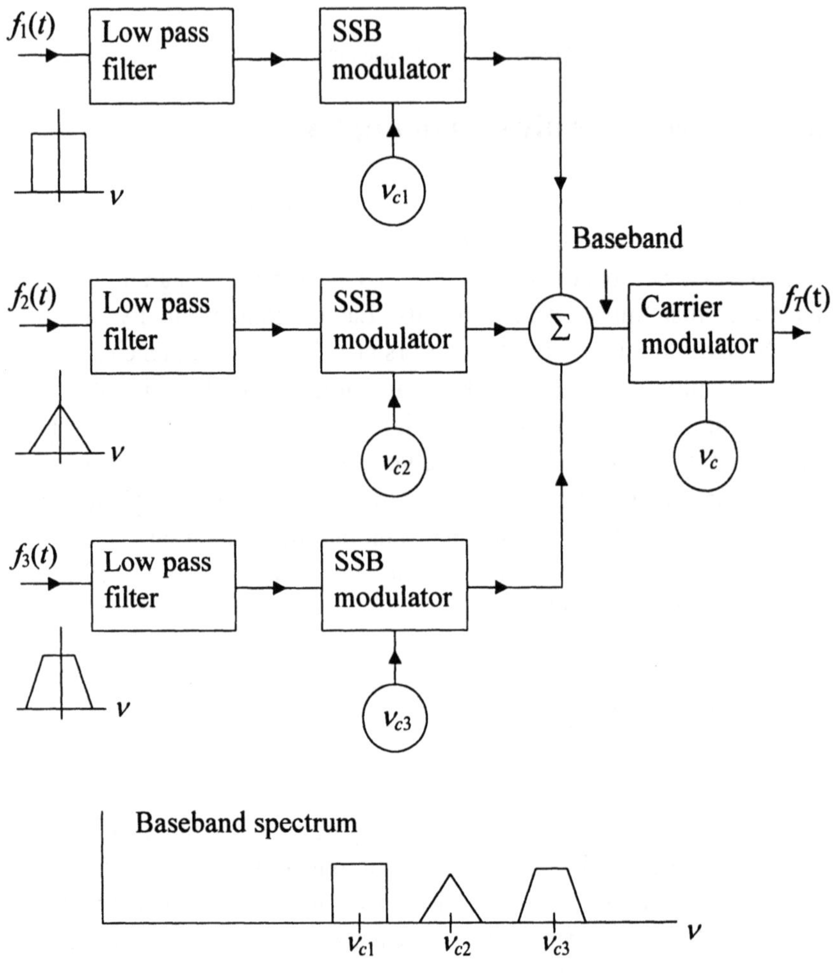

Figure 10. Frequency division multiplexing transmitter

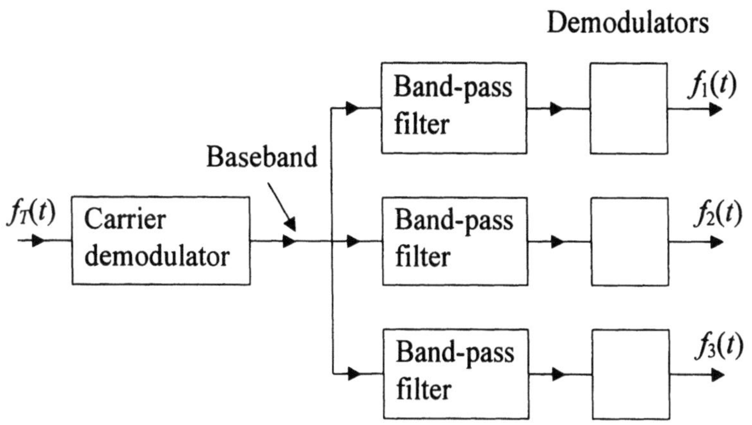

Figure 11. Frequency division multiplexing receiver

Modulation and demodulation can be accomplished using non-linear devices or by direct frequency modulation of sources. Inexpensive integration techniques are important to keep costs sufficiently low.

Frequency modulation would probably be the simplest and best approach for prototype terahertz communications, giving much improved S/N over AM techniques. Frequency division multiplexing would be necessary to make full use of the available bandwidths. Digital communications at terahertz frequencies would require the development of fast switching techniques to achieve the potential information capacity.

The need for terahertz communications depends predominantly on the need for high information rate mobile communication systems. There is every indication that information rate and bandwidth requirements will continue to grow as manufacturers try to provide consumers with ever more sophisticated features.

REFERENCES

[1] Usher, M.J., Guy, C.G., *Information and Communication for Engineers*. Basingstoke: Macmillan Press Ltd, 1997.

[2] Dunlop, J., Smith, D.G., *Telecommunications Engineering*. Wokingham: Van Nostrand Reinhold (UK) Co. Ltd, 1984.

[3] Carlson, A.B., *Communication Systems*. New York: McGraw-Hill Book Company, 1975.

MICROMACHINING FOR TERAHERTZ CIRCUITS AND SYSTEMS

R. E MILES, D. P. STEENSON and J. M. CHAMBERLAIN
School of Electronic and Electrical Engineering
University of Leeds
Leeds LS2 9JT
UK

Abstract

This chapter reviews the circuit techniques available for use in terahertz systems and concludes that there are significant advantages in scaling the existing RF rectangular waveguide structures to operate at these higher frequencies. It goes on to describe how rectangular wave guide with the required dimensions of ~ 100 microns and good surface finish can be fabricated using an ultra-thick photo resist. The electrical properties of this photoresist at THz frequencies are examined in terms of its use in a variety of circuit configurations.

1. Introduction

Existing terahertz systems, as used in for example astronomy, remote sensing and spectroscopy, are based on a combination of quasi optics and conventionally machined waveguide components such as mixer blocks. The resulting circuits are not only bulky but are also very expensive because of the specialised equipment and highly skilled operators required for their manufacture. This state of affairs while, not being a problem for the highly specialised "one-off" applications addressed by THz technology hitherto, becomes a serious issue as THz technology is applied to a wider range of uses. Many of these will require an inexpensive, robust and hence mass produced "integrated circuit" implementation. In the medical field alone the use of THz imaging and sensing in the routine diagnosis and monitoring of skin conditions, wound care and dehydration will require large numbers of portable systems in hospital wards and clinics. They could even become a standard item in a general practitioner's "black bag".

If terahertz technology is to satisfy even a fraction of its promised potential there is a real need for a high volume, low unit cost method of circuit and system manufacture. This inevitable means replacing the skilled machine operator with automatic systems. As will be shown below, at THz frequencies, typical circuit dimensions are measured in microns and therefore it is natural to look towards the highly developed methods used in semiconductor integrated circuit manufacture (i.e. micromachining), for ways of fabricating THz circuits. The successful use of these techniques will automatically satisfy the need for robust, inexpensive and miniaturised systems.

There are two main thrusts to the micromaching approach, namely devices and circuits. Passive semiconductor devices such as the Schottky diode (and now the Heterostructure Barrier Varactor) are already an established part of the technology and steady progress (see elsewhere in this volume) is being made towards the realisation of solid state THz sources (in the form of fundamental THz oscillators and

R.E. Miles et al. (eds.), Terahertz Sources and Systems, 285–297.

as up or down converters). This chapter will however focus on the circuit aspects of THz technology and discuss how micromaching can be put to use in the manufacture of these very high frequency circuits.

2. Terahertz Circuit Requirements

In simple terms we might say that "we already know what we need to know about terahertz circuits". After all, it is only high frequency electromagnetic radiation and a highly developed transmission line technology already exists, albeit at lower frequencies. All that is necessary therefore is to scale down the dimensions of the plethora of well understood waveguide and microstrip based components that have been developed over the years. If you want a THz coupler, take the design of a 10 GHz component and divide the dimensions by 100! (For an example of this see the e-plane filter discussed in section 3.2). The problem is, how do we manufacture these components with the required dimensional accuracy and surface finish to operate efficiently at THz frequencies? In addition to the manufacturing issues there are also considerations in choice of materials. As frequency increases so do the energy losses in a transmission medium. These losses are associated with molecular and atomic vibrations which are not apparent at lower frequencies. (These, of course, are the very phenomena that make THz spectroscopy so attractive a proposition in the identification of chemical signatures).

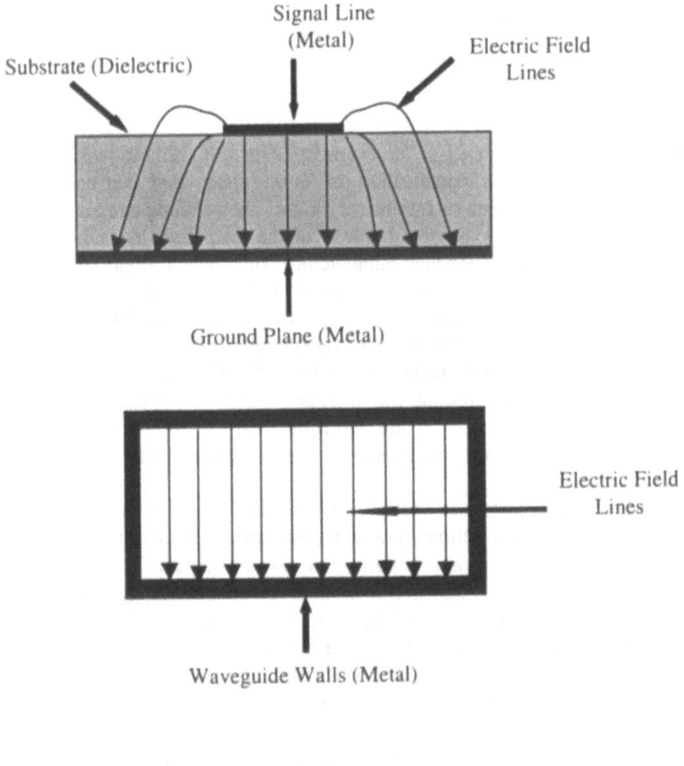

Figure 1.Electric fields in microstrip (top) and rectangular waveguide (bottom)

By the very nature of microstrip circuits, energy is transmitted in the media surrounding the transmission line i.e. air and the dielectric substrate (see Figure 1a). Because of its high dielectric constant, transmission is mostly in the substrate where of course it is subject to energy loss mechanisms. Energy losses also occur in the metal lines. There is therefore an argument which says that the only acceptable transmission medium is a vacuum - although air will do! In rectangular wave guide there is no propagation through a potentially lossy dielectric but losses associated with currents flowing in the metal walls will of course occur (Figure 1b). Rectangular waveguide also has an advantage in that the signal is confined by the metal walls and is not radiated out of the circuit.

In energy loss terms, rectangular waveguide is an improvement on microstrip but is not as lossless as quasi optics. However, as mentioned in the Introduction, quasi optical circuits are bulky and expensive and cannot be considered as serious contenders for mass market THz systems (quasi-optical techniques will of course be used where free space THz transmission is required, in for example communication or imaging applications). For these reasons, this Chapter will be concerned mainly with the fabrication of rectangular waveguide components for THz frequency systems but will also make some mention of micromachined microstrip components.

2.1 Terahertz Rectangular waveguide dimensions

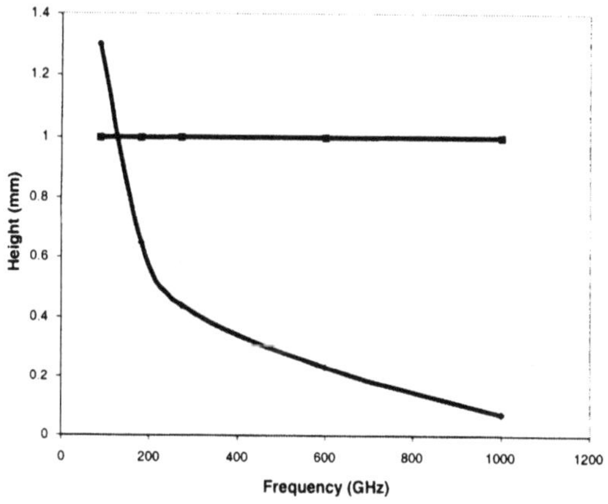

Figure 2. Waveguide height v. frequency

Figure 2 shows the relationship between frequency of operation and the height of rectangular waveguide. Except for the 1 THz point, the dimensions used in Figure 2 are for standard waveguide. The THz figure is calculated for a free space wavelength of 300 microns and taking a height to width ratio of 2:1. It can be seen that, as we enter the THz frequency range at about 300 GHz, the waveguide height drops to below 1 mm. We can start to think in terms of micron dimensions and hence the techniques used in semiconductor device fabrication. However, we need one more ingredient. The heights of the waveguide are still on the large side for conventional photoresist and etching processes. A technique known as LIGA (LIthographie, Galvanoformung und Abformung) can be used to write features in resist layers >1mm thick but this process uses an X-ray source and is therefore expensive and inconvenient to use. The last few years have seen the introduction and widespread use of ultra-thick photoresists that can be patterned using conventional photolithographic equipment. Perhaps the most popular of these is known as SU-8, developed by IBM for inkjet printer technology (a web site detailing the history, uses and properties of SU-8 can be found at http://aveclafaux.freeservers.com/SU-8.html). SU-8 can be patterned in thicknesses > 1 mm to give, where required, aspect ratios of better than 6:1. Figure 3 shows a 170 μm diameter metalised SU-8 post having a height of 700 μm.

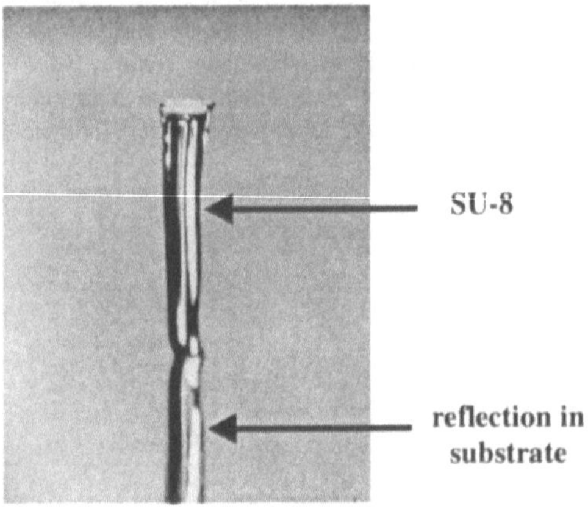

Figure 3. 700 μm × 170 μm SU-8 post

SU-8 can therefore be patterned directly using UV exposure and developer to form waveguide structures without any need for chemical etches as in other micromachining technologies (see Si micromaching later in this chapter). Metallisation is achieved by sputtering/electroplating gold onto the SU-8 formers, to a thickness greater than 3 skin depths, at the frequency of operation in order to minimise ohmic losses.

3. Wave Guide Structures Fabricated using SU-8

The following sections outline the techniques developed at the University of Leeds [1,2] for the fabrication of rectangular waveguide components using SU-8.

3.1 Snap-Together Waveguide

This method of construction, which fabricates the waveguide in two halves simultaneously, was pioneered by Caroline Collins (now McIntosh) at The University of Leeds in 1998 [1]. Figure 4 shows the process sequence and illustrates the principles of the method of fabrication. The waveguide can be fabricated on any flat substrate that can stand the backing temperature of ~ 100 C. Typically a Si wafer or a glass microscope slide are used.

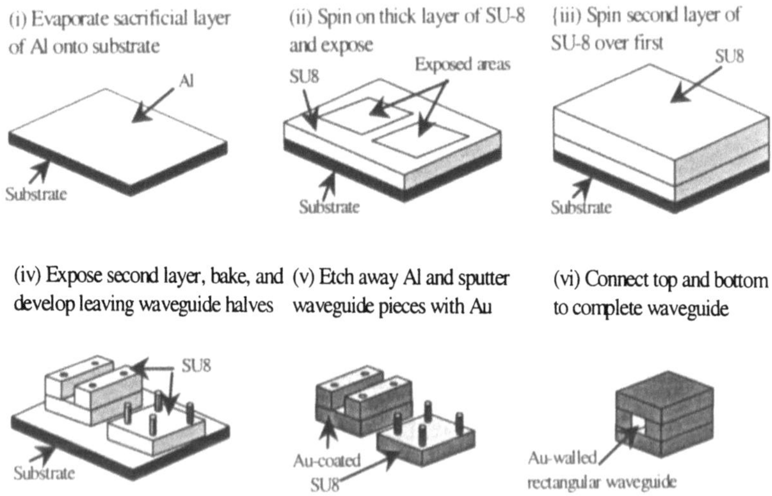

Figure 4. Snap together SU8 wave guide fabrication sequence

The first stage (i) is to evaporate a sacrificial layer of aluminium over a suitable substrate (e.g. glass, Si). This is followed by a layer of SU-8 which is prebaked and exposed through a mask, but not developed, to form the top and bottom layers of the waveguide (ii). A second layer is then spun on top to a thickness equal to ½ of the desired waveguide height (iii). After another prebake, this second layer is patterned to form the side walls of the structure, in the case shown here a straight length of waveguide (iv). The patterning also includes a series of pegs in one section and matching holes in the other. After developing, the structure in the layers is revealed and the resulting 2 halves are metalised by sputtering (and if necessary, electroplating) a layer of gold all over (v). Finally after dissolving away the layer of aluminium in dilute NaOH the two pieces can be separated and "snapped" together to form the final structure (iv). Note that in this process, apart from dissolving away the etch release Al layer, no chemical etching is necessary and because the SU-8 acts only as a former for the waveguide structure it's electrical properties at THz frequencies are immaterial as the signal is shielded from the remaining resist by the metallisation.

Photographs of the finished components are shown in Figure 5 together with an unmetallised assembled section which, because the exposed SU-8 is transparent in the visible, shows the internal details.

The waveguide illustrated was designed to operate at W-band (i.e. 77 – 110 GHz) in order to facilitate electrical characterisation using a commercial network analyser but higher frequency designs are just as easy, if not easier to fabricate. S-parameter measurements were obtained by mounting the waveguide section in a special jig into which conventional w-band connectors had been incorporated. The results of the electrical characterisation are shown in Figure 6 where it can be seen that the wave guide has an acceptably low insertion loss of < 4dB and high return loss of > 18dB over the w-band frequency range. These figures are for a waveguide length of about 20 mm.

Figure 5. Snap-together waveguide sections. The one in the centre is assembled but unmetallised showing the internal structure through the transparent SU-8. The outer pieces are the two metalised halves before assembly.

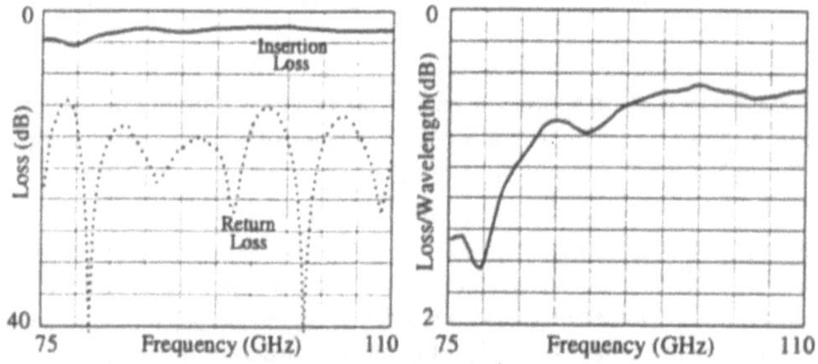

Figure 6. Measured Insertion, Return and Transmission Losses for the snap-together waveguide shown in Figure 5.

It is a simple matter to generalise this process to make other components such as power dividers, couplers, circulators and waveguide horns and mixer blocks (see below).

3.2 E-Plane Filter

The components discussed in the last section form the essential "wiring" of a THz circuit but it is of course necessary to have other, more functional, elements such as filters, detectors, multipliers sources and, when they become available, THz transistors (the 3 terminal RTD?). These functions can be accommodated by mounting devices inside the micromachined waveguide and this will be illustrated here by looking at the realisation of an E-plane filter [2].

The dimensions for this filter were derived from a design by Budimir and Postoyalko [3] intended for operation at X-band, by simply dividing by a factor of 10 to give a W-band device. (W-band was chosen again for ease of characterisation but the process can be easily scaled down to operate at higher frequencies). Figure 7 shows the stages in fabrication which are similar to, but somewhat less complicated than those discussed above for the snap together waveguide.

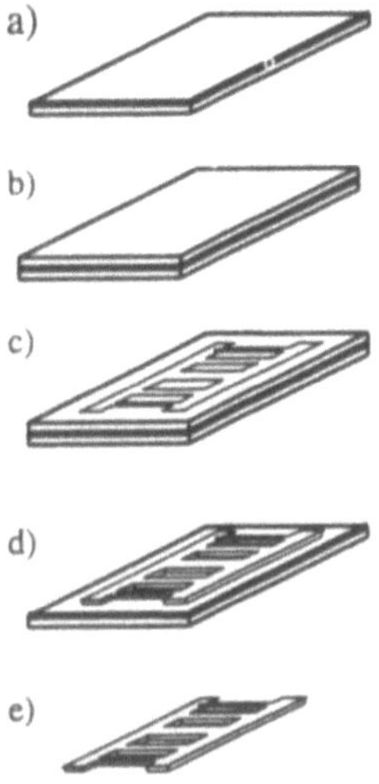

a) Evaporate sacrificial layer of Al onto glass substrate

b) Spin thick layer of SU-8 onto substrate and prebake

c) Expose layer to UV via mask and post expose bake

d) Develop away unexposed areas of SU-8 and hardbake

e) Etch away Al to release filter element from substrate and sputter both sides with Au

Figure 7. Steps in the fabrication of an SU-8 E-plane filter.

Referring to Figure 7; (a & b) A single layer of SU-8 is spun onto a sacrificial layer of aluminium predeposited on a glass substrate and then given a suitable prebake. In this instance the thickness of the SU-8 was ~ 30 μm. (c) The SU-8 is next exposed through a mask using a conventional mask aligner and (d) developed to form the e-plane insert. (e) The insert is then released from the substrate by dissolving the Al away in dilute NaOH and metalised as before (> 3 skin depths), but this time on both sides. The finished e-plane insert is shown in Figure 8.

Figure 8. Completed E-plane insert

For electrical characterisation purposes the insert was clamped between the two sides of a conventionally machined length of W-band waveguide which was split in the vertical direction. The results of the characterisation are given in Figure 9.

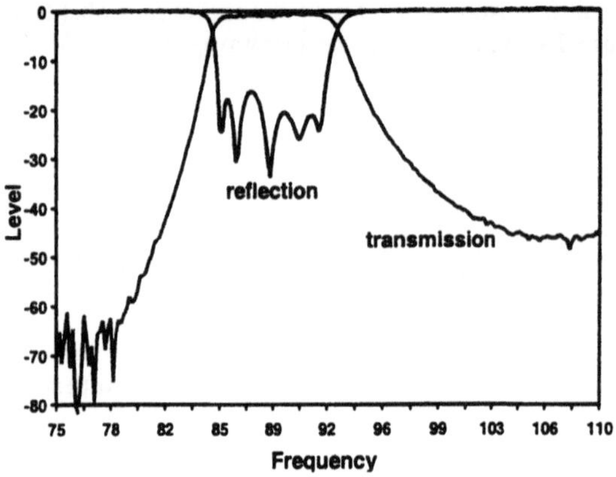

Figure 9. E-Plane Filter Response

The results in Figure 9 show that the filter characteristics are very close, but not quite identical, to those predicted (see Table 1). The reason for the slight discrepancy is not yet understood but it was found that the performance depended critically on the quality of the electrical contact between the e-plane insert and the wave guide walls.

Table 1. Comparison of the measured E-plane pass band characteristics with theoretical predictions.

	WG-16 prototype	EEsof simulation	W-band SU-8 Cu
Insertion loss. dB	0.45	0.1	0.7
Return loss. dB	19	21.1	16.3
Bandwidth. %	6.21	9.05	9.06

E-plane inserts are now being incorporated into the snap-together structures described in 3.1 above. In this case the waveguides will be fabricated on their "edge" such that the joining plane is in the "vertical direction. This is the preferred method of splitting a wave guide along its length because current does not flow across the top of a waveguide. Once more it should be noted that the THz properties of the SU-8 are not relevant to this technology.

3.3 Active Device Mounts in SU-8

Attempts have been made, especially by the EPSRC funded TINTIN consortium in the UK to build on-chip waveguide around semiconductor devices fabricated in the substrate. Difficulties associated with contacting the devices arose in this work which have not as yet been overcome. However, the e-plane insert described in the previous section can be easily modified to form the mount for a semiconductor device in THz micromachined waveguide. Workers at IEMN at Lille in France [4] are perfecting an epitaxial lift-off technology for semiconductor devices which can then be mounted on these inserts.

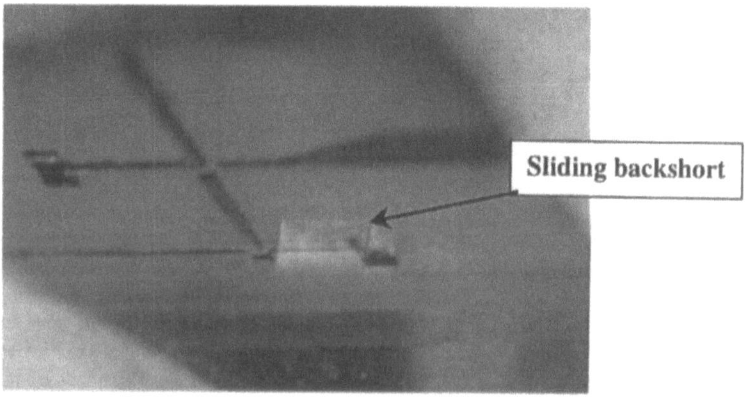

Figure 10. Backshort in 500 GHz SU-8 Mixer block

Alderman et al have fabricated 500 GHz mixer blocks using SU-8 for space borne applications which incorporate a tuned cavity, Schottky Barrier Mixer Diode and LO and IF ports. These blocks are formed in two halves which must be mated together and a problem of SU-8 thickness uniformity is identified. This is overcome by the use of planarisation techniques to ensure that the two halves form an intimate contact. A typical structure is shown in Figure 10. [Courtesy B. Alderman]

Figure 11. Embedded E-plane filter (Alderman et al)

Recent work with SU-8 has been towards developing an embedded mask technology [5] which has the potential of greatly increasing the range of 3-dimensional structures that can be fabricated in SU-8. Figure 11 shows a view looking along a monolithic e-plane filter where it is just possible to make out the vertical columns.

4. Terahertz Properties of SU-8

It has already been pointed out that the high frequency characteristics of the components described above do not depend on the material properties of the SU-8. In most cases the remaining SU-8 resin acts as a metalised former and is shielded from the THz fields. However it will be difficult to produce e-plane active device mounts were SU-8 is not exposed to the electromagnetic environment. Similarly, the performance of microstrip circuits based on SU-8 will be critically dependent on the THz properties of SU-8.

A number of workers have reported measurements of the dielectric properties i.e. the complex dielectric constant of SU-8 over a wide range of frequencies [6], [7] and [8]. Agreement is not always 100% and it has become apparent that the dielectric properties depend very strongly on the preparation conditions. For spin-on purposes SU-8 resin is dissolved in • -butyrolactone (GBL) to an extent depending on the final thickness required. At frequencies ~ 20 GHz SU-8 has a relative permativity ε_r of about 4 and GBL of > 10. The presence of GBL can therefore have a large effect on the measured ε_r and the value is therefore strongly dependent on the backing time and temperature after the SU-8 is spun on i.e. the prebake conditions. Figure 11 is a three dimensional plot of the variation of the real part of the relative permativity of SU-8 as a function of frequency for different prebake times with the sample being held at 90^0C . Figure 12 is a cross-section at a single frequency, 24.62 GHz. These results were obtained from measurements of the resonant properties of microstrip lines deposited on SU-8 which in turn had been spun-on to a GaAs substrate. Due allowance was made for the substrate. A photograph of the lines measured are shown in Figure 13.

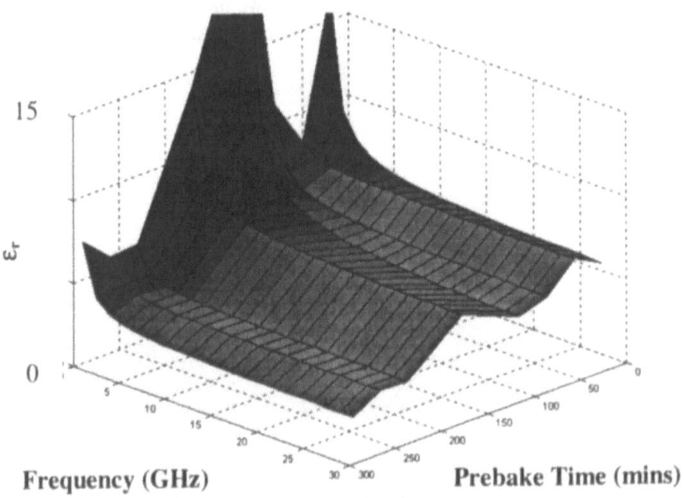

Figure 12. Variation of ε_r with prebake time and frequency

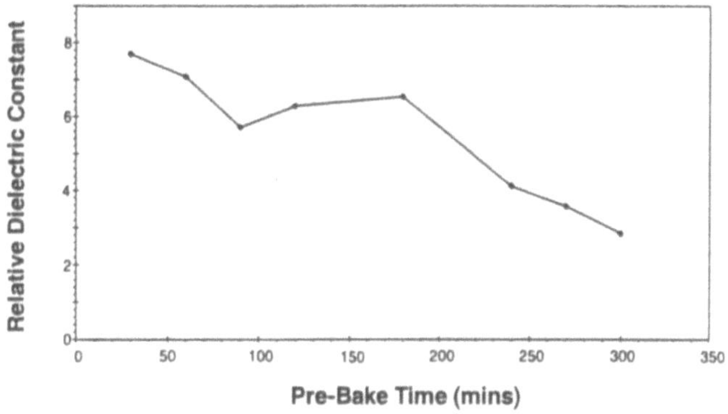

Figure 13. Variation of ε_r with prebake time at 24 GHz

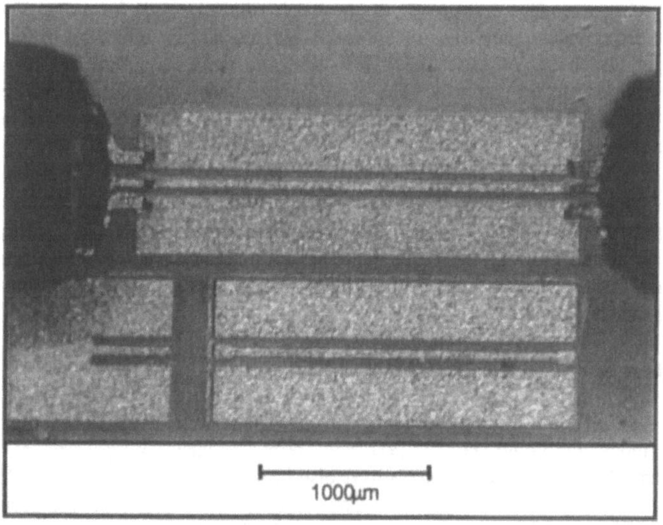

Figure 14. Coplanar stripline structures (and probe tips) as used for measuring the dielectric properties of SU-8.

It is probably not reasonable to assume that all the structure in these plots has a physical meaning. Most likely it is a result of experimental error. However these results agree well with those for long prebake material reported by Arscott et al and reproduced in Figure 15.

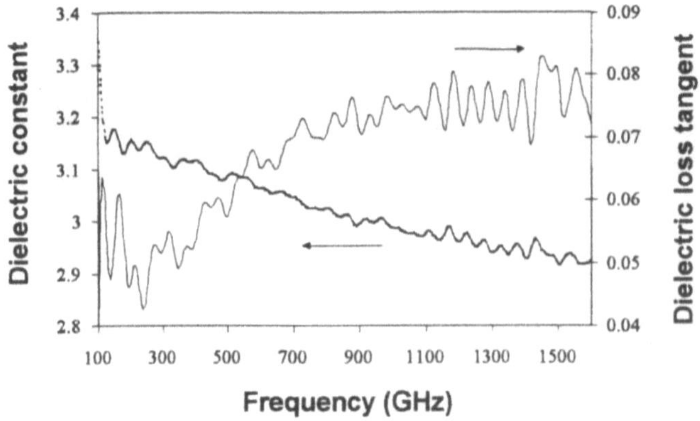

Figure 15. Relative dielectric constant and loss tangent for SU-8. [Arscott et. al.]

The general trend is clear. The relative permativity decreases with frequency as would be expected and it also decreases with increasing prebake time as more and more of the GBL is driven off. The figures show that at in the 10 – 30 GHz region ε_r is only beginning to level out after a prebake time of about 5 hours. This time does however depend on the thickness of the SU-8 which governs the removal rate of GBL. Experiments show that the prebake has the biggest effect on ε_r but small changes also occur on exposure to UV light and during a subsequent post bake. Observations suggest that if the GBL is not removed completely during prebake then it is "locked" in as the resin polymerises on exposure and cannot be removed by even an extensive post bake.

The absorption coefficient of SU-8 as measured in [7] is in the region of 20 to 30 cm^{-1} at 1 THz. This corresponds to a 3 fold reduction in amplitude of the transmitted signal over distance of 400 μm. Losses in SU-8, (as also confirmed in [6]), are not insignificant at THz frequencies and it is therefore not an ideal substrate material for microstrip circuits at these frequencies unless the layer is very thin or the lengths short.

5. Silicon Micromachined Terahertz Circuits

The non-isotropic etching properties of Si have been used to micromachine waveguides (McGrath et. al [9]) and horn antennas Wright et. al. [10]. Micromachined silicon components have also been combined with SU-8 in 585 GHz superheterodyne mixer blocks by Crowe et. al. [11]. An excellent review of micromachined THz circuits based on Si is to be found in [12]. While this area of micromachining will not covered in any detail here, suffice it say that because silicon, even in when undoped is a conductor ($\sigma_{min} \sim 4 \times 10^{-6}\ \Omega^{-1}m^{-1}$) the emphasis is on membrane substrates in order to minimise energy losses.

Acknowledgements

The authors would like to thank Byron Alderman, Tony Champion, Caroline McIntosh and Paul Simpson for their experimental work without which this chapter could not have been written. We are also grateful to the EPSRC and the EU for their financial support.

References

[1] C.E.Collins, J.W.Digby, G.M.Parkhurst, R.E.Miles, R.D.Pollard, J.M. Chamberlain, D.P.Steenson, "A Novel Micromachined Millimetre-Wave and Terahertz Snap-Together Rectangular Waveguide Technology" *IEEE Trans. On Mic.and Guided Wave Lett.*, vol. 9, no. 2, pp. 63-65, Feb., 1999.

[2] R.E.Miles, D.P.Steenson, A.J.Champion, F.Masot, "Micromachined Components for Terahertz Circuits," *Proc. 7th Int. Conf. on Terahertz Electronics*, Nara, Japan, pp 52-55, Nov. 1999.

[3] V.Postoyalko, D.S.Budimir, "Design of Waveguide E-Plane Filters with All Metal Inserts by Equal Ripple Optimization", *IEEE Trans.On Microwave Theory & Techniques*, vol. MTT42 no. 2, pp. 217-222, Feb. 1994.

[4] S.Arscott, L.Duvillaret, P.Mounaix, F.Garet, J-L.Coutaz and D.Lippens, "Micromachining techniques at terahertz frequencies," *Proc. EUROPTO Conference on Terahertz and Applications II*, Munich, Germany, SPIE Vol. 3828 pp 335-341, June 1999.

[5] B.E.J.Alderman, C.M.Mann, M.L.Oldfield, J.M.Chamberlain, "An embedded mask for application in high-frequency integrated circuits using UV positive resist technology," *J. Micromech. Microeng*, Vol. 10, pp. 334-336, 2000.

[6] J.R.Thorpe, D.P.Steenson, R.E.Miles, "High frequency transmission line using micromachined polymer dielectric," *Electron. Lett.* Vol. 34, No. 12, pp 1237-1238, 1998.

[7] C.E.Collins, R.E.Miles, R.D.Pollard, D.P.Steenson, J.W. Digby, G.M.Parkhurst, J.M.Chamberlain, N.J.Cronin, S.R.Davies and J.W.Bowen, "Millimetre-Wave Measurements of the Complex Dielectric Constant of an Advanced Thick Film Photoresist", *J.Elec Materials*, Vol. 27, No. 6, pp L40-L42, June 1998

[8] S.Arscott F.Garet, P.Mounaix, L.Duvillaret, J-L.Coutaz and D.Lippens, "Terahertz time-domain spectroscopy of films fabricated from SU-8," *Electronics Letters*, Vol. 35, No. 3, pp 243-244, Feb. 1999.

[9] W.R.McGrath, C.K.Walker, M.Yap, and Y.-C.Tai, "Silicon micromachined waveguides for millimeter-wave and submillimeter-wave frequencies," *IEEE Microwave Guided Wave Lett.*, vol. 3, pp. 61–63, Jan. 1993.

[10] J.A.Wright, S.Tatic-Lucic, Y.-C.Tai, W.R.McGrath, B.Bumble, and H.LeDuc, "Integrated silicon micromachined waveguide circuits for submillimeter wave applications," in *6th Int. Symp. Space Terahertz Technol.*, Pasadena, CA, pp. 387–396, Mar. 1995.

[11] T.W.Crowe *et al.*, "Inexpensive receiver components for millimeter wavelengths," in *8th Int. Symp. Space Terahertz Technol.*, Cambridge, MA, pp. 377–384, .Mar. 1997.

[12] M.Lubecke, K.Mizuno, G.M.Rebeiz, "Micromachining for terahertz Applications", *IEEE Trans. Microwave Theory and Techniques*, MTT46, Vol. 11, pp. 1821-1831, 1998.

Theme 8

THz Sensing and Spectroscopy

WIDEBAND TERAHERTZ SENSING AND SPECTROSCOPY WITH ELECTRONIC SOURCES AND DETECTORS

D.W. VAN DER WEIDE
University of Wisconsin-Madison
Department of Electrical and Computer Engineering
1415 Engineering Drive
Madison WI 53706 USA

A growing variety of reflection and transmission spectroscopy in the 1-1000 GHz regime can be done with wideband all-electronic terahertz (THz) spectrometers. We have measured gas absorption spectra and energetic material reflection spectra with such instruments using phase-locked microwave sources to drive picosecond GaAs nonlinear transmission lines, enabling measurement of both wideband spectra and single lines with Hertz-level precision, a new mode of operation not readily available with optoelectronic THz techniques. We take two approaches to coherent measurements: (1) spatially combining the freely propagating beams from two coherent picosecond pulse generators (which have discrete spectra ranging from ~ 6 to > 500 GHz), or (2) using a more conventional coherent source/detector arrangement with sampling detectors. The first method employs a dual-source interferometer (DSI) modulating each harmonic of one source with a precisely-offset harmonic from the other source—both sources being driven with stable phase-locked synthesizers—the resultant beat frequency can be low enough for detection by a standard composite bolometer. Room-temperature detection possibilities for the DSI include antenna-coupled Schottky diodes. Finally, we have recently introduced a reflectometer based on serrodyne modulation of a linearized delay line, using a technique that is process-compatible with pulse generator circuits.
Keywords: wideband electronic terahertz techniques, gas spectroscopy, reflection spectroscopy, nonlinear transmission lines, samplers, coherent measurements, dual source interferometer

1. Introduction

There is a strong need for integrated circuit (IC) sensors in the THz regime. Applications include multi-species gas sensing to monitor automotive and factory emissions, particularly in the presence of particulates (e.g. soot from diesels) that scatter near-infrared radiation used in today's

R.E. Miles et al. (eds.), Terahertz Sources and Systems, 301–314.

near-infrared (NIR) gas sensors. Other applications might include measuring reflections with portable, inexpensive equipment for security and location of buried objects. Finally, all of these electronic techniques would benefit from IC-compatible coherent measurement capabilities[1]. Electronic techniques for generating such frequencies fall into narrowband-tunable (e.g. [2]) and wideband categories. Here we discuss wideband (pulsed) generation and detection.

For wavelengths longer than the NIR, Fourier-transform infrared (FTIR) spectroscopy is the dominant method for measuring absorption spectra[3]. While FTIR has been used in the millimeter-wave regime with specialized equipment[4], the relative weakness of the blackbody source below 10 cm^{-1} (300 GHz) requires prohibitively long scans and averaging times.

As an alternative to FTIR, we can employ a dual-source interferometer (DSI) that is not only compact but also free of the need for a coherent detector so that an incoherent (e.g. photoconductive or bolometric) detection mechanism can be used[5, 6]. This eliminates the noise inherent in a sampling detector and establishes a baseline for future fully integrated implementations that would use sampling detectors. In the first section of this paper we describe and analyze the DSI, and demonstrate its use in measuring gas absorption spectra.

We then briefly discuss measurements of reflection spectra of energetic materials using a more conventional sampling detector, and finally review a new means of generating coherent signals from a single source. This approach will play a role in future handheld electronic THz systems.

To generate wideband (discrete) spectral energy in the 1-1000 GHz regime, we use nonlinear transmission line (NLTL) pulse generators coupled to wideband planar antennas[7-9]. The GaAs IC NLTLs used in this work consist of series inductors (or sections of high-impedance transmission line) with varactor diodes periodically placed as shunt elements. On this structure at room temperature a fast (~ 1-2 ps) voltage step develops from a sinusoidal input because the propagation velocity u is modulated by the diode capacitance, $u(V)=1/\sqrt{LC(V)}$, where L is the line inductance and $C(V)$ the sum of the diode and parasitic line capacitance[10, 11]. Limitations of the NLTL arise from its periodic cutoff frequency, waveguide dispersion, interconnect metallization losses, and diode resistive losses. Improvements in NLTL design have resulted in sub-picosecond pulses at room temperature[12].

2. Dual-source interferometer

In this spectrometer we spatially combine the freely propagating beams from two coherent NLTL pulse generators, resulting in a compact, reliable, and easy-to-implement micro- and millimeter-wave source for spectroscopy. Because the DSI modulates each harmonic of one source with a precisely offset harmonic from the other source—both sources being driven with stable phase-locked synthesizers—the resultant beat frequency can be low enough for detection by a standard composite bolometer. Room-temperature detection possibilities for the DSI include antenna-coupled Schottky diodes, which we have under development.

3. Description of DSI

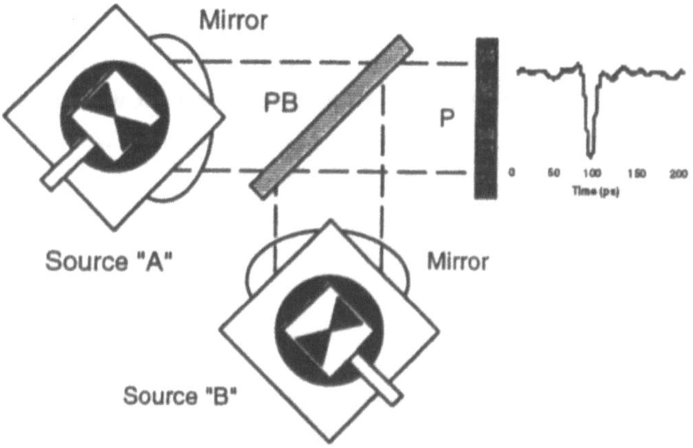

Figure 1. Dual-source interferometer configuration. Each source/antenna combination is at the focus of a paraboloidal mirror and radiates a polarized beam, which is transmitted ("A") or reflected ("B") by the polarizing beamsplitter (PB). The output polarizer (P) selects half the power of each beam. The output waveform is that detected by a bolometer [1, 5, 6]. © IEEE

In the dual-source interferometer, the output of each NLTL feeds an integrated bow-tie antenna (Fig. 1) mounted at the focus of a hyperhemispherical high-resistivity silicon lens[13]. These lenses in turn are mounted at the foci of off-axis paraboloidal mirrors. The beams collimated by the mirrors are either transmitted (Source "A") or reflected (Source "B") by a wire-grid polarizing beamsplitter. Each beam then contributes equally to the final, linearly polarized beam by arrangement of a final wire-grid polarizer mounted at 45° to the beamsplitter. Note that, while the prototype construction is already small (~ 170 mm long, 120 mm wide, and 80 mm high), it would be possible to fabricate two antennas and their circuitry on the same substrate, making the whole system extremely compact[6]. We also note that other workers have described similar ideas using laser-triggered photoconductive switches[14-16], but one advantage of the current system is its all-electronic approach, which offers precise amplitude (0.01 dB) and frequency (1 Hz) control of the fundamental excitation.

Each source is fed by a 100-900 mW sinewave generated by one of two microwave synthesizers, both of which share a common 10 MHz timebase. The output of one synthesizer is offset by $\Delta f \ll f_0$ ($\Delta f \sim 100$ Hz; $f_0 = 3$-10 GHz), and this offset is used as a trigger for a Fast Fourier Transform (FFT) spectrum analyzer or digital lock-in amplifier. While the synthesizers and wideband power amplifiers used in the present demonstration are expensive, they could be replaced by an inexpensive single source combined with a new NLTL-based frequency translator[17, 18].

4. Analysis of DSI

Here we analyze the viability of a detection method based on combining two waveforms of slightly offset frequencies at a power sensor such as a bolometer. Assume first that we have two periodic waveforms $v(t)$ and $w(t)$ of the same frequency $f = 1/T$, so that

$$\begin{aligned} v(t+T) &= v(t), \\ w(t+T) &= w(t). \end{aligned} \tag{1}$$

We combine them and excite a detector with the result. The detector is slow enough that it can only measure the average power of the radiation that impinges upon it. If one of the waveforms is shifted with respect to the other by a we obtain

$$\begin{aligned} P(a) &= \frac{1}{T}\int_0^T [v(t)+w(t+a)]^2\, dt = \frac{1}{T}\int_0^T v^2(t)dt + \frac{1}{T}\int_0^T w^2(t+a)dt + \frac{2}{T}\int_0^T v(t)w(t+a)dt \\ &= \frac{1}{T}\int_0^T \left[v^2(t)+w^2(t)\right]dt + \frac{2}{T}\int_0^T v(t)w(t+a)dt \end{aligned} \tag{2}$$

as the detected power. Clearly, only the last term of the above formula is dependent on the amount of the relative shift a between the two waveforms; the first term is merely a DC offset. In the subsequent calculations we will focus only on the shift-dependent term that we will call $2\widetilde{P}(a)$. Since the waveforms $v(t)$ and $w(t)$ are real functions, we have

$$\widetilde{P}(a) = \frac{1}{T}\int_0^T v(t)w(t+a)dt = \frac{1}{T}\int_0^T v^*(t)w(t+a)dt \tag{3}$$

where the "*" denotes complex conjugation. Since $v(t)$ and $w(t)$ are periodic, we can write them as Fourier series

$$v(t)=\sum_{m=-\infty}^{\infty} v_m \exp\left(\tfrac{2\pi i m}{T} t\right) \tag{4}$$

$$w(t)=\sum_{n=-\infty}^{\infty} w_n \exp\left(\tfrac{2\pi i n}{T} t\right) \tag{5}$$

yielding

$$\begin{aligned}
\int_0^T v^*(t)w(t+a)dt &= \int_0^T \sum_{m=-\infty}^{\infty} v_m^* \exp\left(-\tfrac{2\pi i m}{T} t\right) \sum_{n=-\infty}^{\infty} w_n \exp\left(\tfrac{2\pi i n}{T}(t+a)\right) dt \\
&= \sum_{m=-\infty}^{\infty} v_m^* \sum_{n=-\infty}^{\infty} w_n \int_0^T \exp\left(-\tfrac{2\pi i m}{T} t\right) \exp\left(\tfrac{2\pi i n}{T}(t+a)\right) dt \\
&= \sum_{m=-\infty}^{\infty} \sum_{n=-\infty}^{\infty} v_m^* w_n 2\pi T \delta_{mn} \exp\left(\tfrac{2\pi i n}{T} a\right) \\
&= 2\pi T \sum_{n=-\infty}^{\infty} v_n^* w_n \exp\left(\tfrac{2\pi i n}{T} a\right) = 2\pi T \sum_{n=-\infty}^{\infty} v_n^* w_n \exp\left(\tfrac{2\pi i n}{T} a\right)
\end{aligned} \tag{6}$$

or

$$\widetilde{P}(a)=\frac{1}{T}\int_0^T v(t)w(t+a)dt=\frac{1}{T}\int_0^T v^*(t)w(t+a)dt=2\pi\sum_{n=-\infty}^{\infty} v_n^* w_n \exp\left(\tfrac{2\pi i n}{T} a\right) \tag{7}$$

If the relative shift a between the two waveforms is varied (slowly compared to the frequency $f=1/T$), the detected average power will vary as well, according to the above equation. In particular, if $a=Tf't$, where $f' \ll f$ is the offset frequency, then

$$\widetilde{P}(t)=2\pi\sum_{n=-\infty}^{\infty} v_n^* w_n \exp(2\pi i f' n t)=\sum_{n=-\infty}^{\infty} \widetilde{P}_n \exp(2\pi i f' n t). \tag{8}$$

That is, the detected power $\widetilde{P}(t)$ varies in time with frequency f', $\widetilde{P}(t+1/f')=\widetilde{P}(t)$, and the harmonics of the original waveforms $v(t)$ and

$w(t)$ are encoded in its spectral composition as in the above formula. The attenuation of the n-th harmonic,

$$\left| {v'_n}/{v_n} \right| = \left| {w'_n}/{w_n} \right|, \tag{9}$$

where v'_n (v_n) is the amplitude of the signal with (without) the sample present in the beam path, can be easily recovered from the spectral analysis of the detected power,

$$\left| \frac{v'_n}{v_n} \right| = \left| \frac{w'_n}{w_n} \right| = \sqrt{\left| \frac{\tilde{P}'_n}{\tilde{P}_n} \right|} = \sqrt{\left| \frac{P'_n}{P_n} \right|}, \tag{10}$$

where P'_n (P_n) is the n-th harmonic of the detected power signal with (without) the sample in the beam path.

5. DSI Results

As shown in Fig. 2, the DSI can illuminate a single-pass absorption cell using off-axis paraboloidal mirrors to collimate the radiation and collect it for detection by a bolometer.

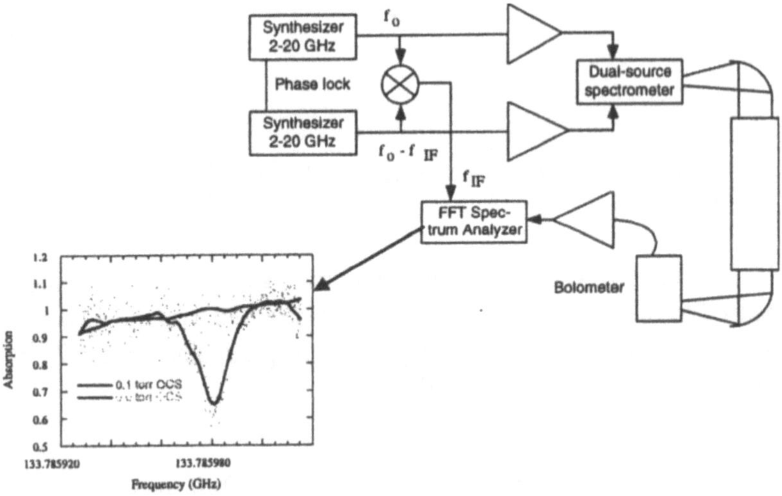

Figure 2. Dual-source spectrometer system configured for gas absorption measurements. Inset shows raw and smoothed data from 0.1 torr OCS vs. vacuum from a 76 cm long cell, uncorrected for background power slope. Horizontal scale is 20 kHz per large division[1].
© IEEE

Using a digital lock-in amplifier, we could measure a single harmonic of the DSI, in this case at 133.785980 GHz, the 20th harmonic of the fundamental, 6.689299 GHz. To measure the data shown, we used a third source to provide the 10 MHz timebase for the microwave synthesizers then varied this 10 MHz signal to continuously sweep the harmonics of the DSI. This enabled the extremely narrow linewidth of 0.1 torr OCS to be resolved, with an equivalent Q ($f_o/\Delta f$) of > 8×10^9. In Fig. 3 we show the effect of pressure broadening on this line.

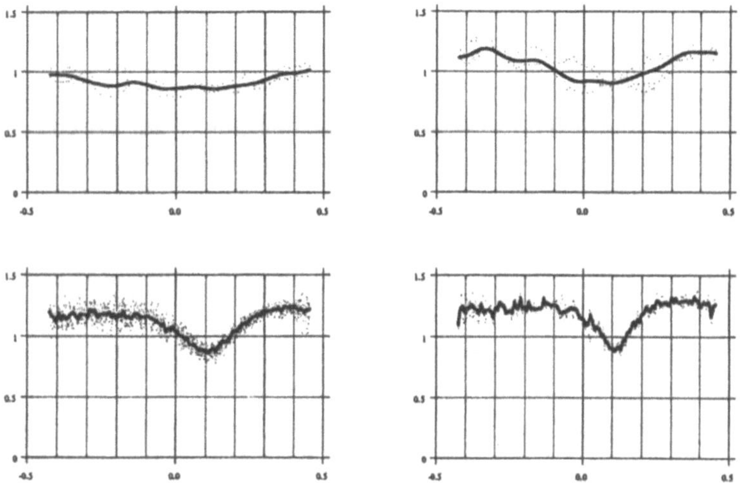

Figure 3. Pressure broadening of OCS line measured by DSI. Center frequency 133.78597 GHz (20th harmonic of 6.6893 GHz), sweep width 11.8 MHz. Pressure, from left to right, top to bottom: 0.5, 0.3, 0.2 and 0.1 torr[19]. © SPIE

6. Reflection Spectroscopy

Increasingly sophisticated weapons and explosives require increasingly sophisticated detection technologies. Non-metallic varieties of these threats are especially important because they elude familiar metal-detecting portals, so they have motivated development of a multi-pronged approach to detection, including residue sniffing and computerized tomography. These techniques, however, have significant drawbacks, suffering from invasiveness, slowness, unfamiliarity to the public, and significant potential for false negatives[20].

Threats like these, however, appear to be readily detectable and even identifiable using a wideband of signals in the sub-THz regime (1-500 GHz), based on experiments reported here. Traditional equipment for generating and detecting these frequencies has, however, been difficult, bulky and expensive. The objectives of this work are to develop and apply all-electronic and monolithically integrated technology for generating and detecting these wideband signals to the problem of imaging the reflection spectra of plastic weapons and explosives on human subjects.

Many of the concepts we employ here are being pursued at lower frequencies for target detection at higher resolutions than traditional narrowband radar allows. This ultrawideband (UWB), carrier-free, impulse, or baseband radar has been rapidly gaining popularity in applications where complex and elusive targets are the norm[21]. UWB radar has benefited from very recent advances in semiconductor technology enabling the production of sub-nanosecond pulses with peak powers of over 1 megawatt but having average powers in the milliwatt regime.

By contrast, the technology we employ—the integrated-circuit nonlinear transmission line (NLTL)—essentially trades power for speed, producing pico- or even sub-picosecond pulses with peak powers less than one watt and average powers in the low microwatt regime. These power levels are non-ionizing and biologically inconsequential, but because we can employ coherent detection, rejecting noise outside the frequencies of interest, we can still measure useful spectra with them.

Baseband pico- and sub-picosecond pulses of freely propagating radiation, usually generated and coherently detected with photoconductive switches and ultrafast lasers[22-24], have been useful for wideband coherent spectroscopy of materials, liquids, and gasses in the THz regime. Such systems have even been used for what could be called scale-model UWB radar[25]. These highly versatile beams of ultra-short electromagnetic pulses can be treated quasi-optically: They are diffracted and focused with mirrors and lenses, and the resultant effects can be readily observed in the time-domain waveform at the detector. Consequently, such beams are singularly useful for spectroscopy in a difficult-to-access spectral regime, and recent reports of spectroscopic imaging with these optoelectronic systems have generated much interest[26].

Using system concepts identical to the optoelectronic THz systems but with electronic components (Fig. 4), we have configured an NLTL-antenna generator and coherent sampling detector in a focused reflection spectroscopy system using off-axis ellipsoidal mirrors.

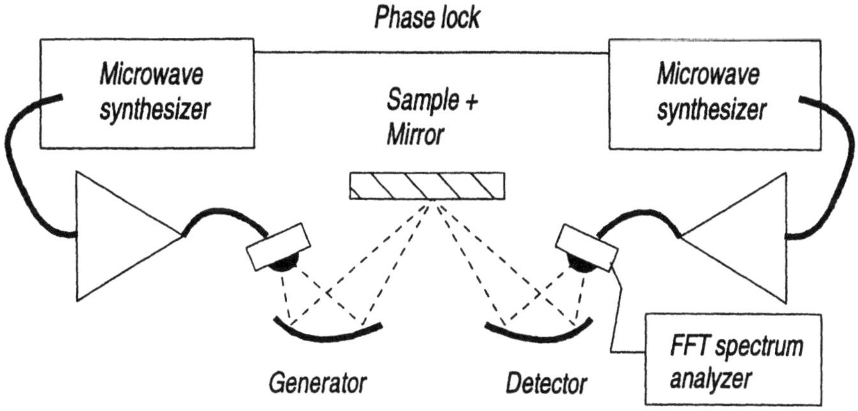

Figure 4. System diagram for electronic THz reflection spectroscopy[19]. © SPIE

Wideband (as opposed to single-wavelength) imaging has the chief advantage of flexibility: if weapons change composition over the years, a single-wavelength or narrowband source may no longer detect the new composition, but having a broad range of frequencies maximizes the opportunity to detect the new threat's signature.

One example of this advantage is shown in the data we present here. C-4 is primarily RDX with a plasticizing binder. As shown in Fig. 5, the wideband reflection spectra out to 450 GHz show clear similarity between C-4 and RDX; both are distinct from HMX, demonstrating specificity in the dielectric response of these targets. The contrast we measure is not fully explained, but probably arises from the granularity of the materials, their dielectric constants and orientation of the sample, as distinct from rotational absorption of chemical bonds.

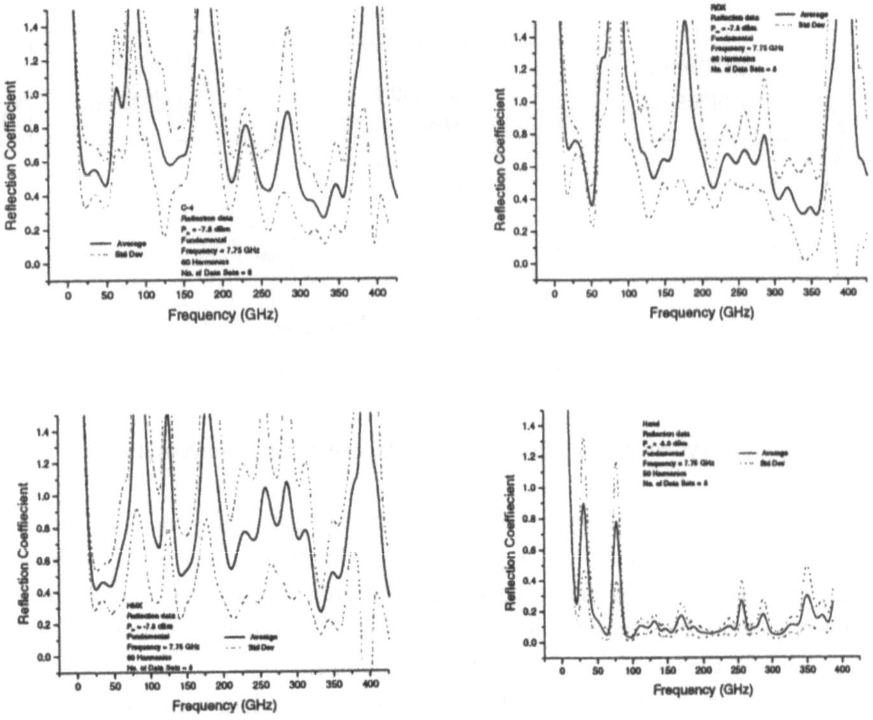

Figure 5. Measured reflection spectra of various energetic materials (C-4, RDX, HMX) compared to a human hand (lower right) [19]. The data are uncorrected for standing waves that give rise to reflection coefficients greater than unity. © SPIE

These spectra each show the average and standard deviation of 8 data sets using 50 harmonics of a 7.75 GHz fundamental with the sample occasionally repositioned to simulate a security screening application. The presence of standing waves in the highly reflective arrangement resulted in sample-to-background ratios exceeding unity at some frequencies.

We have measured reflection spectra from several potential threats, such as HMX, RDX, C-4, TNT, napthalene (with a similar chemical structure to some explosives), ammonium nitrate solids and slurries, and gasoline, as well as common objects like keys, metals, wallets, credit cards, and notebooks. All show unique and repeatable signatures.

The new information gained from these results can enable progress toward developing multi-pixel arrays of these sources and detectors for a new screening technology. Our current focus is on building inexpensive "single pixel" screening technology by pursuing advances in microfabrication of these devices, integrating more functions on a chip, and reducing the cost of these systems.

7. Coherent Signal Generation with Scanned Delay Lines

In this final section, we briefly review a new frequency translator based on a nonlinear transmission line (NLTL) phase shifter. Rather than forming shock waves on the NLTL with large signal excitation, we use its voltage-variable delay together with both amplitude and phase linearization to modulate the phase of a small 0.5-3.0 GHz microwave signal, performance that can be scaled with integration of the circuit. The resultant single sideband modulator exhibits > 45 dBc carrier and spurious suppression. This new approach has significant applications in both instrumentation and sensing, particularly because it offers a clear path toward complete integration of a coherent measurement system.

This is a new solution that can enable complete integration of wideband network analyzers, directly addressing the need for instruments to characterize 100 GHz devices, circuits, and systems, as well as the growing opportunities for sensors in this regime. Combining a NLTL phase shifter with serrodyne (sawtooth) modulation results in a frequency translator that can use an inexpensive–ultimately integrated–microwave source to coherently convert a wideband microwave signal directly to baseband. This invention, coupled with improved directional sampling circuits, could enable high-performance, inexpensive, and field-capable 100 GHz vector network analysis, as well as several other new military and commercial applications which benefit from a monolithic coherent generation/ detection system.

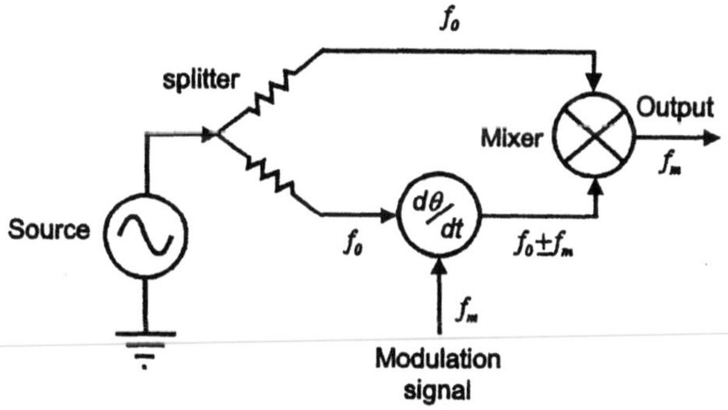

Figure 6. Block diagram of frequency translator application (above). Phase shifter is serrodyne modulated at f_m and output of mixer is a sinusoid at frequency f_m[18]. © IEEE

As shown in Fig. 6, a coherent electronic measurement system need not rely on phase-locked sources, but instead could use a modulated delay line to impart a (single sideband) frequency shift onto one arm of an interferometer. By serrodyne-modulating the delay line, additional sidebands are generated due to the flyback transient; these can be minimized by sufficiently short flyback times or triangle-wave modulation with appropriate Fourier transform reversals at baseband.

We built a demonstration phase shifter on a brass block with 30 abrupt junction diodes ($C_{j0} = 2$ pF) with 4.6 mm interval spacing for the first section and 20 abrupt junction diodes ($C_{j0} = 0.8$ pF) with 0.6 mm spacing for the second. This circuit was able to achieve > 2 ns delay, yet was usable beyond 3 GHz. For comparison, single GaAs IC NLTLs can achieve ~ 160 ps delays[12] with higher usable frequencies; they can also be cascaded for additional delay.

We implemented a phase linearization and amplitude compensation scheme using a gain-controlled amplifier (HP IVA 14208) for amplitude and a 12-bit D/A for phase. We measured the beat note between the original microwave signal and its frequency shifted version using both serrodyne and triangle wave modulation. Both methods gave nearly pure sinusoidal results, but the serrodyne version exhibited a flyback transient, while we could reconstruct the output sinusoid without this transient by using triangle-wave modulation by time-reversing each second cycle of the output.

In spite of the cleaner time-domain appearance of the triangle-wave modulated output, when driven at 0 dBm the serrodyne version exhibited > 45 dB of carrier and harmonic suppression compared to ~ 35 dB for the triangle wave. Fig. 7 shows the downconverted spectrum. In the future, we can address the shortfall of the triangle wave modulation with more careful waveform reconstruction.

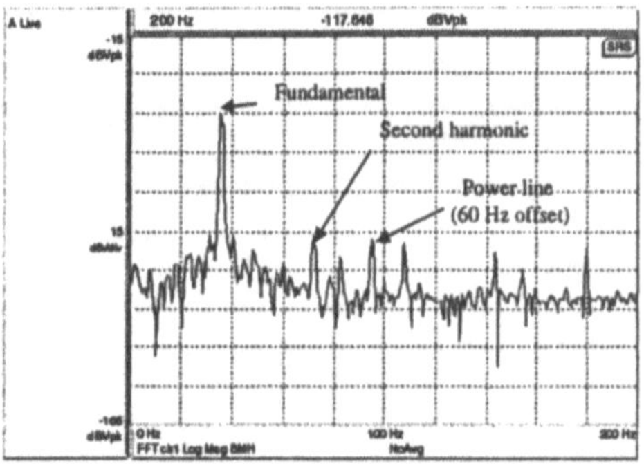

Figure 7. Downconverted spectrum of single-sideband modulator shown in Fig. 6. Harmonics and spurious due to the line frequency and the sawtooth retrace transient are evident[19]. © SPIE

We have shown that modulated NLTL frequency translators are ultimately candidates for integration with NLTL pulse generators and diode sampling bridges, enabling for the first time the foundation of a complete monolithic electronic THz system. We extended this approach to building a 1-3 GHz reflectometer with performance comparable to a commercial network analyzer[17, 18].

8. Conclusions

We have described and demonstrated a new technique for gas absorption spectroscopy using a new all-electronic THz interferometer having no moving parts but rather one whose phase is controlled electronically via a frequency offset between its two identical sources. We have also shown a more conventional coherent source/detector arrangement, but used for reflection spectroscopy of energetic materials. Finally, we presented a new means for eliminating these precision sources for a very compact and low cost approach. While this technique lays a promising foundation for inexpensive coherent microwave instrumentation, it can be further extended to other micro- and millimeter-wave sensors, such as handheld reflectometers operating in the THz regime for applications such as demining as well as sensing gasses, nonmetallic weapons and explosives.

9. Acknowledgments

This work is supported by a Ford University Research Award, DARPA, ONR YIP, NSF PECASE, and FAA. Thanks to J. Murakowski, V. Agrawal P. Akkaraekthalin for analysis, and F. Keilmann and K. McNesby for discussions.

10. References

1. van der Weide, D.W., Murakowski, J., and Keilmann, F. (2000) Gas-absorption spectroscopy with electronic terahertz techniques, *IEEE Transactions on Microwave Theory and Techniques* **48**, 740-743.
2. Eisele, H., Rydberg, A., and Haddad, G. (2000) Recent advances in the performance of InP Gunn devices and GaAs TUNNETT diodes for the 100-300-GHz frequency range and above, *IEEE Transactions on Microwave Theory and Techniques* **48**, 626-631.
3. Chantry, G.W. (1984) *Long-wave optics,* Academic Press, London.
4. Afsar, M.N. (1984) Dielectric measurements of millimeter-wave materials, *IEEE Transactions on Microwave Theory and Techniques* **32**, 1598-1609.
5. van der Weide, D.W. and Keilmann, F. (1996) Picosecond dual-source interferometer extending Fourer-transform spectrometer to microwave regime, *1996 IEEE MTT-S International Microwave Symposium Digest,* IEEE, New York, NY, USA, pp. 1731-1734.
6. van der Weide, D.W. and Keilmann, F. (1998) Coherent periodically pulsed radiation spectrometer, United States patent 5,748,309.
7. van der Weide, D.W., Bostak, J.S., Auld, B.A., and Bloom, D.M. (1991) All-electronic free-space pulse generation and detection, *Electronics Letters* **27**, 1412-1413.

8. van der Weide, D.W., Bostak, J.S., Auld, B.A., and Bloom, D.M. (1993) All-electronic generation of 880 fs, 3.5 V shockwaves and their application to a 3 THz free-space signal generation system, *Applied Physics Letters* **62**, 22-24.
9. Bostak, J.S., van der Weide, D.W., Bloom, D.M., Auld, B.A., and Özbay, E. (1994) All-electronic terahertz spectroscopy system with terahertz free-space pulses, *Journal of the Optical Society of America B* **11**, 2561-2565.
10. Rodwell, M.J.W., Kamegawa, M., Yu, R., Case, M., Carman, E., and Giboney, K.S. (1991) GaAs nonlinear transmission lines for picosecond pulse generation and millimeter-wave sampling, *IEEE Transactions on Microwave Theory and Techniques* **39**, 1194-1204.
11. Rodwell, M.J.W., Allen, S.T., Yu, R.Y., Case, M.G., Bhattacharya, U., Reddy, M., Carman, E., Kamegawa, M., Konishi, Y., Pusl, J., Pullela, R., and Esch, J. (1994) Active and nonlinear wave propagation devices in ultrafast electronics and optoelectronics (and prolog), *Proceedings of the IEEE* **82**, 1035-1059.
12. van der Weide, D.W. (1994) Delta-doped Schottky diode nonlinear transmission lines for 480-fs, 3.5-V transients, *Applied Physics Letters* **65**, 881-883.
13. van der Weide, D.W. (1994) Planar antennas for all-electronic THz systems, *Journal of the Optical Society of America B* **11**, 2553-2560.
14. Greene, B.I., Federici, J.F., Dykaar, D.R., Jones, R.R., and Bucksbaum, P.H. (1991) Interferometric characterization of 160 fs far-infrared light pulses, *Applied Physics Letters* **59**, 893-895.
15. Ralph, S.E. and Grischkowsky, D. (1992) THz spectroscopy and source characterization by optoelectronic interferometry, *Applied Physics Letters* **60**, 1070-1072.
16. Karadi, C., Jauhar, S., Kouwenhoven, L.P., Wald, K., Orenstein, J., and McEuen, P.L. (1994) Dynamic response of a quantum point contact, *Journal of the Optical Society of America B* **11**, 2566-2571.
17. Akkaraekthalin, P., Kee, S., and van der Weide, D.W. (1998) Distributed wideband frequency translator, *1998 IEEE MTT-S International Microwave Symposium Digest*, IEEE, New York, NY, USA, pp. 1431-1434.
18. Akkaraekthalin, P., Kee, S., and van der Weide, D.W. (1998) Distributed wideband frequency translator and its use in a 1-3 GHz coherent reflectometer, *IEEE Transactions on Microwave Theory and Techniques* **46**, 2244-2250.
19. van der Weide, D.W., Murakowski, J., and Keilmann, F. (1999) Spectroscopy with electronic terahertz techniques, presented at Terahertz Spectroscopy and Applications II, Munich.
20. Mann, P. (1996) TWA disaster reopens tough security issues, *Aviation Week & Space Technology*, pp. 23-27.
21. Herskovitz, D. (1995) Wide, Wider, Widest, *Microwave Journal* **38**, 26-40.
22. van Exeter, M. (1989) Terahertz time-domain spectroscopy of water vapor, *Optics Letters* **14**, 1128-1130.
23. Grischkowsky, D., Keiding, S., Exeter, M.v., and Fattinger, C. (1990) Far-infrared time-domain spectroscopy with terahertz beams of dielectrics and semiconductors, *Journal of the Optical Society of America B* **7**, 2006-2015.
24. Nuss, M.C., Goossen, K.W., Gordon, J.P., Mankiewich, P.M., O'Malley, M.L., and Bhusan, M. (1991) Terahertz time-domain measurement of the conductivity and superconducting band gap in niobium, *Journal of Applied Physics* **70**, 2238-2241.
25. Cheville, R.A. and Grischkowsky, D. (1995) Time domain terahertz impulse ranging studies, *Applied Physics Letters* **67**, 1960-1962.
26. Hu, B.B. and Nuss, M.C. (1995) Imaging with terahertz waves, *Optics Letters* **20**, 1716-1718.

TIME-DOMAIN TERAHERTZ SPECTROSCOPY AND SENSING

P. HARING BOLIVAR, M. BRUCHERSEIFER, M. NAGEL,
H.P.M. PELLEMANS AND H. KURZ
Institut für Halbleitertechnik II,
RWTH Aachen, D-52056 Aachen, Germany

1. Introduction

Spectroscopy in the terahertz (THz) region of the electromagnetic spectrum is an important and growing field of research with a broad range of applications. The characteristic properties of THz radiation (1 THz = 10^{12} Hz), localised between the infrared and the microwave frequencies, suggest many applications in the spectroscopy of gases, liquids and condensed matter. As an example, the rotational transitions of small molecules lie in this spectral range, enabling the direct quantitative detection and chemical analysis of gases. Moreover, important semiconductor properties, such as carrier concentration and mobility, lead to resonances in this spectral range, which enables one to derive such information directly from THz-transmission experiments. THz spectroscopy therefore constitutes an important analytical tool complementary to standard characterisation methods which opens up a wide variety of application perspectives.

The terahertz frequency range was traditionally covered by heterodyne or incoherent measurement techniques. However, the small bandwidth of heterodyne systems reduces the range of applications to very narrowly defined problems. This situation changed with the advent of ultrashort pulsed lasers, notably the Kerr-lens mode-locked Ti:sapphire laser (Spence *et al.* 1991), which opened the way for the development of time-resolved terahertz spectroscopy (Smith *et al.* 1988, van Exter and Grischkowsky 1990). This technique allows extremely broadband characterisation from 50 GHz up to 43 THz (Wu and Zhang 1997, Han and Zhang 1998, Leitenstorfer *et al.* 1999) and is far more flexible than conventional approaches. In contrast to incoherent techniques (e.g. bolometric detection) which detect the intensity of the THz radiation, the time-resolved approach is intrinsically coherent and hence allows the detection of amplitude and phase of the radiation, enabling one to derive a wider range of information directly without recourse to complicated Kramers-Kronig analysis. Another advantage is, that coherent detection automatically suppresses the influence of any incoherent noise, and is far more sensitive than conventional bolometric detection methods, enabling the detection of extremely small signals below the thermal background radiation (van Exter and Grischkowsky 1990).

Several general overviews of time-domain THz spectroscopy are available in the literature (e.g. Nuss and Orenstein 1998, Haring Bolivar 1999). This chapter will therefore only briefly recapitulate the general advantages of time-domain THz spectroscopy, before going on to discuss examples of recent progress. We apologise in

R.E. Miles et al. (eds.), Terahertz Sources and Systems, 315–328.

advance for not being able to include all of the interesting activities of the numerous groups involved in time-domain THz spectroscopy and sensing owing to the restricted space in this chapter.

This chapter is structured as follows: the initial part focuses on simple examples of linear THz spectroscopy, before going into non-linear (optical pump - THz probe) THz experiments, modulation THz spectroscopy, and spatially resolved measurements which demonstrate the wide flexibility of THz sensing for the characterisation of THz devices. We will show how glass-fibre-integrated and synchronised THz techniques are especially powerful means of characterising THz systems and components.

2. Time-domain THz spectroscopy

2.1. General approach

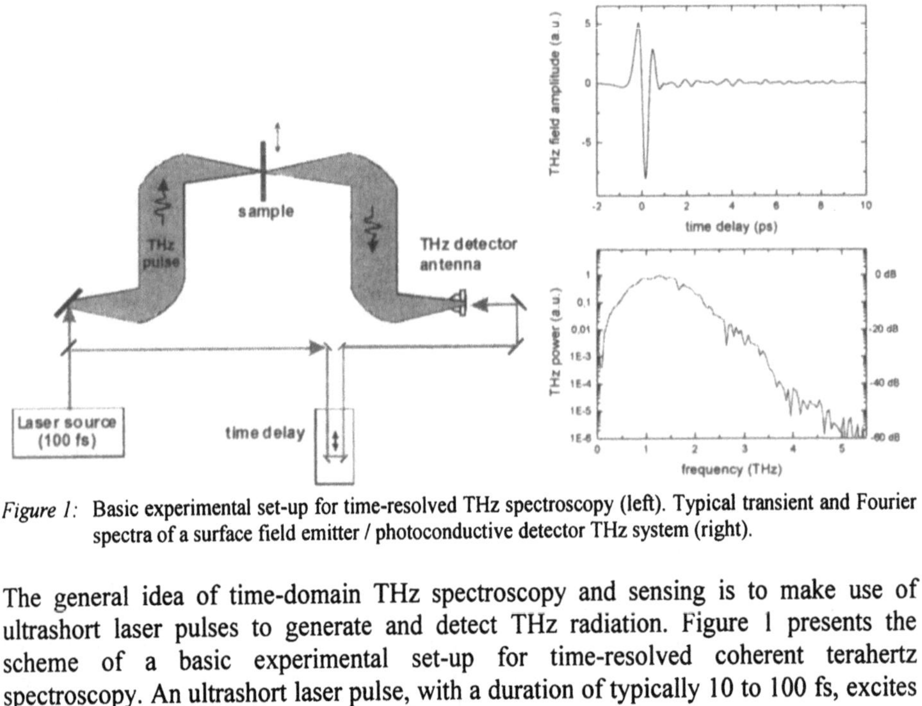

Figure 1: Basic experimental set-up for time-resolved THz spectroscopy (left). Typical transient and Fourier spectra of a surface field emitter / photoconductive detector THz system (right).

The general idea of time-domain THz spectroscopy and sensing is to make use of ultrashort laser pulses to generate and detect THz radiation. Figure 1 presents the scheme of a basic experimental set-up for time-resolved coherent terahertz spectroscopy. An ultrashort laser pulse, with a duration of typically 10 to 100 fs, excites a terahertz emitter. The radiated THz impulse is then collimated and focussed onto a detector, which monitors the momentary THz field at the time it is gated by a second time-delayed laser pulse. The correlation of excitation and detection pulses allows one to monitor directly in the time-domain the evolution of the THz field amplitude. A wide range of alternatives for emitters and detectors has been developed, but the basic operation principles can be grouped into two categories: photoconductive (Smith *et al.* 1988) and electrooptic (Auston and Nuss 1988). The first principle is based on the photogeneration of charge carriers, which are accelerated to emit THz radiation, or used to gate a detection photoswitch ("Auston switch"). The second approach is based on a

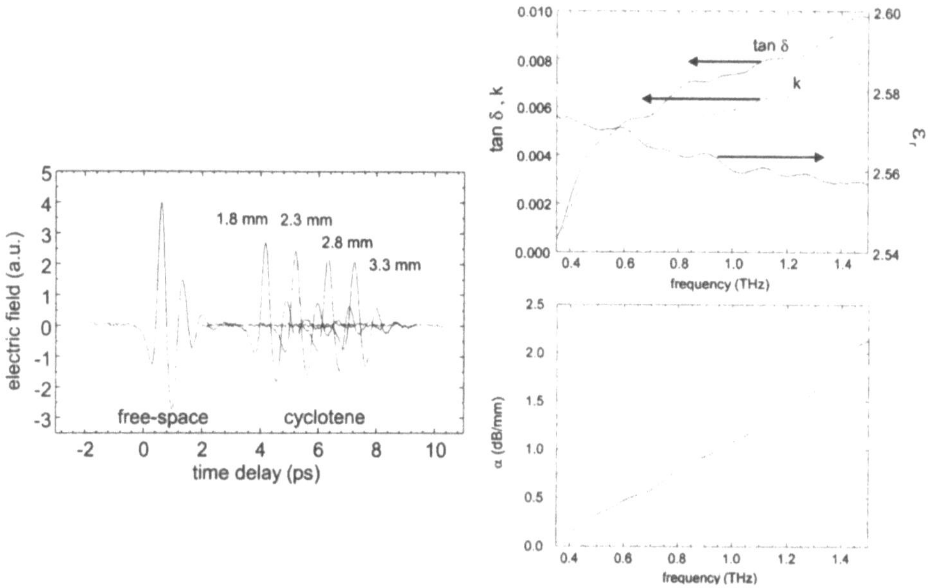

Figure 2: Example of transmission transients taken through air, and through free-standing cyclotene (BCB) films of varying thickness (left). The right figures illustrate technologically relevant material properties (tan d, k, e r and a) which can be extracted from the transients.

purely optical interaction of the light field with the material: Via non-linear $\left(c^{(2)}\right)$ coupling of the electromagnetic field, two optical fields can be mixed to generate THz radiation, or a THz field detected by analysing the polarisation modulation of an optical pulse in a non-linear crystal. In general, the generation process for THz radiation is extremely inefficient with a power conversion ratio of the order of 10^{-5} (Haring Bolivar 1999b) for typical excitation parameters with 100 MHz repetition rate laser systems. Nevertheless, the exceptional sensitivity of the coherent detection process allows very sensitive measurements with a signal to noise ratio (S/N) easily surpassing 10^{6}, as demonstrated by the THz spectrum depicted in Figure 1.

THz spectroscopy is performed by placing an object in the beam path of the THz impulses in different configurations. A usual approach is depicted in Figure 1, where the THz beam is focused onto a sample, and the transmitted pulses are detected. Reference transients are taken without the sample, in order to be able to correct for the spectral dependencies of the experimental setup. By numerical solution of the Fresnel formula (including Fabry-Perot effects) one can then determine directly the complex propagation constants of the analysed object (e_1 and e_2, or n and k), without requiring complicated Kramers-Kronig analysis. Appropriate numerical approaches are described e.g. in (Duvillaret *et al.* 1996, Nuss and Orenstein 1998). Many applications of time-resolved THz spectroscopy have been demonstrated in the past. The power of THz spectroscopy is a result of the frequencies being in resonance with many interesting material properties, such as rotational transitions in gases (for environmental monitoring technologies), strong absorption of polar liquids (for detecting reaction kinetics or for biological analysis), elementary charge carrier excitations in semiconductors, materials,

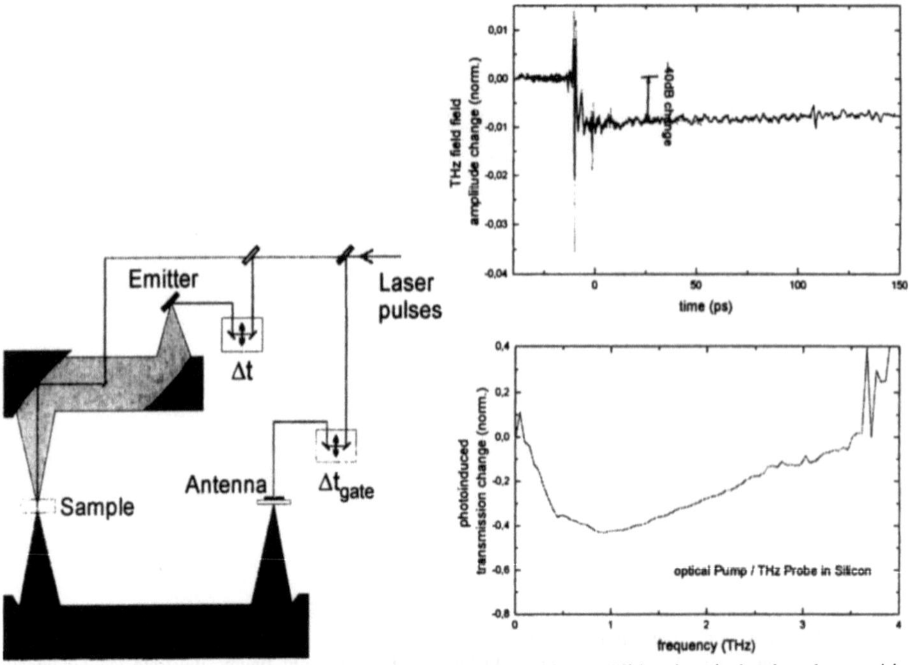

Figure 3: Setup for optical pump – THz probe experiments, with an additional optical pulse photoexciting the sample (left). Examples of the THz transmission after photoexcitation of InGaAs (top right) and silicon (lower right).

such as ceramics or dielectrics are transparent to THz radiation, hence opening up attractive imaging and tomography perspectives. Apart from commercial applications, THz spectroscopy fulfils additionally an important role in fundamental research, as it can deliver detailed information about coherent carrier dynamics (see e.g. Roskos *et al.* 1992, Haring Bolivar 1999c). For examples of linear THz spectroscopy, see the transmission data in Figure 2.

2.2. Time resolved Optical Pump – THz Probe spectroscopy

One of the outstanding features of the time-resolved THz spectroscopy approach which cannot be accessed with standard far-infrared techniques, is that it can not only measure static material properties at THz frequencies, but can also be easily extended to permit dynamic measurements of THz properties, e.g. by monitoring the THz transmission of a sample after exciting it with an additional laser pulse. By changing the time delay of this optical 'space' pump pulse with respect to the THz probing pulse, one can monitor directly in real time the evolution of the THz transmission with a femtosecond time resolution. Examples of such measurements are presented in Figure 3. Such approaches were first applied to the transient monitoring of the carrier mobility after optical excitation (Nuss *et al.* 1987). The examples presented in Figure 3 illustrate the THz transmission after photoexcitation of silicon and InGaAs. While in InGaAs one can detect even small (< 40 dB) dynamic changes of the THz transmission after optical excitation, the silicon signal does not show any transient response. Only static changes

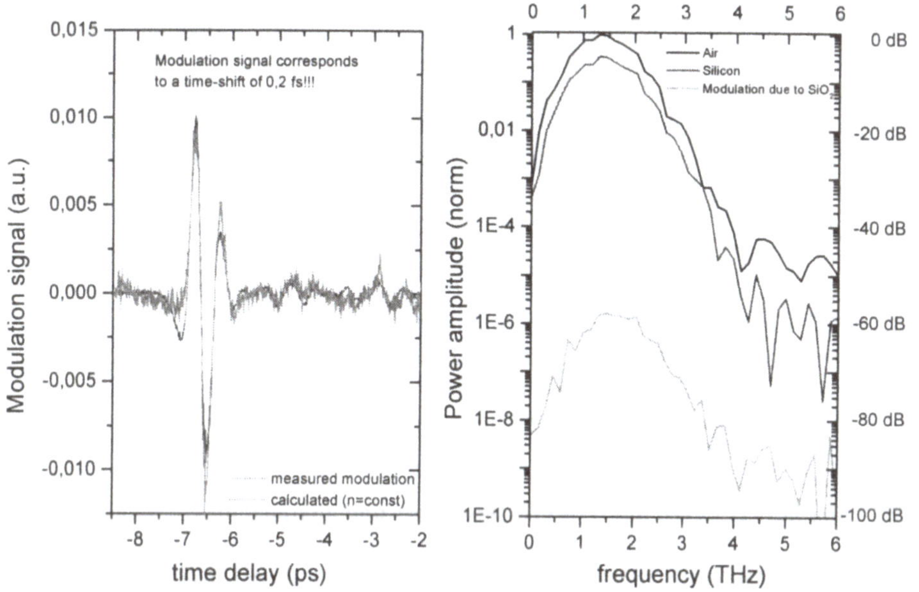

Figure 4: Example of modulation THz spectroscopy. The left plot presents a measured transient of a 100 nm SiO_2 film in comparison to the theoretical expectation (for n = 1.6 and k = 0). The right plot illustrates spectra of the measured modulation THz amplitude in comparison to reference transmission spectra.

of the THz transmission can be deduced from the data. This is due to the accumulation of charge carriers in silicon which inhibits the detection of transient changes.

2.3. Modulation THz spectroscopy

As discussed in the previous sections, time-domain THz spectroscopy has proved to be a very attractive characterisation technique. However, the usual characterisation of materials requires measurement of a transmission transient through a sample and through a reference to extract the desired information. Measurement systems have greatly improved in the recent past, nevertheless due to the remaining intrinsic instabilities of the system (e.g. pointing instability of lasers), it is extremely difficult to detect small changes of THz transients typically below $\lambda T/T \approx 10^{-6}$. (e.g. induced by thin films). With a recently proposed approach (Brucherseifer *et al.* 2000, Jiang *et al.* 2000, Jiang *et al.* 2000b, Brucherseifer *et al.* 2000b) it is possible to reach detection levels of the order of $\lambda T/T \approx 10^{-9}$ (Brucherseifer *et al.* 2000b). The basic idea to monitor small THz transmission changes is to modulate the position of the sample (e.g. with a shaker at typically 20 – 100 Hz), thereby measuring alternatively the THz transmission through the sample and reference, and monitoring only the difference signal with a lock-in amplifier. By exploiting this simple idea, one can readily reach extremely high sensitivities, which make the characterisation of sub-μm thickness films possible. This is especially attractive for biological applications where large material quantities are difficult to obtain. Figure 4 shows an example of the characterisation of a 100 nm SiO_2 film on silicon. The film is partially etched off, for comparison (modulation) with the transmission without the SiO_2 film. The left figure illustrates a modulation THz transient of the transmission through the film. The measured transient agrees well with a simple theoretical expectation assuming a frequency independent constant refractive index of n = 1.6 and disregarding absorption losses (k = 0). The right hand figure illustrates the spectra of this modulation signal in comparison to reference signals taken through air or through the silicon substrate. It can be seen that while the S/N of the reference signal is "only" of the order of 10^5 to 10^6 (50-60dB), the modulation signal reaches much lower noise levels of about 10^9 (90dB). This remarkable S/N permits the characterisation of the 100 nm thin film, in order to extract for example its frequency-dependent complex refractive index (n(n),k(n)) (Brucherseifer *et al.* 2000b). We expect that further refinements of this approach will open up the possibility of analysing even thinner films in the range of 10 - 30 nm.

3. Spatially resolved analysis of THz devices and components

THz sensing can easily be expanded to allow spatially resolved measurements of THz properties. This idea has given rise to THz imaging applications, which are described in more detail in the chapter by X.-C. Zhang. The focus of the following sections is to demonstrate alternative spatially resolved approaches developed especially for the characterisation of THz devices and components (Pfeiffer *et al.* 1996).

Basically two fundamental alternatives exist for the analysis of THz devices: One method uses femtosecond laser pulses to generate and detect THz fields which interact with the analysed device. This "impulsively excited" method permits extremely broadband analysis in order to e.g. extract the THz propagation characteristics through a THz component. In many cases it is desirable to detect the THz field generated by a free-running THz component. This can also be achieved with time-domain THz techniques, by a second "synchronised approach", where a femtosecond laser is synchronised to the free-running device in order to allow the phase-locked coherent detection of the THz fields generated by the analysed device. In the next section we will present examples of both approaches, showing that in both cases it is possible to make spatially and temporally resolved measurements to characterise THz devices, components and systems in detail.

3.1. Impulsive spatio-temporal mapping for the analysis of THz components

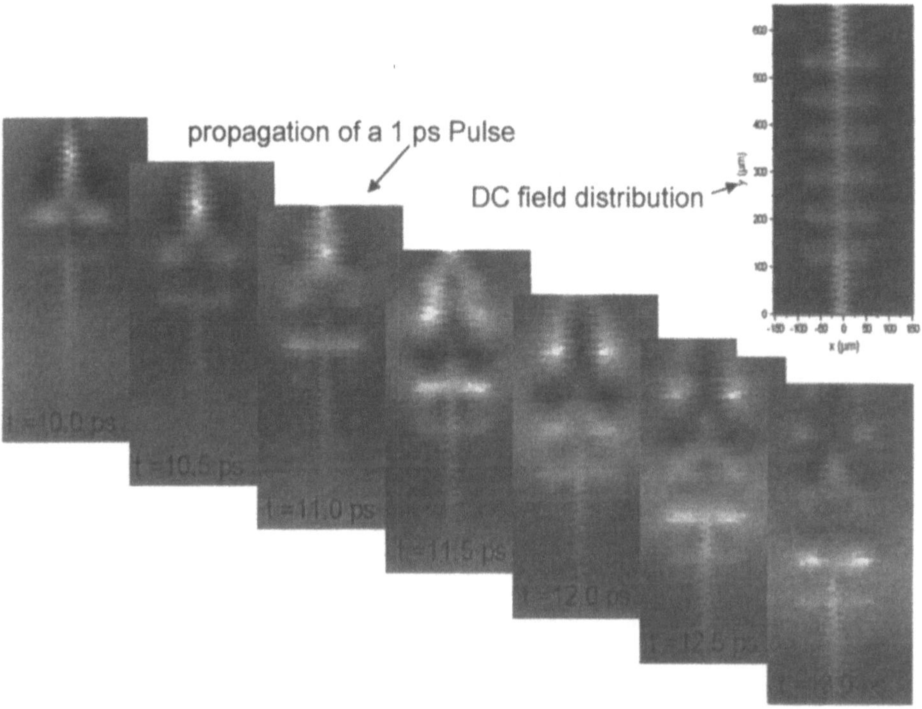

Figure 5: Example of the spatio-temporal characterisation of a planar 350 GHz filter structure. The inset in the upper right illustrates the DC field distribution, which illustrates the geometry of the structure. The remaining field distributions reflect the time-resolved images of the propagating THz pulse at different time delays.

The following section presents two examples of the impulsive spatio-temporal analysis to illustrate the flexibility of time-domain techniques for the characterisation of THz devices. The first example illustrates the near-field analysis of a THz component; the second the far-field characterisation of a THz antenna.

The first example shows the characterisation of a planar 300 GHz bandstop filter structure integrated into a microstrip THz waveguide structure (Nagel *et al.* 1999). A diagram of the filter geometry is depicted in Figure 5. For the measurements, a broadband THz pulse is injected into the waveguide by the photoexcitation of an Auston switch. An electrooptic detector is then used to detect the near-field of the device as a function of the time delay with respect to the excitation laser pulse as described in (Pfeiffer *et al.* 1996). By moving the sample with respect to the probing head, and by changing the time delay between the excitation and probing laser pulses, one can take complete spatio-temporal images of the propagation of the THz pulse along the filter structure. Such spatially resolved maps of the THz field at different time delays are depicted in Figure 5. The maps show, in detail, the propagation of the THz pulse along the filter structure, which both allows the extraction of the propagation characteristics, (as depicted in Figure 6) and gives a detailed knowledge of the internal field distribution of the device.

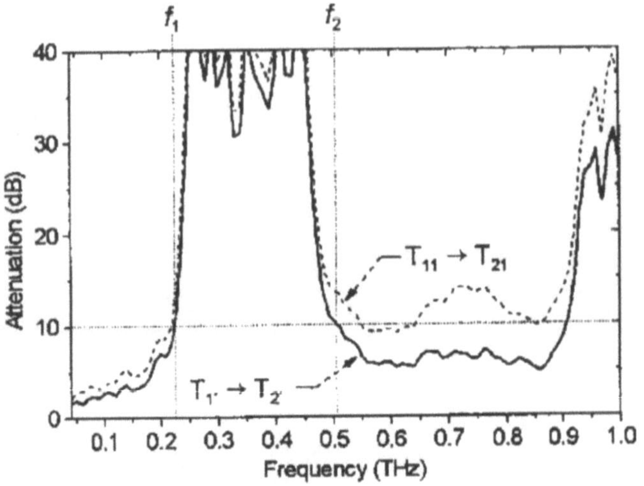

Figure 6: Attenuation constant for the propagation through the filter structure depicted in Figure 5. The $T_{11} \rightarrow T_{21}$ transmission coefficient includes the propagation through the waveguide, $T_{1'} \rightarrow T_{2'}$ in-cludes only the transmission through the filter. f_1 and f_2 are lower and upper design frequencies for the bandstop filter structure.

The previous example illustrates the power of time-domain characterisation techniques to analyse passive electronic devices. Apart from its exceptional bandwidth, time-domain techniques using electrooptic detection schemes to make possible the analysis of signals within THz components directly, without being restricted to defined in/output ports or to any probing contact patches.

The above spatially resolved measurement was made possible by scanning the filter structure with an x-y translation stage below the electrooptic probing head in order to avoid having to move optical components associated with the laser pulse guiding. By using glass fibres, one can realise far more flexible THz analytic systems. This use of glass fibres for the analysis of THz components is illustrated in the following example (Pellemans *et al.* 2000).

Figure 7 illustrates a setup designed to detect the far-field radiation pattern of THz components. A broadband THz pulse is generated in an InGaAs surface field emitter and coupled into a THz waveguide. The probing laser pulse is coupled into a glass fibre, which is fixed to the probing THz antenna in order to facilitate the free spatial movement of the detection antenna without having to readjust optical components. In order to compensate for the dispersion in the glass fibre and maintain short optical gate-pulses, (and hence maintain a high THz bandwidth), one has to pre-chirp the laser pulses. This can easily be attained for the optical pulses of typically 100 fs by using a grating pair.

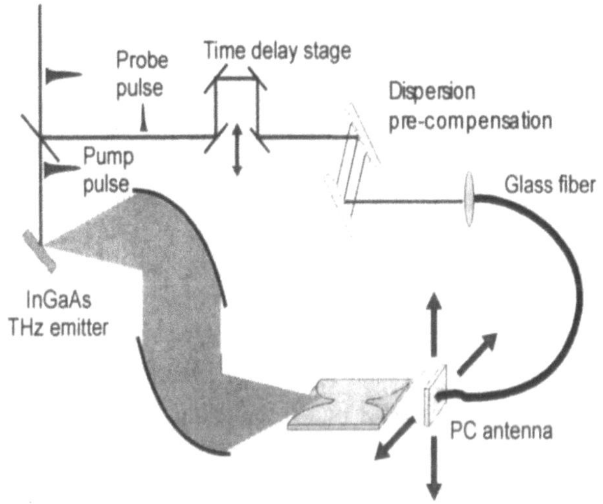

Figure 7: Scheme of a glass fibre integrated setup to detect the emission characteristics of a THz antenna.

Using such a setup, it is possible to detect the spatio-temporal THz emission characteristics of devices. Most analyses have been performed on THz devices of collaborating partners (Digby *et al.* 1997). Figure 8 illustrates a simple example of the THz emission from a corrugated horn with a design wavelength of 2.5 THz. The left graph illustrates the spatial dependency of the THz emission power. The right plot illustrates the Fourier decomposition of a THz transient detected at one point, which allows us to resolve the frequency composition of the THz field radiated by the horn. The low frequency cut-off is mainly due to the cut-off of the waveguide employed for feeding the corrugated horn antenna.

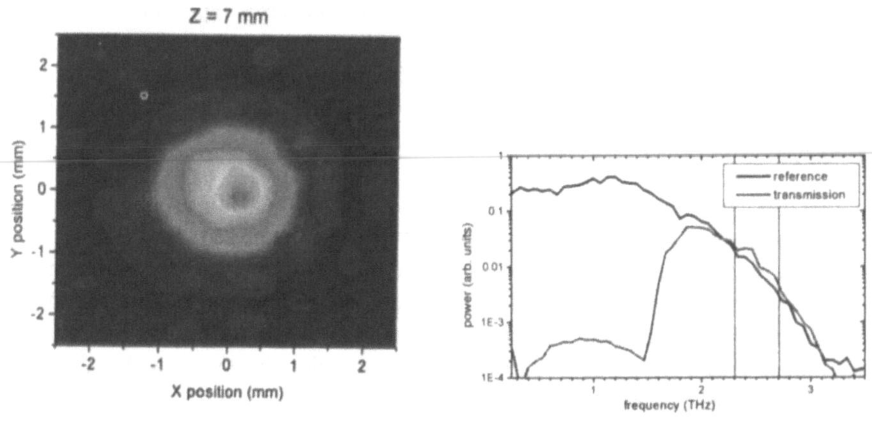

Figure 8: Spatially resolved plot of the THz emission power of the corrugated horn structure after a propagation distance of 7mm (left). The right plot illustrates the frequency resolved emitted power at one spatial point, in comparison to the input THz pulse.

3.2. Synchronised spatially resolved measurements of GHz and THz devices

The above examples illustrate the wide flexibility of impulsive THz probing of THz devices. Nevertheless, in many cases it is not only desirable to test the passive response of a THz component to an input THz field, but necessary also to test a free-running THz device and monitor, say, it's internal dynamics or it's THz emission. Free-running devices can be characterised by time-domain THz techniques, as long as a synchronisation of the probing femtosecond laser to the free-running device is guaranteed. This can easily be achieved in a photoconductive switch used for mixing the field generated by the free-running device with the laser repetition frequency. As described in more detail in (Löffler *et al.* 1996) one can then use this mixing signal of the device frequency f_D and a higher harmonic of the laser frequency nf_D to lock the probing laser to the device, and hence enable coherent probing of the device. Due to the 1/f noise dependence of the probing laser system it is advantageous to lock to a non-zero difference frequency $\Delta f = f_D - n \quad f_D$ to attain a more stable locking and better S/N ratio. Phase locking can be achieved by, for example, using the difference frequency signal Δf as the error signal in a phase-locked loop, where the laser repetition rate is varied to maintain phase locking. The signal is thus downconverted from f_D to Δf, where it can easily be detected with standard electronic components, (lock-in amplifier or fast digitalising cards). The choice of difference frequency is usually made to give the frequency response of standard electronic components in the setup – typically we use a Δf on the order of 20 – 200 kHz.

Examples of synchronised measurements are presented to illustrate the near- and far-field monitoring of THz device characteristics in order to exemplify the flexible capabilities of time-domain THz techniques for analysing THz components. The examples will focus on sub-100 GHz devices, as free-running THz devices are not readily available. The principles presented are nevertheless applicable to higher frequency components. The detection systems and mixing components clearly have THz bandwidths, as demonstrated by the experiments presented in previous sections. The only critical aspect is the synchronisation: as the laser repetition frequency of "only" 100 MHz and the difference frequency Δf on the order of 100 kHz are many orders of magnitude lower than the device frequency, a high degree of frequency stability for the free-running device is required to enable a stable locking of device and laser source.

The first measurement example demonstrates the near-field electrooptic probing of internal dynamics within a high-frequency device. The analysed device is a circular resonator which is an integral part of a planar mixer for free-space communications. Figure 9 presents the near field at different time delays. One can readily see the time evolution of the internal field, which is proportional to the waveguide potential. The measurement hence allows us to directly monitor the electric signals within the device. In the measurement presented, one observes a rotating component in the electric field pattern. This turned out to be the consequence of a second undesired mode in the resonator (a good single-mode resonator should give a standing wave pattern) which inhibited the proper operation of the device. Time-domain analysis of the internal dynamics was the only technique capable of isolating this design error, which was not predicted by simulations or observed by measurements at the in/output ports of the mixer circuit. Hence this measurement exemplifies the relevance and capabilities of time-domain techniques for monitoring the internal dynamics of devices under operation.

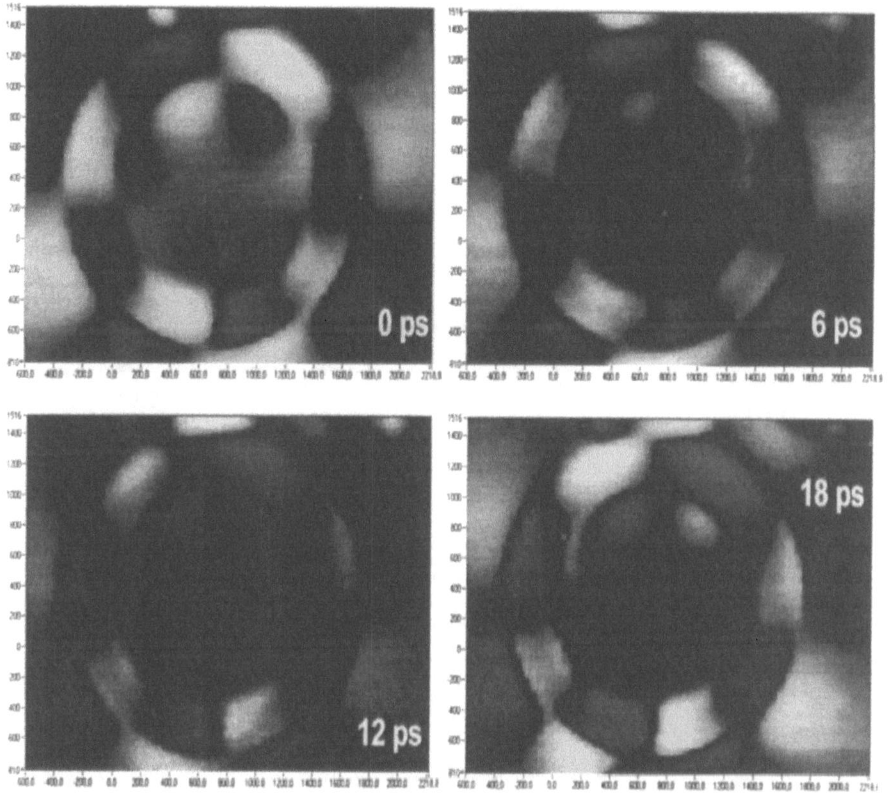

Figure 9: Spatially resolved plot of the electromagnetic field on top of a circular resonator structure. The x and y dimensions are given in μm.

A second example of a synchronised measurement is the analysis of the free-space emission of a radome-covered 94 GHz antenna system for high-resolution radar (Pellemans *et al.* 2000b). Phase-locked synchronisation of the laser repetition rate with the 94 GHz source is attained as described above. The detection part of the measurement setup is in principle the same as the one depicted in Figure 7. The free-space radiation is detected using a resonant dipole antenna with a built-in LT-GaAs photoconductive switch. The antenna is fabricated on a thin polymer substrate to increase the coupling efficiency to free-space radiation and lower the reflections from the detector. A 2 m long glass fibre is used to couple the laser pulses to the photoconductive switch, making the detector antenna freely positionable within the large space, which is necessary in view of the large size of the analysed antenna. With this setup it is then possible to monitor the amplitude and phase of the free-space radiation of the antenna, with a freely positionable detector. The fibre has the important advantage of maintaining a known phase relationship between the antenna and detector, even when moving the detector and scanning different emission patterns.

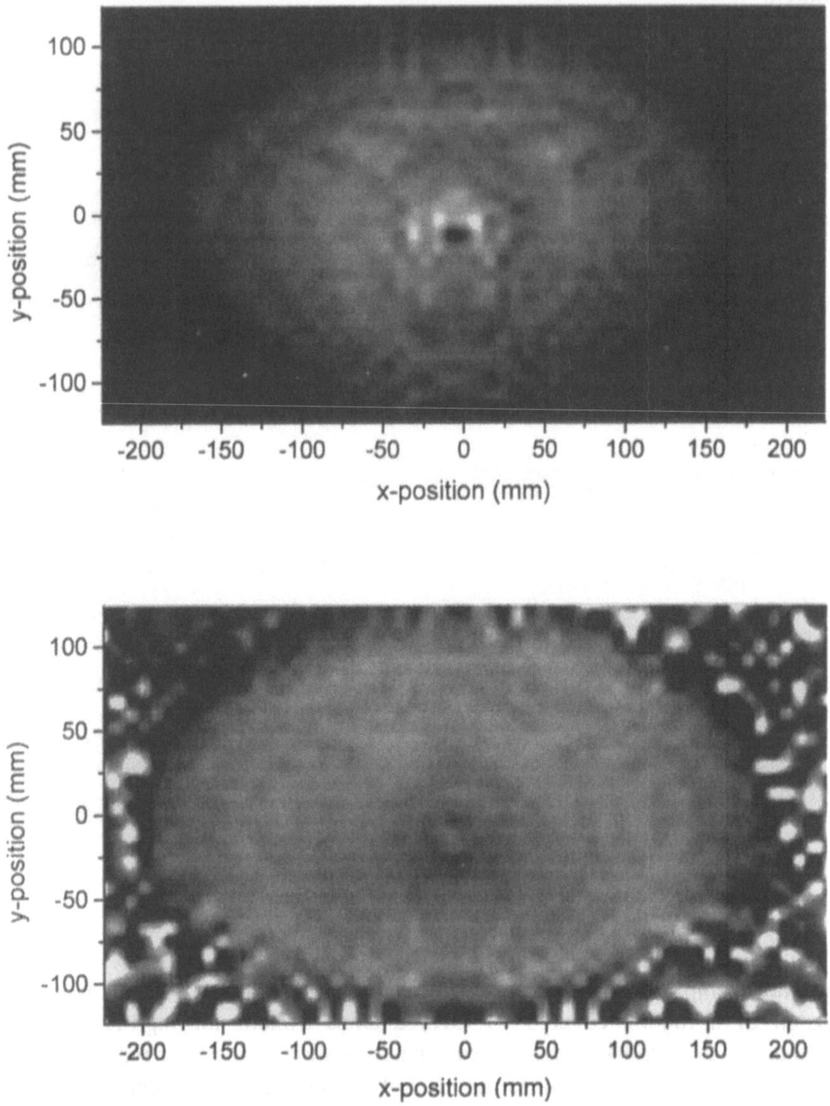

Figure 10: Near-field aperture scan of a 94 GHz antenna, showing the electric-field amplitude (top) and phase (bottom). The spatial resolution is 8 mm, measured at a distance of 25 mm from the radome.

Figure 10 shows the electric field amplitude and phase of a representative scan of 45 25 cm, with a spatial resolution of 8 mm, measured at 25 mm from the radome.

from the environment which deform the emitted radiation pattern significantly. Despite the low source power available, we can obtain a signal-to-noise level of 40 dB, which is limited only by residual scattering from the anechoic chamber placed around the setup. This 40 dB limit corresponds to a power level detected by the antenna (effective area 1 mm^2) of approximately 5 pW, which is indicative of the excellent sensitivity of time-domain detection techniques. The phase is measured with very high accuracy (3 degrees), and is reproducible on a time scale of several hours, demonstrating the stable phase-locking of laser and device.

4. Conclusions

Time-domain THz spectroscopy and sensing have developed into a wide field of research with numerous applications. The flexible variability and wide capabilities of pulsed optical techniques provide the means for wide ranging THz applications extending from basic research of elementary excitations in semiconductors, superconductors and other attractive material systems, to the characterisation of complete THz systems and components. Many applications such as environmental sensing of gases, benefit from the unsurpassed bandwidth and sensitivity of time-resolved techniques. Another advantage of time-domain techniques is their intrinsic capability for monitoring dynamic processes. Integration into glass fibre technologies has enhanced the versatility of THz sensing enormously in the recent past. Additionally, the development of synchronisation techniques overcomes the earlier impossibility of analysing active THz systems and components. In the future, further developments with regard to integration of the analytic systems, ease of operation and enhanced capabilities will support further the breakthrough of time-domain THz techniques, to become a standard analytic approach in the THz frequency range.

5. Acknowledgements

We appreciate the close collaboration and fruitful discussions with S. Hadjiloucas, J.W. Bowen, C.M. Mann, J.M. Chamberlain, D. Lippens, A. Stahl, Daimler-Chrysler Aerospace, and Bosch/Marconi Telecommunications, and financial support by the *Deutsche Forschungsgemeinschaft*, by the *Ministerium für Schule, Wissenschaft und Forschung des Landes Nordrhein-Westfalen* and by the European TMR program *Interact.*

References

Auston, D.H. and Nuss, M.C. (1998) IEEE J. of Quant. Electr. **24**, 184.

Brucherseifer, M., Pellemans, H.P.M., Haring Bolivar, P., Kurz, H. (2000) , in *Conference on Lasers and Electro-Optics*, OSA Technical Digest (Optical Society of America, Washington, USA), pp. 553-554.

Brucherseifer, M., Pellemans, H.P.M., Haring Bolivar, P., Kurz, H. (2000b), submitted to Appl. Phys. Lett.

Cai, Y., Brener, I., Lopata, A.J., Wynn, J., Pfeiffer, L., Stark, J.B., Wu, Q., Zhang, X.-C., Federici, J.F. (1998) Appl. Phys. Lett. **73**, 444.

Digby, J.W., Collins, C.E., Towlson, B.M., Karatzas, L.S., Parkhurst, G.M., Chamberlain, J.M., Bowen, J.W., Pollard, R.D., Miles, R.E., Steenson, D.P., Brown, D.A., Cronin, N.J (1997) IEEE MTT-S Int. Micr. Symp. **2**, 561.

Duvillaret, L., Garet F. and Coutaz, J.-L. (1996) IEEE J. of Selected Topics in Quantum Electronics **2**, 739.

Han, P.Y. and Zhang, X.-C. (1998) Appl. Phys. Lett. **73**, 3049.

Haring Bolivar, P. (1999), in Semiconductor Quantum Optoelectronics: From Quantum Physics to Smart Devices ed. by A. Miller, M. Ebrahimzahdeh and D.M. Finlayson, ISBN 0-7503-0628-9, (Institute of Physics Publishing, Bristol), pp. 151-192.

Haring Bolivar, P., Martini, R. and Kurz, H. (1999b), Proc. SPIE **3828**, p. 228-233, Terahertz Spectroscopy and Applications II, J. Martyn Chamberlain; Ed.

Haring Bolivar, P., Dekorsy T. and Kurz, H. (1999c), Semiconductors and Semimetals **66**, ed. by F. Capasso and H.C. Liu (Springer, Heidelberg), chapter 9., pp. 187-217.

Jiang, Z., Li, M., Zhang, X-C. (2000), Ultrafast Phenomena XII, MF30.

Jiang, Z., Li, M., Zhang, X-C. (2000b), Appl. Phys. Lett.**76**, 3221-3.

Leitenstorfer, A., Hunsche, S., Shah, J., Nuss, M. C. and Knox W. H. (1999) Appl. Phys. Lett. **74**, 1516.

Löffler, T., Pfeifer, T., Roskos, H.G., Kurz, H., van der Weide, D.W. (1996) Microel. Eng. **31**, 397.

Nagel, M., Dekorsy, T. Kurz, H. (1999) SPIE-Int. Soc. Opt. Eng. Proceedings of Spie - the International Society for Optical Engineering **3828**, 263-5.

Nuss, M.C., Auston, D.H. and Capasso, F. (1987) Phys. Rev. Lett. **58**, 2355.

Nuss, M.C. and Orenstein, J. (1998) in *Millimeter and Sub-Millimeter Waves*, ed. by G. Gruener (Springer-Verlag, Heidelberg), pp. 7-50.

Pellemans, H.P.M., Haring Bolivar, P., Brucherseifer, M., Kurz, H., Mann, C.M., Bowen, J.W. and Hadjiloucas, S. (2000), in *Conference on Lasers and Electro-Optics*, OSA Technical Digest (Optical Society of America, Washington, USA), pp. 557-558.

Pellemans, H.P.M., Nagel, M., Brucherseifer, M., Haring Bolívar, P., Kurz, H., and Schindler, G. (2000b), in *Conference on Lasers and Electro-Optics*, OSA Technical Digest (Optical Society of America, Washington, USA), p. 361.

Pfeifer, T., Heiliger, H.-M., Loffler, T., Ohlhoff, C., Meyer, C., Lupke, G. , Roskos, H.G. , Kurz, H. (1996) IEEE Journal of Selected Topics in Quantum Electronics **2**, 586.

Roskos, H.G., Nuss M.C., Shah J., Leo K., Miller D.A.B., Fox A.M., Schmitt-Rink S. and Köhler, K. (1992) Phys. Rev. Lett. **68**, 2216.

Smith, P.R., Auston D.H. and Nuss, M.C. (1988) IEEE J. of Quant. Electr. **24**, 255.

Spence, D. E., Kean P. N. and Sibbet, W. (1991) Opt. Lett. **16**, 42.

van Exter, M. and Grischkowsky, D.R. (1990) IEEE Trans. MTT **38**, 1684.

Wu, Q. and Zhang, X.-C. (1997) Appl. Phys. Lett. **71**, 1285.

Overview

TERAHERTZ TECHNOLOGY, QUO VADIS?

W.Th. Wenckebach
Dept. of Applied Physics, Delft University of Technology
P.O.Box 5046, 2600 GA Delft, The Netherlands
wenckebach@tn.tudelft.nl

Keywords: Terahertz Technology, Applications, Technological issues

Abstract At the end of the NATO-ARW the participants were invited to present their opinions on what they believed will be the main application areas of terahertz technology, the necessary technological requirements for these applications and how we are to meet these requirements. The present article highlights some major issues raised in these discussions.

1. INTRODUCTION

An attempt was made to identify crucial issues which need to be resolved in order for terahertz technology to be applied successfully. To this end discussion groups were formed and asked to formulate answers on such questions as: which applications in the terahertz frequency range may be considered to be useful or even essential, what kind of terahertz technology needs to be developed for these applications, and which key scientific problems still need to be solved. Subsequently, the groups' conclusions and recommendations were discussed in a general forum. Some of the major issues raised in these discussions are presented here. Hopefully these issues will be helpful in defining future directions for research on terahertz technology.

The reader is warned that what follows is rather the view that emerged from the ARW's discussion than the result of a thorough investigation by the author. To avoid false impressions references are restricted to the contributions of the participants in the present proceedings.

2. APPLICATIONS

One might argue that development in the terahertz frequency range is still in many respects being pushed by technology rather than pulled by applica-

R.E. Miles et al. (eds.), Terahertz Sources and Systems, 331–336.

tion. However, in the end applications determine which components need to be developed and what must be their specifications. Therefore we will start by presenting the most important potential applications mentioned during the discussions and give an overview of their technological needs. Note that we will *not* assess whether terahertz technology would be able to compete successfully with other methods.

2.1. SENSING

Many molecules, particularly in gases, have an absorption spectrum in the range from 0.5 - 10.0 THz which is a precise fingerprint of the species. Moreover, terahertz radiation penetrates many substances that are opaque to visible or near infrared light enabling observations in otherwise inpenetrable locations. This allows terahertz spectrometers to be used for sensing of environmental pollution, exhaust gases, monitoring processes in industry, security, etc.

A terahertz sensing system would basically be the same instrument as a terahertz spectrometer as is presently used for scientific research. Thus it consists of a source, a frequency discriminator and a detection system. Also to reach the object under study and to retrieve the returning signal one will certainly need a variety of passive components.

2.2. IMAGING

A terahertz imaging system is nothing other than a terahertz spectrometer with added spatial resolution. Thus such a system allows us to determine the location of specific chemical species in an object. This extra facility may be interesting for some sensing applications, e.g. in industrial processing and security. However, the ability to determine position is vital for medical diagnosis.

A terahertz imaging system would contain the same components as a terahertz sensing system, except that significant data reduction is essential to produce usefull images for e.g. medical diagnosis.

2.3. COMMUNICATION

The terahertz frequency range may be of interest for mobile communications because it adds extra bandwidth. E.g., the 40 GHz wide window at 400 GHz provides several orders of magnitude more bandwidth than presently available for mobile telephony. However, apart from sources, detectors and passive elements, many other devices such as amplifiers, modulators, multiplexers and isolators are also needed.

3. TECHNOLOGY

Given the present technology, three possible lines can be envisaged for developing a terahertz system. In the lower frequency range from 0.1 to 2 THz, one could resort to microwave technology extended to the terahertz frequency range. Above 2 THz a purely optical system can be envisaged. The whole frequency range is however covered by pulsed optical terahertz generation systems (see figure 1).

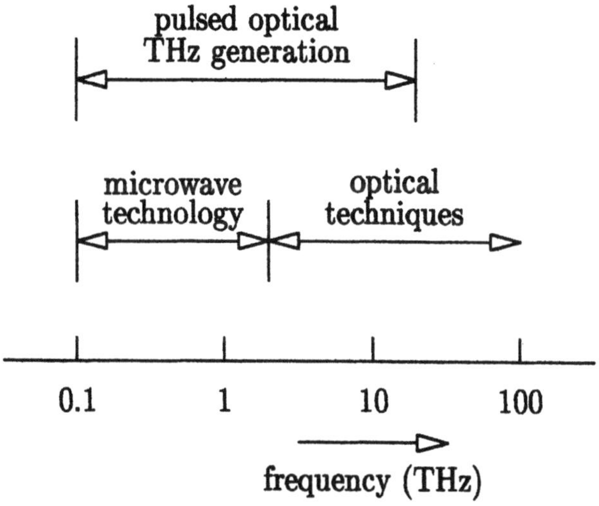

Figure 1 The frequency range of some terahertz technologies.

It is interesting to note that terahertz communication systems are envisaged at the lower frequency end where microwave technology seems most appropriate. Thus the whole range of devices from sources, detectors, passive elements, to amplifiers, modulators, multiplexers and isolators will be needed

On the other hand sensing and imaging applications need the whole terahertz frequency range. However, in this case a more limited range of devices, comprising sources, detectors and passive elements, is probably sufficient.

3.1. KEY TECHNOLOGICAL ISSUES FOR 'MICROWAVE' TECHNOLOGY

The amount of development work still needed varies wildly from one class of device to another. To name some examples:

- Schottky diode detectors can be produced to satisfy the requirements of any of the applications mentioned above.

- The use of Heterostructure Barrier Varactor (HBV) triplers as sources up to 1 THz seems to be within reach. However, still further work to increase frequency, bandwidth, power and efficiency is required. New materials and diamond heat sinks are expected to improve performance [1].
- Also within reach are microfabricated components, in particular frequency multiplexers, maybe using photonic crystal structures. Special design work may still be needed to reduce transmission losses in waveguides [2]. As materials properties define fundamental limits to the frequency range of semiconductor devices, renewed interest arises in vacuum tube devices, possibly to be realised using microfabrication techniques.
- It was the participants' view that no clear path towards amplifiers, modulators and isolators has yet been established.

Hence, using presently available technology it appears feasible to realise terahertz sensing and imaging systems by extending microwave technology to a higher frequency. However, the critical path towards terhertz communication systems, seems to be dependent on the development of a terahertz amplifier, which can possibly also serve as an isolator and a modulator.

3.2. KEY TECHNOLOGICAL ISSUES FOR OPTICAL SYSTEMS

Passive devices are readily available and Schottky diode detectors are capable of operating satisfactorily over the whole terahertz frequency range. Hence, the key issue is the development of sources. Many types of CW THz sources have been proposed and some of them have been realised. However, each of them has its disadvantages. To name a few:

- Optically pumped gas lasers have been around for a long time, but they remain large and more importantly, they cannot be tuned.
- The development of p-Germanium hot hole lasers still continues and CW operation might one day be possible. They can be made smaller in size than optically pumped gas lasers and they are tuneable. However, presently they only operate at cryogenic temperatures.
- Quantum cascade lasers are small, operate at room temperature and may be tuned. However, presently they only operate above 10 THz and the key issue is to lower their frequency [3].
- Bloch oscillators appear to be very interesting on paper, but presently they will not operate in the terahertz range because of domain formation.

3.3. KEY TECHNOLOGICAL ISSUES FOR PULSED SYSTEMS

Terahertz sensing and imaging has been demonstrated successfully using pulsed optical terahertz generation. However, the time needed to obtain useful images is still three orders of magnitude too long to be of practical use. However, the technology still allows for significant improvement, e.g.:

- Devices for rectifying femtosecond laser pulses into terahertz pulses as well as those for detecting terahertz pulses using femrosecond laser pulses are still far from optimal. Research is needed to increase efficiency of these devices.

- Efficient data reduction related to the features observed in the spectra, has barely been tried. It is known from other imaging techniques that such methods may yield orders of magnitude improvement in image acquisition time.

- Terahertz frequencies correspond to wavelengths of the order of 100 μm or more and the resolution of terahertz images is therefore similar. If a better resolution is needed special techniques need to be developed.

4. CONCLUSION

For sensing applications, solutions based on microwave technology or using pulsed terahertz generation seem to be within reach. However, an all optical system would need the development of a room temperature tuneable solid state source, which is still not available.

Imaging applications – using the whole terahertz frequency band – are also within reach, provided pulsed optical terahertz generation is considered as the technology of choice. Again, an all optical system needs the development of sources before it can be realised. If one is allowed to restrict oneself to the lower frequency region, then systems based on microwave technology may also be feasible.

Terahertz communication systems – though envisaged at relatively low frequencies – are probably best realised using microwave technology. However, amplifiers, modulators and isolators are still lacking.

Thus, significant efforts are needed to develop sources, amplifiers and devices for the creation and detection of terahertz pulsed in pulsed terahertz generation. When microwave systems are envisaged then microfabrication of passive components and embedding of active components needs to be further developed. It is interesting to note that in all these cases materials issues seem to play a dominant role.

References

[1] Kolberg, E., This book, page 29, Lippens, D., This book, page 56
[2] Miles, R.E., This book, page 279
[3] Faist, J., This book, page 93, Tredicucci, A., This book, page 106

LIST OF SPEAKERS

Professor S J Allen
Quantum Institute
University of California at Santa Barbara
California 93106
USA

Dr D Arnone
Toshiba Research Europe Ltd
Cambridge Research Centre
260 Science Park
Cambridge
UK

Dr J W Bowen
Department of Cybernetics
The University of Reading
Whiteknights
Reading RG6 6AY
UK

Professor J M Chamberlain
School of Electronic & Electrical Engineering
University of Leeds
Leeds LS2 9JT
UK

Dr T Crowe
University of Virginia
Charlottesville
VA 22903
USA

Professor M Dyakonov
Laborotoire de Physique Mathématique
cc 070
Université Montpellier II
34095 Montpellier
France

Dr H Eisele
Solid-State Electronics Laboratory
University of Michigan
1301 Beal Avenue
Ann Arbor
Michigan 48109-2122
USA

Professor J Faist
University of Neuchâtel
Rue A.L. Brèguet 1
CH-2000 Neuchâtel
Switzerland

Dr P Haring Bolivar
Institut für Halbleitertechnik II
RWTH- Aachen
D-52056 Aachen
Germany

Professor M Koch
TU Braunschweig
Schleinitzstr.22
38106 Braunechweig
Germany

Professor E Kollberg
Microwave Electronics Laboratory
Chalmers University of Technology
SE-412 96 Göteborg
Sweden

Professor D Lippens
Institut d'Electronique et de Micróelectronique du Nord
Université des Sciences de Lille 1
France

Dr C Mann
Rutherford Appleton Laboratory
Chilton
Didcot
Oxfordshire OX11 0QX
UK

Dr R E Miles
School of Electronic & Electrical Engineering
University of Leeds
Leeds LS2 9JT
UK

Professor D Rutledge
California Institute of Technology
Department of Electrical Engineering
MS 136-93
Pasadena
CA91125
USA

Professor E Schomburg
Institute of Applied Physics
University of Regensburg
D-93040 Regensburg
Germany

Professor M Sherwin
University of California at Santa Barbara
Department of Physics
CA 93106
USA

Professor M S Shur
Rensselaer Polytechnic Institute
CII-9017, CIEEM
110-8th Street
Troy NY 121180
USA

Dr K Siebert
Physikalisches Institut
der J W Goethe-Universitaet
Robert-Mayer-Str. 2-4
60054 Frankfurt (M)
Germany

Dr A Tredicucci
Scuola Normale Superiore
Piazza Del Cavalieri 7
I-56126
Pisa
Italy

Dr K Unterrainer
Inst F Solid State Electronics
Floragasse 7
A-1040 Vienna
Austria

Professor D van der Weide
University of Wisconsin-Madison
1415 Engineering Drive
Madison WI 53706
USA

Dr S Verghese
MIT Lincoln Laboratory
Room E-124F
244 Wood Street
Lexington
MA 02420-9108
USA

Professor T Wenckebach
Delft University of Technology
P O Box 5046
2600 G A Delft
The Netherlands

Professor X.-C. Zhang
Rensselaer Polytechnic Institute
Troy
NY 12180-3590
United States of America

Professor N N Zinovev
School of Electronic & Electrical Engineering
University of Leeds
Leeds LS2 9JT
UK

LIST OF PARTICIPANTS

S Barbieri
Laboratorio Europeo di Spettroscopie Non-Lineari
Università di Firenze
Larfo E. Fermi
2-50125 Firenze
Italy

D Burns
US Air Force
223/231 Old Marylebone Road
London NW1 5TH
UK

L Chusseau
Université Montpellier II
34095 Montpellier
France

S Davies
Department of Physics
University of Bath
Bath BA2 7AY
UK

L Dillner
Department of Microelectronics
Chalmers University of Technology
SE-41296 Göteborg
Sweden

A Dmitriev
A F Ioffe Physical-Technical Institute
Politechnicheskaya 26
194021 St Petersburg
Russia

L Duvillaret
LAHC
Universite de Savoie
Savoie Technolac
73376 Le Bourget-du-lac cedex
France

M Feiginov
Institute of Radioengineering & Electronics
Russian Academy of Sciences
Mokhovaya St. 11
Moscow 103907
Russia

Y Gousev
Radioelectronics Lab
Electrum 229
164 40 Stockholm
Sweden

P Harrison
School of Electronic & Electrical Engineering
University of Leeds
Leeds LS2 9JT
UK

Q Hu
Massachusetts Institute of Technology
MIT 36-465
Cambridge MA 02139
US

M Ingvarson
Microwave Electronics Lab
Chalmers Institute of Technology
Rannvagen 6
SE-412 96 Göteborg
Sweden

M Johnston
University of Cambridge
Cavendish Laboratory
Madingley Road
CAMBRIDGE CB3 0HE
UK

V Kacharovskii
A.F. Ioffe Physical-Technical Institute
Politechnicheskaya 26
194021 St Petersburg
Russia

T Klaassen
Delft University of Technology
P O Box 5046
2600 GA Delft
The Netherlands

A Krotkus
Semiconductor Physics Institute
Agostauto II
Vilnius 2600
Lithuania

H Pellemans
IHT-2, RWTH Aachen
Sommerfeldstraße 24
D-52074 Aachen
Germany

P Planken
Delft University of Technology
P O Box 5046
2600 G A Delft
The Netherlands

I Scherbatko
School of Electronic & Electrical Engineering
University of Leeds
Leeds LS2 9JT
UK

R Scheurer
University of Regensburg
Universitätsstr 39
93053 Regensburg
Germany

J Scott
University of California, Santa Barbara
Santa Barbara, CA 93106
US

P Shiktorov
Semiconductor Physics Institute
Agostauto II
Vilnius 2600
Lithuania

K A Shore
University of Wales, Bangor
School of Informatics
Dean Street
Bangor LL57 1UT
Wales, UK

J Stake
Chalmers Institute of Technology
SE-41296 Göteborg
Sweden

E Starikov
Semiconductor Physics Institute
Agostauto II
Vilnius 2600
Lithuania

D P Steenson
School of Electronic & Electrical Engineering
University of Leeds
Leeds LS2 9JT
UK

A Suziedelis
Semiconductor Physics Institute
Agostauto II
Vilnius 2600
Lithuania

W Truscott
UMIST
Department of Electrical Engineering &
Electronics
P O Box 88
Manchester M60 1QD
UK

J Ulrich
FKE, Technical University of Wien
Floragasse 7
A-1050 Wien
Austria

D Woolard
US Army Research Laboratory
Army Research Office
P O Box 12211
RTP, NC 27709
US

1. Michael JOHNSTON
2. Algirdas SUZIEDELIS
3. Dave RUTLEDGE
4. Bill TRUSCOTT
5. Paul STEENSON
6. David BURNS
7. Jevgenij STARIKOV
8. Pavel SHIKTOROV
9. Arunus KROTKUS
10. Daniel van der Weide
11. Simon VERGHESE
12. Tjeerd KLAASSEN
13. Laurent CHUSSEAU
14. Michael FEIGINOV
15. Chris MANN
16. Lionel DUVILLARET
17. Karsten SIEBERT
18. Alessandro TREDICUCCI
19. Steven DAVIES
20. Paul PLANKEN
21. Yuri GOUSEV
22. Henricus PELLEMANS
23. Tom WENCKEBACH
24. Mark SHERWIN
25. Jochen ULRICH
26. Heribert EISELE
27. Igor SCHERBATKO
28. John BOWEN
29. Alan SHORE
30. Michael DYAKONOV
31. Roland SCHEUERER
32. Dwight WOOLARD
33. Stefano BARBIERI
34. Jeff SCOTT
35. Qing HU
36. Jerome FAIST
37. Peter HARING BOLIVAR
38. Karl UNTERRAINER
39. Ekkehard SCHOMBURG
40. Paul HARRISON
41. Lars DILLNER
42. Martyn CHAMBERLAIN
43. X.-C ZHANG
44. Don ARNONE
45. Erik KOLLBERG
46. Mattias INGVARSON
47. Valentine KATCHOROVSKI
48. Ulla KOLLBERG
49. Fariba FERDOS
50. Anna FERDOS
51. Jan STAKE
52. Didier LIPPENS
53. Nick Zinovev
54. Jim ALLEN
55. Bob MILES
56. Lise LINDQUIST

INDEX

MIX
Papier aus verantwortungsvollen Quellen
Paper from responsible sources
FSC® C105338

If you have any concerns about our products,
you can contact us on
ProductSafety@springernature.com

In case Publisher is established outside the EU,
the EU authorized representative is:
Springer Nature Customer Service Center GmbH
Europaplatz 3, 69115 Heidelberg, Germany

Printed by Libri Plureos GmbH
in Hamburg, Germany